D1298062

International Telecommunications

by Phyllis W. Bernt, Ph.D.
Martin B. H. Weiss, Ph.D.

PROFESSIONAL REFERENCE SERIES

A Division of Prentice Hall Computer Publishing
11711 North College, Carmel, Indiana 46032 USA

Phyllis Bernt: In loving memory of Josef Weinroth, Chuma Reiskind Weinroth, and Samuel Moskovits with gratitude for teaching me why it is important to overcome the boundaries that separate nations and people.

Martin Weiss: I dedicate this book to Ginny and Evan, in lieu of time I would otherwise have spent with them. This book would not have been possible without their loving support.

FIRST EDITION
FIRST PRINTING—1993

International Standard Book Number: 0-672-30119-9

Library of Congress Catalog Card Number: 92-83920

96 95 94 93 4 3 2 1

Interpretation of the printing code: the rightmost double-digit number is the year of the book's printing; the rightmost single-digit, the number of the book's printing. For example, a printing code of 93-1 shows that the first printing of the book occurred in 1993.

Composed in Palatino and MCPdigital by Prentice Hall Computer Publishing

Printed in the United States of America

Trademarks

Publisher
Richard K. Swadley

Acquisitions Manager
Jordan Gold

Acquisitions Editor
Gregg Bushyeager

Development Editor
Ken Sochats

Senior Editor
Grant Fairchild

Production Editor
Hugh Vandivier

Copy Editors
Sandy Doell
Mary Corder

Editorial Coordinators
Rebecca S. Freeman
Bill Whitmer

Editorial Assistants
Rosemarie Graham
Sharon Cox

Technical Editor
Joe Pelton

Cover Designer
Dan Armstrong

Director of Production and Manufacturing
Jeff Valler

Production Manager
Corinne Walls

Imprint Manager
Matthew Morrill

Book Designer
Michele Laseau

Production Analyst
Mary Beth Wakefield

Proofreading/Indexing Coordinator
Joelynn Gifford

Graphics Image Specialists
Jerry Ellis
Dennis Sheehan
Sue VandeWalle

Production
Katy Bodenmiller,
Christine Cook, Lisa Daugherty,
Carla Hall-Batton, Howard Jones,
John Kane, Sean Medlock,
Tim Montgomery, Angela Pozdol,
Roger Morgan, Susan Shepherd,
Greg Simsic, Alyssa Yesh

Indexer
Loren Malloy
Suzanne Snyder

Contents

Foreword

This book is the first publication in the Sams Professional Series in Telecommunications. The selection of international telecommunications to launch this series is no accident. Internationalization is an important social, political, economic, and technical trend in the world today. We stand at the beginning of this trend. The forces that brought about internationalization are still present and, if anything, increasing in intensity and scope.

This book gives telecommunications professionals an overview of the field of international telecommunications, provides insights into the many facets of designing and operating an international telecommunications network, and supplies the reader with the skills and knowledge necessary to analyze the telecommunications environment in any country.

The authors of this work, Phyllis Bernt and Martin Weiss, are widely published and well-regarded researchers and theoreticians in the area of international standards. This book represents another of their important contributions to this area. We found Phyllis and Martin easy to work with, highly competent, and dedicated. The high quality of this book lies largely in their untiring efforts to produce a timely and insightful publication.

The Sams Professional Series in Telecommunications, which we edit, is a continuing effort to produce books on leading edge topics in the area of telecommunications. These books are directed to a professional audience composed mainly of telecommunications practitioners. Although this book is the first in our series, several other books stand at various stages of development. We hope to develop a portfolio of books that offer coverage of all of the important emerging technologies, practices, and issues in telecommunications.

Ken Sochats
Jim Williams
Pittsburgh, Pennsylvania
December, 1992

Foreword

Preface

This book is intended as a primer for those approaching international telecommunications for the first time. It is written with the practicing telecommunications professional in mind, although it can also be used in undergraduate and graduate courses in international telecommunications with some supplementation. The book is organized into three major parts.

The first part of the book (Chapters 1 through 5) provides a basic overview of the important issues surrounding international telecommunications. After introducing the challenges inherent with global telecommunications (Chapter 1), we discuss the major regulatory themes (Chapter 2), the economic issues (Chapter 3), the organizations that play a major role in the international arena (Chapter 4), and the technical aspects of telecommunications (Chapter 5). As a rule, these chapters are relatively independent of each other, so readers don't need to read them all or read them sequentially to benefit from this book. Clearly, each of these chapters deserves a book in its own right, so readers who are expert in certain areas might find them lacking in detail and completeness.

Chapter 6 presents an assessment methodology. The purpose of this methodology is to provide an organized, structured approach to gaining an understanding of a country's telecommunications environment. You would presume too much to imagine that any methodology could provide a complete and all-encompassing approach to analysis; hence, you must understand that this is an approach to gaining an initial understanding. You can only obtain more detailed information from close observation, interviews with people doing business in the target country, and so on. The methodology, then, is an approach to gathering and organizing public domain data in such a way that it becomes useful information. You can apply this methodology to both developed and developing nations, and it is meant to serve as a guide for determining the regulatory climate, the state of the infrastructure, and the range of expected services.

Chapter 7 consists of case studies where the telecommunications environments of nine nations are presented. These case studies represent countries that have been in the forefront of change during the past decade (Japan, the United Kingdom and the United States); and those that have been slower to

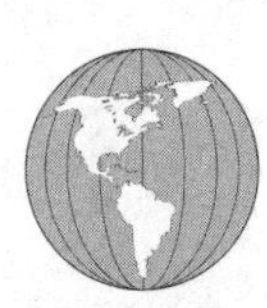

transform their approach to telecommunications (Germany). The nations selected also represent developments in Eastern Europe, Latin America, Asia, Africa, and the Pacific. In addition to an overview of conditions in each country, this chapter presents applications of the assessment methodology explained in Chapter 6. A complete country-by-country survey of telecommunications would fill a book in its own right and would likely be out of date as it was being printed.

The book ends with a brief look toward likely developments in the future given today's trends. Gazing into a crystal ball is not without its risks, particularly in a field as dynamic as this. Our hope is that history does not render our predictions as it has the prediction of Thomas Watson, Sr., former president of IBM, when he anticipated that perhaps a dozen computers could handle the entire data processing needs of the world. Forecasting is indeed a risky business.

International telecommunications is complex and constantly changing. We hope that this book sheds light on some complexities and brings some order to a perpetually shifting landscape. We would like to thank in particular the people who have stimulated our interests and have been the major influences on our thoughts in this area. They include: Marvin A. Sirbu, Eli Noam, Heather E. Hudson, Joseph Pelton, Charles E. Steinfeld, Meheroo Jusawalla, Eugene Newman, and the many participants at the annual Telecommunications Policy Research Conferences, who have tackled and discussed these issues. Their work in this and related fields has been outstanding. We would also like to thank the editors of the Sams Telecommunications Series, James G. Williams and Kenneth M. Sochats, for first stimulating this effort and providing their support and suggestions as we progressed. We would also like to thank the students of the University of Pittsburgh and Ohio University, whose lively discussions and interesting questions over the years have helped us crystalize our thoughts.

Phyllis Bernt
Athens, Ohio
bernt@ouvaxa.cats.ohiou.edu

Martin Weiss
Pittsburgh, Pennsylvania
mbw@pitt.edu

November 1992

Overview of Telecommunications

CHAPTER 1

No modern organization can function efficiently without telecommunications. Telecommunications makes it possible to transfer massive amounts of information between computers and workstations; to link a variety of speakers together in conference calls that span cities, countries, and continents; to carry videoconferencing images across oceans; and to deliver pages of facsimile documents on demand. Because the value of information often decays with time, the central focus of organizational success is to reduce *information float*—the time lag between the creation of information and its use. Timely information is more valuable than information that arrives later. Telecommunications is the medium that shrinks the time lag. The globalization of business has made telecommunications an even more essential resource and has created much more complexity in the management of this important resource.

In many ways, globalization and telecommunications build upon one another. Because telecommunications makes it possible to overcome many limitations of time and space, globalization becomes an attainable reality. Organizations extend their reach to new markets or new labor pools as efficient and effective linking to these distant operations becomes a viable option. Improvements in technology, infrastructure, and pricing spur this extension, which further spurs the development of infrastructure. Thus, the process feeds upon itself. As globalization increases, people put greater demands on telecommunications to move more information quickly and cheaply.

The telecommunications industry has responded to these demands by creating international services and partnerships. Telecommunications managers in organizations that use global telecommunications networks face the tasks of installing, managing, and operating these networks. Coordinating telecommunications facilities and services has always been a complex undertaking. Ensuring that the proper facilities are ordered and installed, that equipment is compatible, and that organizations are ordering the most efficient assortment of services requires a great deal of coordination.

When telecommunications becomes global, the complexity of coordination increases. Instead of dealing with one telephone system, many exist. Instead of meeting one country's technical standards, one must worry about diverse standards. Instead of working with one set of prices, similar services can operate under several different sets of prices. Differences in language, legal systems, currencies, cultures, and customs compound these technical complexities. For those who coordinate international telecommunications networks or even for those who use those networks, the entire area of international telecommunications can seem confusing and intimidating.

This book tries to eliminate some of that confusion and provide its readers with a basic understanding of the key issues they face when extending a telecommunications network across national boundaries. Rather than provide a detailed review of telecommunications on a country-by-country basis, this book provides an overview of the central issues, identifies some of the most important organizations involved in international telecommunications, and supplies an analysis methodology to help you assess the important features of telecommunications in every country. Although a country-by-country analysis would no doubt be interesting, such an examination would be obsolete by the time this book came out because telecommunications is changing so rapidly, both technologically and organizationally. Therefore, this book focuses more on identifying trends and changes in the international telecommunications environment.

Ideally, to eliminate bias in the examination of any country's telecommunications system, we'd need to recruit a person from Mars with no experience in Earth's international markets. However, we, the authors, developed our understanding of telecommunications in general, and international telecommunications in particular, from a U.S. perspective. Although we try to be objective in this book, our U.S. biases may often appear. For this, we seek the indulgence of our non-U.S. readers and apologize in advance for any additional burdens this perspective might create.

The Telecommunications Problem

The basic concern of telecommunications is to move information from one location to another. Normally, this movement takes place over a band-limited channel that has well-defined, technical properties. The design of the channel and the equipment used to implement that design determine how much information, the type of information (voice, data, video), and the speed at which the information is carried. Telecommunications networks consist of many such channels, both in parallel and in tandem.

Almost every channel unavoidably distorts the signal it carries and adds noise to it. If the channels in a network vary greatly, the total distortion can be significant. If one user or organization completely controls the channel, that entity can do much to improve the performance of the system. For example, the user or organization can use careful signal conditioning and transmission codes. In fact, Claude Shannon proved that an arbitrarily low error rate was possible over any channel, so long as the capacity of the channel wasn't exceeded. For Shannon, capacity is a formal relationship between the theoretical capacity of the channel (the bandwith), the degree of the signal's distortion, and the noise of the channel. This, however, requires coordination on all parts of the transmission line. This coordination is difficult to achieve when information passes outside an organization's control; it is even more difficult to achieve on an international scale.

The key factor that emerges, from a practical perspective, is who has control over the line or channel. For a local telecommunications network, one can reasonably assume that the user, or a single organizational entity, controls all technical parameters of the line. The user deals with a local telephone service provider and specifies the type of line or channel that is required. As networks expand to cover greater geographic areas, more entities almost inevitably become involved in designing and providing channels. When a telecommunications network is national in scope, several entities might provide different parts of the network.

The issue concerning the locus of organizational control is inextricably bound to the way telecommunications services are provided, and that locus is not the same in all countries. In some countries, a single administration is responsible for all aspects of telecommunications. In other countries, one administration is responsible for local access, and another administration is

responsible for long distance communications. In an extreme case, such as the United States, many private companies are involved in the establishment and provision of the links that comprise a telecommunications channel. Each of these companies must cooperate in order to establish a channel capable of moving information from one desired location to the other with minimum distortion.

When a telecommunications network crosses national borders, the issues become more complicated. Coordination must take place among telecommunications providers in each of the countries involved. Not only are more organizations involved, no single authority can adjudicate disputes and differences. In the fragmented, market-based system of the United States, a regulatory system (the Federal Communications Commission at the national level and the Public Utilities Commissions of the individual states) prevents chaos, and a legal system helps settle disputes. In the international arena, agreements and the settling of disputes are often matters of diplomacy and treaty accords. Often, differences in language and cultural tradition make negotiations for providing telecommunications channels exasperating and difficult for all parties involved.

In addition to these organizational and political issues, potential technical differences also arise. As discussed earlier, certain design choices must be made during the development of equipment in order to provide telecommunications channels. Different countries or different vendors often make different decisions. The problem then becomes how to make these differing channels or systems operate together, or interoperate, and how to determine who bears the cost of the equipment needed to enable the interoperation of systems. The establishment of technical standards provides the solution to this problem of different channel designs. If uniformly implemented, standards eliminate the need for conversion devices for interconnection among different national networks.

Once systems can interoperate and service is provided, the various entities that provide the service must be paid. It then becomes necessary to establish a system for determining what the payments should be, to whom they should be paid, and how these payments should be allocated among the various service providers. Individual nations have their own methods of distributing payment among service providers. In a nation where one administration provides all services, the matter is obviously simple. In a nation where multiple providers are involved, a method of compensating each provider must be determined. In the United States, for example, a long distance call from one end of the country to the other involves both local telephone companies and a long distance company. By arrangement, long distance companies pay

local companies for the use of their facilities. The long distance companies then add the amounts they pay local companies to the long distance rates they charge customers.

The picture at the international level is much more complicated. Many countries are involved, and no central clearinghouse exists to determine how telecommunications providers in each country are compensated for the portions of international service they supply. Instead, a fairly laborious process of individually negotiated settlements occurs among countries.

Clearly, the international telecommunications environment is quite different from a strictly national one. In practice, individual service providers in each country carry much of the additional burden created by the international environment, and occasional international callers are usually unaware of these complexities. However, as their networks expand across national boundaries, private network operators are likely to face many of the same problems encountered by service providers. These private network operators need to understand the potential problems involved in establishing an operational network.

Similar problems arise in the provision of international data communications capability . Because this industry is somewhat younger, it isn't burdened with the same weight of tradition as the voice communications industry. With these networks, many countries first began to experiment with competition and alternative industry structures, so situations that hold for voice communications might not hold for data communications. Users of data communications must also be more sensitive to the transborder data flow regulations among countries. These rules often apply most strictly to data communications networks.

Elements of Telecommunications

Many influences have shaped telecommunications. Because policymakers have traditionally regarded it as a natural monopoly, some form of regulation or government ownership has always played a role. Even in countries that are experimenting with the competitive provision of telecommunications services, regulation remains an influential factor. The characteristics of international telecommunications result in substantial peer-to-peer coordination

of standards, electromagnetic frequency assignment, and other issues. Several organizations have formed to facilitate these varied and necessary interactions. Some are treaty organizations, such as the International Telecommunications Union (ITU) and Intelsat (International Satellite Organization). Others, such as the International Standards Organization (ISO), are voluntary.

These organizations are useful for facilitating telecommunications, but the basic driver is economics. A demand exists for international telecommunications. The economic aspects of international telecommunications range from the settlements issue, discussed in the previous section, to transborder data flow and economic development issues. All are particularly relevant to private network operators. Technological factors are a major element in shaping telecommunications. Changes in technology have frequently spawned new organizations. Intelsat wasn't needed until the introduction of communications satellites. Changes in technology also change the economics of telecommunications. For example, fiber-optic undersea cables altered the cost and nature of international telecommunications circuits. Technological changes also affect regulation. As technological advances provide new services, new regulatory approaches are developed to deal with those services. This chapter will explore each of these areas.

Regulatory Issues

Conflicting beliefs cause an underlying tension in the regulatory treatment of telecommunications services. That tension results from the differing opinions of those who believe that telecommunications should be treated as a monopoly and those who see competition as the best way to deploy telecommunications services. These two approaches pose vastly different implications for how services are provided, the range of services provided, and who is allowed to provide them.

Proponents of the monopoly approach regard telecommunications as a natural monopoly: the cost of building and maintaining a network to support the services is so high that it represents a formidable barrier to entry. Moreover, the economies of scale and scope provided by one network are such that these greater efficiencies (and the resultant lower unit prices) are sufficient to justify a monopoly provider of services. Proponents of competition contend that allowing a variety of providers for a wide range of services enhances innovation, forces each provider to be more efficient, and encourages lower prices.

Nations have structured their telecommunications industries in a variety of ways. Most countries regard telecommunications as a monopoly, usually government-owned. In most nations, the government agency responsible for the postal system and the telegraph also handles telecommunications. For example, in Germany the Postal Authority (Deutsche Bundespost) handles the responsibility of providing telephone circuits to German citizens. Only recently has the telephone provision been spun off. (It is now DBP Telekom.) With government ownership, the creation of a regulatory body hasn't been deemed necessary. After all, why have one unit of government regulating another? The pattern of government ownership and monopoly is beginning to break down as nations observe the U.S. example of competition. The United Kingdom and Japan are classic examples of this, although other nations, such as Canada, are also beginning to allow competition in their telecommunications sectors.

The history of telecommunications in the United States has actually been marked by a pendular swing between competition and monopoly. In its early years, the telephone industry was marked by competition, with hundreds of providers building telephone systems in communities throughout the nation. The inefficiencies, duplication of facilities, and chaos of having several independent networks eventually led to the creation of franchise monopolies and regulatory agencies. In return for providing a specified level of service, a telephone company gained the right to be the exclusive service provider for a designated geographic area. To link these geographic areas, American Telephone and Telegraph (AT&T) became the monopoly provider of long distance services.

The American example is interesting in many regards. Unlike most other nations, telecommunications in the United States has historically been privately owned. The government, through regulatory bodies and legislation, exercised policy-making authority over the building of a telecommunications infrastructure. In the earliest stages of the industry, private entrepreneurs built networks at will. A unified national network didn't become a reality until the United States government, at both the state and federal levels, established an elaborate monopoly. Indeed, building a national network of affordable telephone service for all who wanted it became a federal mandate with the creation of the Communication Act of 1934. This universal service mandate continues to be a driving force behind much U.S. telecommunications policy-making.

The regulated monopoly approach resulted in the construction of a national infrastructure. Until the 1960s, one unified system, comprised of the Bell system and a host of independent, local telephone companies, held a monopoly

in the United States over customer premise equipment (CPE), local service, long distance service, and private or dedicated lines. The monopoly structure, however, couldn't withstand technological advances and the demands of large business customers for lower rates and more advanced services. The development of the computer industry presented U.S. regulators with services beyond their regulatory purview. Increasing pressure from CPE manufacturers and potential long distance providers, such as MCI, eventually broke down the end-to-end network monopoly. The result has been enhanced CPE products and lower long distance rates, along with the growth of hundreds of new enterprises offering new services and enhancing competition for existing service.

Many other nations regard this move by the U.S. toward increasing competition in telecommunications as a success and have begun efforts to duplicate some aspects of the U.S. approach. Many nations, the United Kingdom for example, have privatized their formerly government-owned telecommunications enterprise. Some have introduced competition into portions of their telecommunications industries. Japan and the United Kingdom, for example, have had competition for some time in the provision of long distance services.

How far a nation goes in privatizing its telecommunications network profoundly affects the money available to modernize the network. Governments are often loath to raise taxes for purposes of network modernization; private companies can borrow money or issue stock. How far a nation goes in introducing competition affects the kinds of services available and who provides those services.

A secondary theme of the monopoly-versus-competition argument concerns the ultimate goal of the adopted telecommunications policy. As the U.S. example shows, a monopoly approach lends itself to the attainment of universal service. In a monopoly, network rates can be averaged across a large number of users. Because the cost of providing service to each customer varies, the result is cross-subsidization. In the United States, subsidization has been an explicit policy, so business customers pay more for local service than residential customers, and long distance rates have traditionally been priced to provide a subsidy for keeping local rates low. As the U.S. example also shows, those who pay the subsidies eventually become disenchanted and ask for lower rates and more innovative services. Because those who pay the subsidies are usually large users with the financial ability to build their own networks if they can't get what they want from the public network, they are often successful in their demands.

Analyzing a nation's policy approach can be highly instructive when ascertaining whether the major goal is to attain universal service or to meet the demands of large and sophisticated users. Both policies possess real implications for the range of services available and for the pricing of those services.

Organizational Issues

As national networks began to expand across national boundaries, network providers soon discovered some additional issues. On a technical level, provider networks needed to be interconnected so that signals could move freely from one system to another. On an administrative level, service providers needed a mechanism for resolving disputes. On an economic level, providers needed some assurance of receiving fair compensation for the portion of international service that they provided. As the technology matured to include radio, a need emerged to coordinate the use of radio frequencies to prevent or minimize interference, particularly in border regions. As radio technology expanded to include satellites, it seemed reasonable to designate a single agency to coordinate satellite transmissions for recovering the cost of satellites more efficiently.

Today, a wide variety of organizations exists to facilitate international telecommunications. Some of these organizations are dedicated to telecommunications alone, and others include telecommunications within their purview. Some organizations are specifically designed to meet the needs of a region (Europe, for example), and others are of a worldwide scope. A substantial number of these organizations are dedicated to the technical problems of interconnection. In many senses, these are the easier problems to solve on an international scale. Other problems have more profound political and economic implications, making solutions much more difficult to attain.

Economic Issues

Closely linked with organizational and regulatory issues are economic issues. Although some can say that economics underlies everything, isolating some economic issues for closer study can be useful. Such economic issues include international settlements, pricing concerns, trade in services issues, and economic development matters. Settlement and pricing issues relate to the flow and allocation of funds associated with completed calls and the pricing relationships among various services. The other issues are largely ones of national

policy that proceed onto the international stage. Standards issues are also of this type. The economic role of standards is evident in the attempt to develop international standards for color television in the 1960s and in the development of high definition television (HDTV) standards today. These are largely issues of industrial policy in individual countries, although their resolution must necessarily occur on an international stage so that the market is sufficiently large for mass-production economics to dominate.

Many of these economic issues bear upon the cost of using or providing international telecommunications services. The relationship between the prices charged for services and the underlying costs of those services is not always clear. Within a country, some services may subsidize others for policy reasons. For example, the provision of local services in the United States was subsidized by long distance services to meet the explicit policy objective of universal service. Similarly, the logic of including telecommunications in the postal organization was to subsidize a socially desirable, but often unprofitable, service (the delivery of letters and packages) with a profitable one (telecommunications). One effect of competition is often to drive prices closer to cost, making it more difficult to maintain a system of subsidies.

When services are international, the issues of pricing become more complex, and the issue of settlements arises. Service providers should base the price of an international service, at least theoretically, on the cost of providing service in each country and on the cost of the international link. What complicates this process is the manner in which the service providers in the respective countries are compensated for the costs they incur. The revenues available to cover the costs involved in an international call come from the money the caller pays to the service provider in Country A for placing the call to Country B. The amount of money shared is determined by the allocation amount (or settlement) agreed upon for the route between Countries A and B. What further complicates this process is that the settlement amounts are not related to underlying costs, such as the price charged a customer for an international call. No way of enforcing symmetry on international routes exists. In other words, the price of a call in one direction may differ from the price of a call going the other direction. All this results in asymmetrical rates and rates that may not be related to underlying costs.

Clearly, such asymmetry has ramifications for all involved. Customers adjust their calling patterns to minimize costs, creating asymmetrical calling patterns. As a result, more dollars flow from one nation to another through the settlement payments mentioned previously, which often results in trade imbalance issues. While diplomats argue over trade issues, those who

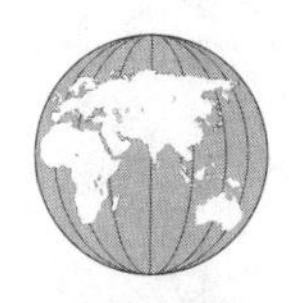

manage international networks, in order to make optimal use of available services, face the added complexity of juggling the asymmetrical international traffic that results from rate asymmetries.

The economic impact of standards and infrastructure issues is not quite as evident as that of pricing issues. Standards, as discussed earlier, can become tools of national industrial policy; in a complementary role they can become a barrier to international trade. In some countries, such as the United States, compliance with many telecommunications standards is voluntary. In a private network, an organization is free to select any vendors and standards, as long as the network works. For example, a firm can choose standard or nonstandard modems, depending on specific needs. Other countries, however, are not quite so liberal. Historically, countries such as Germany and France, which have telecommunications infrastructure operated by the government, restricted the sale of CPE to the telecommunications authority. This was also true in the United States until the early 1980s. Even if private provision is allowed, many countries mandate the use of products that comply with CCITT or ISO standards. (CCITT is the French abbreviation for the International Telegraph and Telephone Consultative Committee.) Such controls not only constrain the equipment used within a country, but also the equipment used on international circuits. Special considerations and pieces of equipment are necessary when interconnecting North American digital networks, which operate under a regional set of standards, with, for example, European networks. These considerations require additional planning by network managers and often require additional costs.

Infrastructure issues are in many ways even more central to the ability to provide an international service than standards issues are. In many regions of the world, a ubiquitous telecommunications infrastructure can't be taken for granted. When locating a facility, a network operator must be aware of the supporting infrastructure or the laws governing private networks. A weak infrastructure can affect the availability of communications facilities, the time needed to provide a telecommunications service, and the quality of the provided service. Where possible, many managers of international networks are compelled to construct private facilities for their networks because of infrastructure problems. The ability to construct private networks often depends upon the approval and cooperation of telecommunications administrations. A weak infrastructure also has an impact on other business practices; if employees or suppliers don't have telephones or can't easily be reached by telephone, other forms of communication become necessary, which increases the management problem.

Technical Issues

Finally, technical issues are important in international telecommunications. The telecommunications industry is characterized by rapid technological change. As equipment at the user's location changes, so do the requirements of an international network. For example, when slow-speed, terminal-oriented devices dominate a data communications network, X.25-based packet equipment is often a good choice. As terminal devices are enhanced to handle higher speeds, other technologies, such as Frame Relay, are utilized.

Technological change occurs in many aspects of telecommunications networks. Generally speaking, telecommunications systems consist of end-user equipment, switching equipment, and transmission equipment. Technological change occurs in each of these areas. When it does, it often has an impact on the other areas as well. A good example of this is the introduction of digital transmission technology in the early 1960s. Digital signaling provides a superior transmission format for noise reduction. With the emergence of integrated circuits, it also became cheaper to fabricate. As digital transmission was introduced into the network, the desire for digital switching increased for cost and efficiency reasons. Once the network became substantially digital (in developed countries), this change began to propagate to end-user equipment. At the same time, certain pieces of end-user equipment, such as private branch exchanges (PBX) became digital and merged with the rapidly proliferating computers. Thus, the service providers felt pressure to make digital and data services available to their customers. Beginning in the 1970s, this motivated the standards committees to develop the Integrated Services Digital Network (ISDN).

Transmission technologies have also changed, partly in response to other technological changes. The emergence of practical, cost-effective fiber-optic telecommunications networks has made cable-based systems formidable competitors of satellite systems for many international applications. The technologies chosen depend not only upon technical capabilities but also on economic issues and the regulatory environment. As the regulatory environment changes, the underlying network economics also change. Thus, a network manager must constantly reconfigure the network to make it as cost-effective as possible.

Regulatory Issues

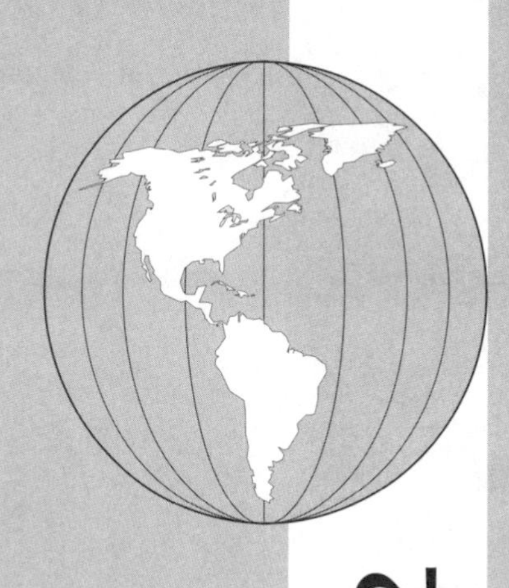

CHAPTER 2

International telecommunications involves many issues that can roughly be considered regulatory. This chapter examines the major ways telecommunications services are regulated in various countries. The approach we take has a taxonomic emphasis although it is unlikely that a pure exemplar of any single category exists. Generally speaking, the telecommunications provider exists somewhere along an axis that you could characterize as ownership. One can envision the provision environment as an axis that might be considered market structure. At one extreme, private providers operate in a competitive marketplace; at the other extreme, government agencies operate in a monopoly environment. Figure 2.1 offers a diagram illustrating this.

Because the telecommunications industry is critical to the economic, industrial, social, and political infrastructure of a nation, some form of regulation or government control is almost certain to exist. Although mechanisms vary from country to country (and sometimes from service to service), control generally consists of some combination of the following:

- Tariffs for the provision of services
- Service quality requirements
- Information policy
- Transborder data flow regulations and policies
- A series of technical policies such as controlling the landing rights for cable and satellite systems

This chapter examines each of these issues in more detail.

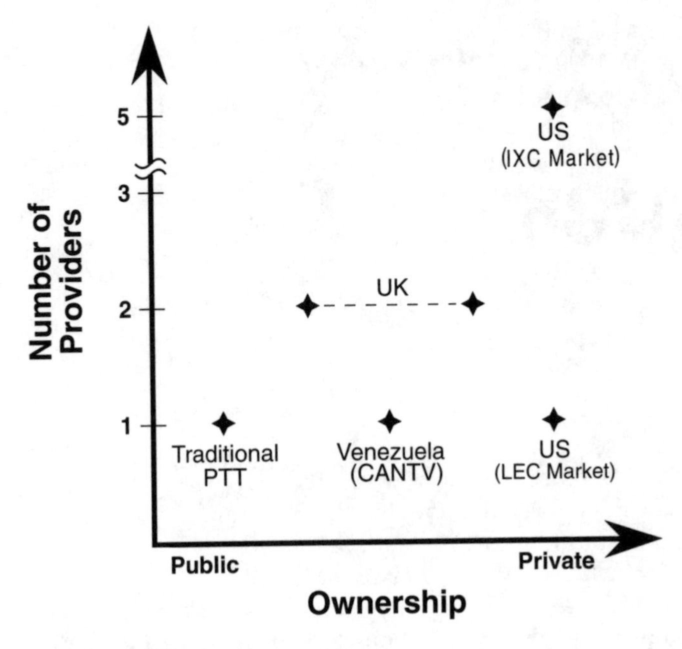

Figure 2.1. The structure of service provision.

Privatization

A major international trend in telecommunications during the past decade has been the move toward privatization. *Privatization* is the transfer of public functions and resources to the private sector. The transfer can entail the operation, management, or actual ownership of publicly owned facilities. Privatization doesn't necessarily include liberalization, which deals with issues of market entry and competition. Privatization without liberalization is possible: a public monopoly merely becomes a private one. In most instances of telecommunications privatization, however, some element of liberalization is involved.

Before the mid-1980s, most nations provided telecommunications services through government-owned monopolies. Some nations, notably those in Latin America, where the telecommunications resource was privately owned,

nationalized telecommunications in the 1960s. During the 1980s and continuing into the 1990s, however, more countries have privatized telecommunications services to some extent. From the sale of company stock to the formation of joint ventures and the drafting of so-called build-operate-and-turnover (BOT) agreements, nations all over the globe are seeking ways to place the administration of networks and the provision of an increasing range of services into private hands.

The success of the private ownership model in the United States provides one reason for this trend. Unlike most nations, the United States approached the development and provision of telecommunications networks and services as private concerns. The telecommunications network, telephones, PBXs, toll switches, and value added services are all owned, controlled, and sold by privately held entities. Government involvement consists of regulatory oversight to ensure that companies meet universal service goals and avoid anticompetitive behavior. The result is a highly successful telecommunications market. As the U.S. economy becomes increasingly dependent on information services, U.S. telecommunications resources can respond to that challenge.

Many observers find it easy to assume that an integral part of U.S. success is its privatization policy and that other nations could benefit from such a policy. G. John Ikenberry refers to this emulation as "policy bandwagoning" and lists it as one of the causes for the spread of such policies.[1] Other causes listed include external pressures from international lending agencies, such as the World Bank and the International Monetary Fund (for developing nations), and a process of "social learning," in which knowledge about a policy and its benefits spread through a nation's decision makers.[2]

Whatever the specific reason, other nations are embracing a privatization strategy. These nations hope that privatization will help them prepare for an information-based economy, to support the demands of a growing number of sophisticated international telecommunications users and keep pace with the continued technological development of networks and services around the world. The degree of privatization and the specific methods adopted by different nations vary according to the state of their capital markets, economic conditions, and social and cultural norms.

The Advantages and Disadvantages of Public Ownership

Nations provide telecommunications through government-owned and operated monopolies so they may reap the following benefits:

- A monopoly presents the opportunity to capitalize on the economies of scale and scope inherent in a telecommunications network.
- Governments can cross-subsidize and support less profitable telecommunications services, such as local and rural service, with profits from long distance and international calling. Moreover, governments can use telecommunications profits to subsidize other government services.
- Governments can retain control of the infrastructure for political, social, economic, and defense purposes.[3]

Such benefits are by no means minor. Government monopolies reap the benefits of increased traffic and added services through greater revenue streams, which are then used in a variety of ways.

On one hand, governments use more profitable services to subsidize less profitable services, thus enhancing efforts to extend service to residential customers and less populated areas. On the other hand, telecommunications revenues are often used to support other services as well. In many nations, the telecommunications administration is, or has been, part of a government entity also responsible for the postal system; hence the term Postal, Telephone and Telegraph (PTT). Telecommunications often subsidize the postal system. Indeed, in many developing nations, the revenue stream from telecommunications services represents a major source of government funding. In Brazil, the telecommunications administration contributed almost $800 million, or 31 percent of gross government revenues in 1986; the Moroccan telecommunications department is expected to transfer $2 billion to general government funds between 1987 and 1994.[4]

For many nations, a public telecommunications administration is a valuable political, social, and economic tool. Experts have predicted, for example, that African nations will be relatively slow to privatize because public enterprises fulfill such important roles. These public enterprises are used as a vehicle for political patronage, and they provide a mechanism by which governments can maintain some control over economic and social conditions.[5]

Although public enterprise offers some benefits, it hasn't always been the most effective vehicle for dealing with technological change and the growing demands of citizens for expanded services. If revenues from telecommunications services are used for subsidizing other government entities, these revenues aren't being reinvested in the telecommunications infrastructure. If a major investment is needed to build up infrastructure, the most obvious recourse a public enterprise has is to raise taxes—a move that most governments want to avoid. In a governmental structure, telecommunications represents only one area of concern. Telecommunications infrastructure issues must compete for funding with such other items as health care, education, and defense. Indeed, a reason often cited for the lack of telecommunications funding in some developing nations is the government's perception that other items are of more immediate importance.

The lack of access to ready capital for building an infrastructure is only one drawback mentioned in regards to public ownership. N.C. Lerner commented that public agencies are often unable to operate efficiently because they lack the following:

- Financial and administrative autonomy
- Incentives for efficient operation and planning
- Management authority to attract and retain personnel
- Management authority to reduce staffing levels
- Pricing and service flexibility
- Access to capital markets
- Political and planning independence[6]

Decisions made by government-run telecommunications administrations are often constrained by political concerns. Legislative action frequently decides issues of pricing and service provision. If a telecommunications administration is expected to generate revenues to fulfill public policy goals or to subsidize other agencies, it has few incentives to offer new services that might jeopardize existing revenue streams. Civil service employment policies handle personnel and staffing matters. The goal of staffing can be to provide jobs rather than to attain optimal efficiency levels. This is borne out by the following statistic: developed countries typically have staffing levels of 7-12 people per 1,000 main lines, "middle income" countries have 20-40 people per 1,000 main lines, and "poor" countries have 60-90 people per 1,000 main lines.[7]

Lack of autonomy in telecommunications administrations often results in inadequate service. Sophisticated services are rarely available; long waiting lines exist for service installation and upgrades; the infrastructure is outdated. This type of situation in the late 1970s, compounded by government deficits that made it highly unlikely for massive tax increases to be passed for infrastructure development, led the British to privatize their telecommunications administration, British Telecom.

The Benefits of Privatization

Nations that privatize former government-run agencies do so because they expect to realize a range of benefits. Most of their expectations center on the infusion of capital and on increased efficiencies that result from the adoption of a market orientation. Specifically, these benefits include

- Raising cash to lower national debt
- Raising foreign exchange, which can be used to extinguish foreign debt or purchase needed imports
- Raising domestic currency in order to expand services or facilities
- Curbing losses incurred by state-owned enterprises and controlling subsidy flows to those enterprises by moving into new business ventures and opportunities
- Increasing operating efficiencies through more liberal personnel policies and more market-oriented procedures
- Spreading corporate ownership more widely through the sale of stock
- Gaining control of, or improvements in, public policy objectives through the creation of regulatory bodies[8]

Rather than losing control of the public policy agenda, some have argued that the British government, through the creation of the Office of Telecommunications (OFTEL), actually has maintained or even increased public control over the direction of telecommunications policy[9]

In addition to seeking the capital infusion needed to rebuild eastern Germany, Germany's Post and Telecommunications Minister, Christian Schwarz-Schilling, is considering privatization of the state-owned Deutsche Bundespost

Telekom to make it more competitive in foreign business markets. The German constitution currently limits DBP Telekom's foreign activities to "functions ancillary to provision of its existing international services to and from Germany." As a result, the state-owned company is effectively barred from participating in joint ventures or in joining with other enterprises in bidding for foreign operating companies.[10] DBP Telekom can't take advantage of a potential valuable source of additional revenue and a means for positioning itself in the international telecommunications market.

Privatization changes an organization's incentives. If the revenues raised by a state-owned entity are used to subsidize other enterprises, few incentives exist for the entity to minimize costs and maximize revenue. A profit orientation, with excess revenues kept by the enterprise, can change those incentives. With profits as a goal, an organization becomes more aware of costs, which are then reduced. One of the steps that the newly privatized British Telecom had to take, for example, was the creation of a management information system to provide data needed to support decision-making.

Telecommunications benefits greatly from privatization. The building of a telecommunications infrastructure requires a great deal of capital. Just building a digital overlay network for Bangkok, Thailand, will cost over $3 billion. Due to politics or economics, a public enterprise often can't raise the needed capital. Governments are loath to raise taxes, and funding for other sectors of the economy is perceived as more critical. Privatization efforts have little success if private entities aren't willing to participate. A nation's telecommunications can, however, prove enticing to private enterprises. A great demand for service exists in many nations. For example, in 1987 over 400,000 people waited for telephone service in Indonesia.[11] Meeting a demand that great can result in significant revenues.

The level of pent-up demand for telecommunications services can make the operation of the public network particularly attractive to private entities. Technologies such as digital switching and fiber optics promise increasing economies of scale and scope in meeting the demand for service. The high levels of demand and the possible savings can result in high levels of profit. Many telecommunications administrations are taking advantage of those economies, as witnessed by the revenue streams generated by many PTTs before privatization. The German Monopoly Commission, for example, found that the German PTT had a profit of DM 5 billion in 1980.[12]

Various Approaches to Privatization

No single formula for privatization exists. Depending on their social and economic situations, nations take a variety of approaches in privatizing their telecommunications industries. Situations range from total ownership by a private entity to the granting of service contracts to private companies. Most approaches fall somewhere between these two extremes. Of course, at one extreme is the United States where almost all telecommunications enterprises are privately owned. A few small communities own and operate local telephone systems, but these cases are rare. In some small cooperatives, customers hold ownership shares in the undertaking. These, too, are rare. Large companies, such as AT&T, the Bell Operating Companies, GTE, and MCI are the norm and are completely owned by stockholders who buy equity shares on the open market. Government's only involvement is through such regulatory agencies as state commissions and the FCC, taxing authorities, and the courts.

However, the U.S. approach isn't the only way to place at least some aspects of telecommunications into private hands. Some others include

- Formation of a state enterprise (or a parastatal enterprise) that has a greater degree of autonomy than a governmental department
- Formation of a state-owned company with operational autonomy but is still subject to some governmental restrictions
- Formation of a mixed-ownership enterprise with either the state or the private sector owning a majority of shares
- Formation of a mixed-ownership enterprise that begins as a wholly state-owned entity but is gradually sold to private interests through the sale of shares
- Collaboration between a public entity and a foreign company or consortium, often with the consortium assuming management control
- Formation of a joint venture with a foreign company or consortium
- Formulation of build-operate-and-turnover agreements (BOTs) in which a private entity builds facilities, operates them for a specific term, and turns them over to the government or telecommunications administration
- Use of fixed price sales in which an enterprise is sold to private interests for a specified sum

- Use of soft loans, in which private investors are repaid through a share of revenues generated
- Use of a management contract with a private entity given operational control over the enterprise[13]

A country can achieve privatization in other ways as well. Some countries allow the construction of private networks, which is a form of privatization. Selling an enterprise to its employees is also a privatization move.

Privatization Approaches

I. Private Ownership

   A. Ownership by one entity

   B. Ownership by multiple stockholders purchasing publicly traded equity shares

   C. Ownership by employees purchasing shares

   D. Cooperatives

II. Government Ownership

   A. A state enterprise rather than a governmental department

   B. A state-owned company with operational autonomy

III. Mixed Ownership

   A. Shares owned by both government and private entities

IV. Consortia and Joint Ventures

   A. Significant issues of management control

   B. Issues of ultimate ownership of facilities

V. Build-Operate-Turnover Arrangements

   A. Management in private hands through concession or franchise

   B. Ultimate ownership in public rather than private hands

VI. Management Contracts

   A. Privatization of day-to-day management rather than privatization of ownership

All approaches are evident in today's telecommunications market. The United Kingdom, for example, is second only to the United States in privatization efforts, having privatized its telecommunications administration through a mixed-ownership arrangement.

The British Telecommunications Administration (BT) was privatized in 1984 when slightly more than 50 percent was sold through a public stock offering. The Thatcher government knew that they would need a large amount of capital to build a modern telecommunications infrastructure. BT wasn't operating efficiently in the early 1980s. According to one estimate, while the French PTT served 125 subscribers for each employee and Bell Atlantic in the United States served 181, BT's subscriber to employee ratio was 83 to 1.[14]

Rules drafted for the newly created private organization contained provisions designed to keep a broad ownership base along with continued government participation. Shareholders were limited to owning no more than 15 percent of the shares; the chief operating officer had to be a British citizen; two of the nonexecutive directors had to be government appointed; and the government held a "golden share" in the organization. This golden share assured that the government could prevent any unwanted changes in the purpose and major undertakings of the company. After a public relations effort to interest potential investors, shares were offered for sale in 1984. Over two million shareholders, including 96 percent of BT's employees, provided over four billion pounds to the British government.[15] The privatized BT continues to provide revenues to the government through dividends on the government's remaining ownership interest and through taxes on BT profits.

The British could privatize through the sale of equity shares because they had a strong capital market. Nations without such an asset usually privatize through foreign capital. The Venezuelan government recently sold a 40 percent equity share in its telephone administration, Compañía Anónima Nacional Teléfonos de Venezuela (CANTV), to a consortium led by GTE Corporation. In addition to a 40 percent equity share, the consortium, which includes Telefonica Internacional de España S.A. and AT&T, also received management control of the enterprise. The Venezuelan government retained a 49 percent equity share and reserved an 11 percent share for the employees of CANTV. The government named four members of CANTV's Board of Directors, and the consortium named five.[16] The Venezuelan government retains a major equity share in the enterprise and a voice in its governance. In addition to capital infusion, the Venezuelan government gained management expertise to run the new operation.

Government-owned enterprises can privatize without selling any equity interests to private concerns. An example of such an arrangement is a BOT. In Thailand, a consortium named TelecomAsia has formulated a BOT arrangement to build a two million line digital overlay network for Bangkok. This network will connect with both the telecommunications administration's domestic network and the country's international gateway. As segments of the overlay network are completed, ownership will be transferred to the Thai telecommunications administration. TelecomAsia will recoup its investment through revenues generated by a 25-year operating and maintenance agreement.

Many varied examples of privatization activities can be cited, but an underlying issue of each is management control. For nations with developed capital markets, such as the United Kingdom, the privatization process can include limitations on foreign ownership and control. Industrialized nations have a domestic work force skilled in modern managerial, financial, and marketing techniques. This is often not the case with less developed nations, for whom the prospect of privatization can present some pitfalls, as well as benefits.

Impact on Developing Nations

The domestic needs of many developing nations do not agree with the interests of a totally profit-oriented enterprise. Several developing countries are far from approaching universal service. In many developing nations, government—including the telecommunications administration—represents a major source of employment for the population. These nations can enhance their telecommunications infrastructure by turning to foreign investors for capital and management expertise, but they run the risk of not fulfilling other policy goals.

In 1987, four of the Association of Southeast Asian Nations (ASEAN) countries (Indonesia, Malaysia, Thailand, and Singapore) employed more than 104,000 people on their combined telecommunications staffs.[17] During this same period, the four nations reported fewer than 3.7 million main lines (about 36 lines per employee). Although such a large number might reflect overstaffing,[18] cutting staffs could cause problems for the domestic economy, especially if alternative job opportunities aren't available. In its privatization efforts, Sri Lanka sought "large-scale training programs" from a new network operator.[19] Without such provisions, foreign network operators are more likely to import labor than to retrain domestic workers.

Recent developments in Puerto Rico illustrate that the interests of foreign investors in newly privatized telecommunications enterprises might not coincide with governmental goals. Hoping to raise $2 billion for education and infrastructure building, the Puerto Rican governor sought to sell the government-owned Puerto Rico Telephone Company, along with its associated long distance, cellular, and telecommunications equipment manufacturing concerns. Although buyers were interested in the long distance, cellular, and equipment manufacturing organizations, few were interested in the local telephone company. Terms imposed by the Puerto Rican government caused this lack of interest. In addition to paying a minimum of $2 billion, the potential foreign buyer was expected to assume another $1 billion in debt, assure that no employees would lose jobs or retirement benefits, and agree not to raise basic service rates for at least three years.[20] What the Puerto Rican governor regarded as desirable policy goals, potential investors regarded as detriments to profitability and, therefore, deterrents to a sale.

The benefits and pitfalls of seeking foreign ownership have not escaped Peter Scherer, an official of the World Bank. Scherer praises the Latin American countries for securing foreign capital, attracting experienced operators, getting new technology into the area, and introducing a new business orientation into the operation of their telecommunications administrations. In Latin America, nations have followed the pattern of selling controlling interests in their telecommunications administrations to foreign operating companies and selling remaining shares to a large number of investors. The World Bank official also points out the following issues and concerns, which have been raised about the process:

- Most new operators were granted monopoly privileges, so no competitive forces are at work to encourage efficiency
- Some concerns arise that foreign operators might be more interested in short-term profits than in long-term gains
- Substantial tariff increases were required to make sales attractive to the new operators
- So far, a relatively small share of proceeds from the sales have actually been reinvested in the telecommunications companies, which must depend on self-generated funds for future expansion
- Building up effective regulatory agencies to oversee the new operators is a long and difficult task[21]

These sales have resulted in higher service rates, diversion of funds away from the infrastructure, and concern about the long-term commitment of the new operators. (These sales have produced these effects, but not all sales turn out this way.)

For some developing nations, privatization doesn't present a necessarily positive alternative; however, many have no choice but to pursue some form of privatization. Users continue to ask for new advanced services. International companies require a high level of service from their host nations. Developments in industrialized nations continue to raise technological standards. These factors contribute to developing nations' needs for capital and management expertise. To accomplish these goals, developing countries must continue to pursue privatization.

Steps to Privatization

Privatization can be a long and complex process. As a first step, nations must determine how much they want to privatize and what approach they are willing to adopt. Some necessary basic actions include:

- At a minimum, making the telecommunications administration a separate entity if it is now part of another function, such as the post office
- Determining the form a new telecommunications administration will adopt (state-owned, mixed-ownership, totally private)
- Passing legislation to make the transformation possible
- Creating a regulatory entity to oversee the new operation

In many ways, the last step can be the most crucial. In moving from a government-owned enterprise that is the embodiment of government policy to a more private operation with more tenuous links to policy goals, a nation needs a regulatory body to ease the transition.

A regulatory body can oversee pricing, terms and conditions of service, quality of service, and industry structure. Regulatory entities can enforce franchise requirements or facilitate competitive entry. Because no country as yet has a fully competitive, market-driven telecommunications marketplace, regulators guard against such abuses as monopoly pricing and unfair competitive practices. A regulatory agency can balance the sometimes diverging interests of customers and service providers.

Regulation of a private enterprise works only if the enterprise perceives it as fair. In Jamaica, for example, fairness is implemented through an operating license to Telecommunications of Jamaica (TOJ), a joint venture between the Government of Jamaica and Cable and Wireless PLC. The license substantially limits the discretion of the government and TOJ's ability to determine pricing. The former is necessary to protect the company from arbitrary and politically motivated directives from the government, and the latter requirement protects the people from the pricing decisions of a monopoly provider. And if necessary, disputes can be referred to courts in the United Kingdom.[22]

Privatization isn't a panacea that ensures a robust telecommunications industry, a sophisticated infrastructure, and a wealth of advanced services and features; however, countries that embrace some form of privatization are trying to accomplish those very things. Most nations are engaging in some privatization of their telecommunications markets. Of the twelve European Community (EC) countries, only four—Germany, France, Belgium, and Luxembourg—still had publicly owned telecommunications administrations in early 1992. In the Caribbean, a host of privatization efforts have been underway. In Barbados, Belize, Grenada, Jamaica, and Trinidad and Tobago, the telecommunications enterprises have been partially privatized. Costa Rica and Guatemala have undertaken another form of privatization, contracting out services or providing concessions to private companies, for the cellular and teleport services.[23]

The telecommunications manager faces a patchwork of varying privatization efforts and strategies. You should understand that a variety of arrangements are possible, what these arrangements are, and how they function. A nation's approach to privatization affects the industry structure, the orientation of the telecommunications enterprise, the entities involved—both commercial and regulatory—and the range of facilities and services available.

Competition Versus Monopoly

An increasingly important theme in both national and international telecommunications is the growing tension between the traditional monopoly and emerging competitive modes of service provision. For most of its history, telecommunications has been regarded as a natural monopoly with one

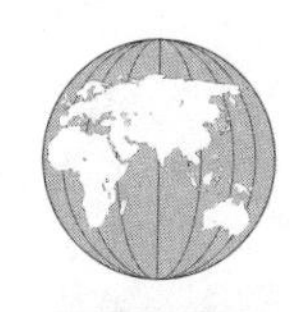

government-owned or privately held but closely regulated provider offering voice service over an analog network. In recent decades, however, irreversible forces have begun to place pressure on that simpler approach.

Technology has opened the door to new services and has enabled alternative providers to offer those services economically. As business has come to recognize the importance of telecommunications as a strategic tool, business users have become more sophisticated and more demanding in their requests for services. Both technological forces and customer demand are pushing toward a competitive approach to the provision of services, an approach that could have significant consequences for the continued viability of public switched networks (PSNs), universal service, and industry structure.

Not surprisingly, early telephone networks were regarded as natural monopolies, especially after the idea of a universally accessible network was articulated. The investment required to build central offices, to install wire in the network and customer's premises, and to purchase, install, and maintain switching equipment for a national network represented a significant barrier to entry. Small companies might be able to interconnect a few locations and establish a few isolated networks, but the problems of coordinating the interconnection of many small, parallel networks suggested that true economies of scale existed in limiting the number of network providers. Certainly the notion of several networks presented the prospects of confusion and duplication of resources.

The benefits associated with a large telephone network reinforced the notion that true economies were available through consolidation: as the number of people on a network increases, the value of connection to that network increases because more people and businesses can be called.[24] One large network therefore offered greater value as a whole and to each user. One large network also presented the opportunity to establish universal service. If universal service is based on offering rates low enough for virtually all customers to afford while generating sufficient revenues to build an infrastructure capable of serving virtually all customers, a large, monopoly-based network offers the opportunity to accomplish that goal.

Pricing Issues

The key to building universal service is pricing. In their analysis of the swing between competition and monopoly in U.S. telecommunications policymaking, Oettinger and Weinhaus note that, "On one side lie monopoly, universal connectability, and averaging of prices; on the other side lie competition, specialized use, and deaveraging of prices."[25] It isn't at all surprising that pricing issues weigh heavily on both sides of the debate.

Monopoly pricing of PSN services isn't concerned with matching price and cost. Rather, pricing strategies are based on averaging rates and creating cross-subsidies designed to ensure affordable connection to as large a number of subscribers as possible. Through geographic averaging, rates are the same for calling on all routes of equal length, despite the variable traffic levels among those routes. In this way, connections to remote areas are as affordable as connections to densely populated areas. By creating cross-subsidies between residential and business services, monopoly providers can maintain more affordable residential rates.

The monopoly approach has definite beneficiaries, especially rural areas and residential customers. But not everyone benefits from the system. Long distance, urban, and business customers pay relatively higher prices for their services. Many of those customers, especially larger business customers, have begun to seek alternatives to highly priced services. As larger users find alternatives and bypass the PSN, the monopoly system of pricing begins to fail. When the customers who pay the subsidies leave the network, rates for remaining customers increase. In response to this situation, monopoly providers are beginning to lower subsidies and move prices closer to cost. As a result, formerly subsidized rates (such as local service costs) increase, and competitive service rates (such as long distance rates) decrease. So, price pressure from larger customers is moving PSN providers away from pure monopoly pricing patterns.

Technology and Innovation

Monopoly pricing has not only given large business users an incentive to seek alternatives to PSN, it has also made potential competitors eager to provide those alternatives. Services priced well above cost offer an attractive margin. Competitors can undercut PSN prices and still reap a generous profit.

Alternative providers don't need to compete solely on price. The monopoly approach to PSN is based not only on averaged prices but to some extent on averaged services. Proponents of competition cite the tendency to spur innovation as one of its main advantages, and evidence does support that contention. For example, in the United States, customer premise equipment (CPE) wasn't feature rich before deregulation. Features such as redial, abbreviated

dialing, and message recording weren't available before deregulation. With no installed base of equipment, alternative providers are poised to move into new services and markets quickly. PSN providers might not be as quick to deploy new technologies, such as fiber-optic facilities, because they have a large installed base of copper facilities. They need to recover the last of those existing facilities. New alternative access providers in the United States, such as Teleport and Metropolitan Fiber Systems, face no such problem and have established all-fiber facilities.

Traditionally, a dichotomy existed between voice and data services because they have very different requirements. With new technology, these services are converging, which encourages competition. Most public data networks used separate switching facilities, called *packet switches*, to handle data traffic and different types, called *circuit switches*, to handle voice traffic. (See Chapter 5, "Technical Issues," for further discussion of these technologies.)

Many policymakers felt voice and data merited different regulatory treatment, partly because voice was the traditional domain of regulation, and data was not. Voice, after all, was often associated with certain social objectives, such as universal service requirements and regulation. In contrast, data services were almost always limited to commercial ventures and hence outside the regulators' traditional scope of authority. In addition, early data services were provided over private networks, which were usually unregulated. When public data networks emerged in the 1970s and 1980s, regulators were generally inclined to treat them differently than voice networks and, to the extent that an interest in competition developed, to experiment with competition in the public data networks.

As more fully discussed in Chapter 5, the technological distinctions between voice and data are blurry. Voice networks, once exclusively analog, now consist of digital switches and fiber-optic transmission facilities. Telephone switches are effectively specialized, software-driven computer systems. Data users now move large amounts of data over packet-switched and circuit-switched networks. In most cases, voice and data networks actually occupy the same physical facilities. For example, ISDN provides the mechanism for delivering voice and data to a customer's location over one set of wires.

If the distinction between voice and data has become fuzzy, so too are distinctions between the providers of those services. In many countries, the PSN operator provides services in competition with value-added network (VAN) providers. Both public and private packet-switched networks exist. Any distinction between providing basic service (merely transmitting a message) and providing an enhanced service (somehow augmenting the message) is difficult to determine. Because of this convergence of voice and data, questions

invariably arise regarding what constitutes the PSN, what type of competitors are possible for PSN services, and how viable such competition can and should be.

Industry Structure

As PSN operators begin to deliver services once provided by VANs, such as electronic mail and database access, competitors seek to move into services that once were the exclusive territory of the PSN operator. For the sake of simplicity, we can say that the basic PSN comprises six components:

- The terminal equipment (telephone) at the premises, also called customer premises equipment (CPE)
- The wiring at the premises
- The local loop connecting the premise to a local switching facility
- The local switching facility itself
- Transmission facilities linking switches
- Switches higher up in the switching hierarchy (Refer to Chapter 5 for more detail.)

Figure 2.2 illustrates these six basic components. As a result of technological advances and customer demand, competition can, and in many countries does, occur in each of these components. The result is a drastic change in the structure of the telecommunications industry.

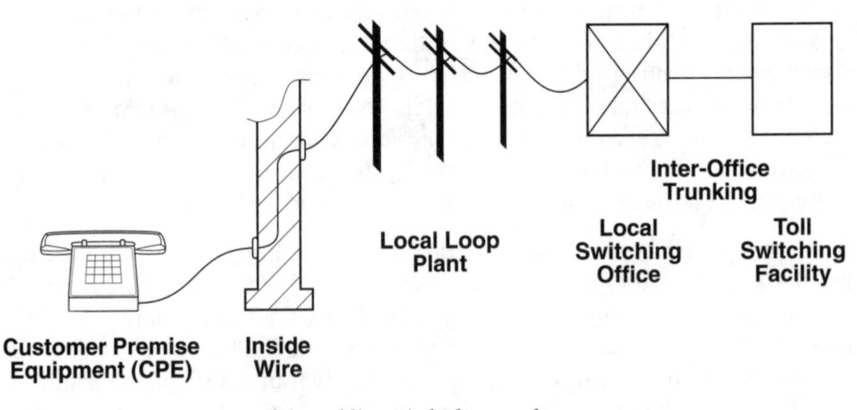

Figure 2.2. Six components of the public-switched network.

The United States has been at the forefront of competitive inroads. Competition began in the United States at the peripheries of the PSN, in the CPE and long distance markets. In the CPE area, the FCC stopped early efforts to do something as simple as attaching a noise-reducing, plastic, cuplike device to a telephone for fear of damage to the PSN.[26] However, after it created rules specifying equipment requirements and requiring that all hard-wired connections be replaced with modular connections, the FCC deregulated CPE and premise wiring. Deregulation of this peripheral market resulted in the creation of many CPE manufacturers and vendors.

After a lengthy series of court cases and FCC proceedings, competition moved further into the network. Today, long distance services in the United States are provided by several interexchange carriers (IXC), including AT&T. The deployment of software-driven, local, switching facilities made it possible for local telephone companies to route long distance calls to the customer's selected IXC. This development has resulted in the creation of hundreds of IXCs and a strong distinction between local telephone companies and long distance carriers. Local telephone companies provide local service and access to long distance companies, or IXCs; long distance carriers haul traffic between local telephone companies but don't provide local service themselves.

In effect, CPE and long distance competition have redefined the PSN and made it a smaller part of the telephone system. The traditional, regulated PSN now consists of the loop, the local switching office, and the transmission facilities that link the local switching office to other such offices and to all IXC facilities. This shrinkage of the PSN has resulted in a smaller scope of universal service, which no longer includes a telephone set and which, for all practical purposes, includes *access* to an IXC, not necessarily IXC service.[27]

Perhaps inevitably, competition is emerging in the remaining portions of the PSN. After competition is allowed at the peripheries of the network and connection at those peripheries causes no noticeable harm to the PSN, the boundary beyond which competition is prohibited becomes less distinct. Alternative local access providers are seeking to compete in the transmission areas of the PSN, both for the facilities that link local switches to IXCs and for loop facilities that link customer premises to local switching facilities. This final competitive thrust into the smaller PSN presents some interesting questions about the continued viability of a traditional, monopoly-based PSN. Figure 2.3 illustrates the provision of competitive local access.

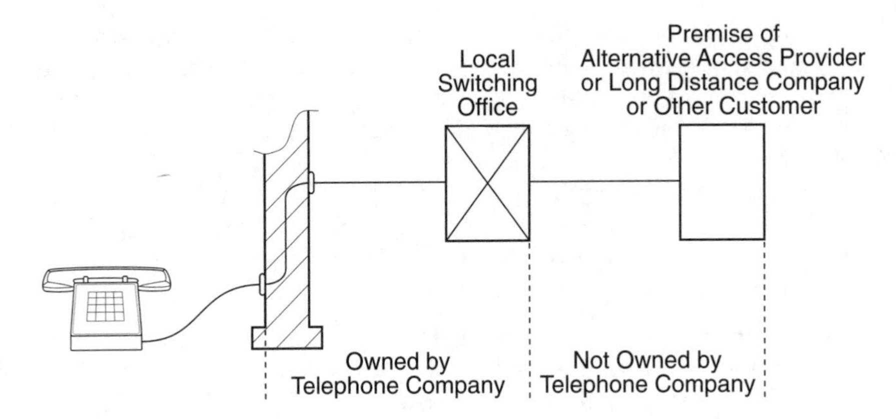

Figure 2.3. *Provision of competitive access service.*

As the U.S. experience illustrates, competition is difficult to prevent when technologies exist that interconnect the distinct facilities of several competitive service providers. The new competitive environment requires willing competitors and a regulatory approach that regards market-based competition as a superior method of regulation. The U.S. experience also illustrates continuing tension between the respective benefits of monopoly and competition. Although competition has introduced feature rich CPE, lower long distance rates, and all-fiber, alternative local access facilities, growing competition has put more pressure on local rates and on the concept of a universal, unified network.

Alternative access providers in effect bypass one portion of the PSN. Telecommunications users can bypass the PSN completely. In the United States, virtually no restrictions are placed on bypassing the PSN through the construction of private networks. Those entities building private networks may need to get right-of-way and zoning permission; however, permission from regulatory bodies such as the FCC or state commissions is usually not necessary. Companies that already have multiple locations and have already taken care of right-of-way problems—railroads are an excellent example—can install their own telecommunications network to connect their own locations. In the United States, when these private networks want to connect with the PSN in order to communicate with locations outside of their network, they can. This capability of connecting with the PSN enhances the attractiveness of private networks. The company can communicate over its own network at virtually no expense to meets its internal communication needs and can access the PSN to meet its external needs. If connection of private networks

to the PSN is not allowed, as is the case in many nations, the attractiveness of bypass through private networks is reduced.

Bypass is a form of competition to the PSN because it provides alternatives to PSN services. This form of competition has also placed pressure on the PSN provider to lower rates and offer more services in order to keep users on the PSN.

Many nations regard the U.S. introduction of competition to the telecommunications arena as a successful model to emulate.[28] Others have taken a different approach in defining the line between monopoly services and competitive offerings. The U.S. model creates a dichotomy between basic local service and long distance. As a result, competition was first allowed in a specific sector of the industry. Although a nucleus of monopoly is maintained (at least for now) at the local level, with local companies still holding exclusive franchises, the FCC and some state commissions have allowed competition to take place at the other side of the local switching facility. Japan follows this model to an extent. Nippon Telephone and Telegraph (NTT) provides its customers with local service along with access to competitive long distance providers, called new common carriers (NCCs).

The EC appears to adopt this same dichotomy between basic services and competitive offerings in the Green Paper, the plans for telecommunications in the EC. One of the Green Paper's provisions is that a monopoly should be maintained only for "a limited number of basic services," with other services open to competition.[29] Through a series of directives, the EC Commission has sought to:

- Open the terminal equipment market to competition
- Establish an EC-wide telecommunications market through the creation of open network provisioning (ONP) arrangements designed to assure access and interconnection to the PSN in all member nations
- Open data-transmission, mobile, and cellular markets to competitive entry
- Encourage PTTs to establish regulatory entities separate from network operators

At the core of the network, however, are basic services that could retain their monopoly status. Even this area of the network is being considered for further liberalization by the EC Commission, though this examination is being met with substantial resistance. Indeed, all efforts of the EC to liberalize the telecommunications market have met with some resistance by member nations.[30]

The United Kingdom, in its early experiments with competition, chose a very different model from the central monopoly of basic services. Rather than dividing the industry into local and long distance, the United Kingdom divided it between two service providers, BT and Mercury. The implications of such an approach are interesting. BT retained a universal service obligation so that, unlike the U.S. experience, the range of universal service is fairly broad. However, the British experiment hasn't fulfilled one of its basic objectives. Duopoly hasn't provided enough impetus to improve BT efficiency and innovation, so the U.K. government has decided to replace duopoly with broader competition.

According to Kenneth Baker, British Minister for Information Technology, in establishing a policy of "fixed links duopoly," the British sought to "avoid market disruption" and provide "an adequate opportunity to adjust...to the competitive environment."[31] Only BT and Mercury were licensed as providers of basic telecommunications services. Other entities couldn't use PSN circuits, leased from BT or Mercury, to resell switched services. BT was expected to provide sufficient interconnection to Mercury so that customers could reach any party regardless of its network provider. Despite these provisions, however, BT retained significant market dominance. In 1989, BT's share of revenues from switched services had declined to 96.2 percent from a total share of 99.8 percent in 1980, a drop of only 3.6 percentage points.[32] Although the core network was kept in the hands of two service providers, value added services and CPE were opened to competition. As an interesting side note, Australia is now pursuing a duopoly approach as a transitional step in its move toward competition and privatization.

Other nations draw the boundary between monopoly and competition farther out in the peripheries. France allows very little competition, most of it from VANs. The German telecommunications network has remained, for the most part, a monopoly although efforts have been made to separate regulatory and oversight functions from operational activities. Competition is allowed in such areas as radio paging, mobile radio, and cellular services. Private satellite systems are also allowed. Private entities can, through the use of leased circuits, offer services to third parties. However, these services can't include voice communications.

Farthest from the U.S. model are those nations that maintain a full monopoly. An example is Singapore, where, according to Mark A. Hukill and Meheroo Jussawalla, "It is not likely that services, even nonbasic services such as paging, will be opened to competition in the next several years."[33] Singapore's reluctance to open its telecommunications industry to competition might stem

from its ability to meet business needs under a monopoly scheme. Unlike most other nations, Singapore has no rural areas and can concentrate on providing service to a geographically small location with a dense population.

The Continuing Debate

At the base of any discussion of monopoly and competition lies the tension between universal service concerns and the benefits of innovation and efficiency promised by competition. Monopoly provides average rates and services for the many; competition provides innovation, lower prices, and specialized services.

When deciding where along the continuum between total monopoly and total market-based competition a nation wishes to place its telecommunications industry, policy makers face many questions. Prominent among them is whether, and how, to maintain universal service and a viable PSN. If the continuation of subsidies is no longer possible through monopoly pricing, nations can levy universal service charges on competitive service providers.

Nations face the concern that PSN providers, as they offer both competitive and noncompetitive services, might take advantage of their monopoly profits to underwrite competitive ventures. For this reason, many countries require PSN providers to offer competitive services through separate subsidiaries or to provide some other means of assurance that a barrier is maintained between monopoly and competitive revenue streams.

The decision that is made has a profound effect on the telecommunications structure in that country, the services available, who offers those services, and service pricing. These are all issues of concern to those who manage international networks. Because uniformity of decisions isn't likely among nations, the whole issue of competition and monopoly presents even more complexities to the international telecommunications manager.

Tariff Structures

Globalization has made tariffs for telecommunications an international issue. Tariffs are approved by regulatory agencies and determine which services are available from the regulated entity, the terms and conditions for ordering and receiving those services, the pricing structure, and amounts charged

for service. Tariffs also identify, by their absence, the services not commonly offered, at least by the regulated entity. In effect, tariffs are a menu offered by PSNs to guide a telecommunications manager through hundreds of choices. Tariffs are usually a published document listing the services the regulator will allow the network provider to offer, the prices which the regulator will let the provider change, the duties and responsibilities of the customer for using the service, and the duties and responsibilities of the provider in offering those services. If tariffs were similar across national boundaries, the process of designing and managing an international network would be much simpler. However, national tariffs aren't similar in pricing structure, the range of services offered, or price, making the telecommunications manager's job much more difficult and complex.

Tariffs do have underlying similarities. The same types of services are usually offered. Though pricing structures differ, they usually follow some standard formats. The differences among tariffs are usually related to price levels, flexibility in accommodating customers' special needs, and the range of services available. All these similarities and differences present telecommunications managers with strategic questions about the most efficient and cost-effective services in each country, the effectiveness of looking for bypass opportunities, and the optimal routes to choose for a network. Tariffs are one aspect of telecommunications that a manager can't afford to ignore.

In most instances, tariffs are required from regulated PSN operators. Some non-PSN services can also be subject to tariffs.[34] In most cases, however, tariffs apply to services offered through public networks, including public packet-switched data networks, and to nonswitched facilities like leased lines.

Tariffs are designed to accomplish several purposes. They direct what a regulated entity can offer customers, on a nondiscriminatory basis, at a specific price. They specify the obligations of both customer and service provider. They also define the parameters of the services offered; for example, the bandwidth and grade of service of a specific type of leased circuit.

As one major purpose, tariffs generate adequate revenues to assure that the PSN remain a viable entity. This makes tariff pricing a highly significant issue. As the CCITT recognizes, the primary objective of tariffs is to accomplish the following:

> *In principle, the prime objective in setting charges for the national telephone system is to recover the cost of providing the service, including running costs, depreciation and a suitable return on the capital investment. The return on capital is usually that agreed with or allowed by a Regulatory Body, normally the Government. This basic principle can*

equitably be applied to individual components of the national telephone tariff, although in practice economic and political structures generally preclude such an absolute approach and cross-subsidy between individual components of the tariff almost invariably applies.[35]

Cross-subsidies, and their impact on specific services, have significant implications for those who order and use services that are subject to tariffs.

Overview of Services and Price Structures

Tariffs present a catalog of what is available, on a common basis, to all customers of the PSN operator. Tariffs also indicate what isn't available through the regulated entity. The customer must then determine whether such services are available through alternate providers. Tariffs offer the customer a starting point from which to determine whether services subject to tariffs are sufficient, or whether other means should be explored to meet the customer's telecommunications needs.

Telecommunications services fall into several basic categories, including:

- Local service
- National long distance
- International long distance
- Leased lines
- Packet switching
- Value added services
- Mobile telephone services
- Customer premise (or terminal) equipment

Another type of telecommunications service has developed since the introduction of competition. As new companies are allowed to provide such services as long distance or cellular communications, they need connection to the PSN in order to reach their customers or complete their service. Arrangements, either through contracts or tariffs, need to be made so that interconnection is provided and compensation is made to the PSN operator.

General Telecommunications Service Categories

I. Local Service: provision of a line between the customer and a switching facility for placing calls inside a defined local calling area

II. National Long Distance: calling outside the local calling area but inside national boundaries

III. International Long Distance: calling beyond national boundaries

IV. Leased Lines: the provision of dedicated lines for use by a specific customer

V. Packet Switching: efficient method for transmitting data from several customers over one facility

VI. Value Added Services: enhanced services, usually provided by entities other than a telephone company, that offer more than just the transmission of a message

VII. Mobile Telephone Services: telephone-type services provided on a mobile basis, either through radio or cellular technology

VIII. Customer Premise Equipment: the terminal device at the end of a circuit or loop

IX. Access or Interconnection Charges: a type of compensation arrangement in which carrier A pays carrier B for the use of B's facilities

Some of these services aren't subject to tariffs. The tariffs applied to others may be widely divergent. For example, the treatment of CPE varies widely. In some nations, all CPE is deregulated, or not subject to tariff; in other nations, CPE is provided solely by the telecommunications administration.

Even though CPE may be offered on a nontariff basis in some nations, it must undergo certain approval processes before it can be marketed in a specific country. Because the manufacture and sale of CPE represents a potent revenue source for many nations, liberalization of the CPE market hasn't been instantaneous. Most nations have significant incentive to buy domestically manufactured CPE rather than consider bids from a variety of providers. When PTTs were the monopoly operators of government-owned networks, their usual approach was to adopt procurement policies limiting purchases to nationally produced equipment.

Even nations that have moved away from this policy maintain a process of type approval in which CPE must meet specific technical conditions to be certified for use. With the separation of regulatory and operational entities, this type approval process is no longer under the control of the network provider that purchases the equipment. For example, type approval in Germany is done by a department of the Ministry of Posts and Telecommunications, which is separate from Deutsche Telekom. In the United States, which is a very liberal telecommunications marketplace, type approval consists of meeting the rules and regulations set forth by the FCC in its Part 68 Rules. A major initiative of the EC is the adoption of directives that liberalize the type approval process among EC nations and open procurement procedures so that equipment from all EC member nations is treated equally across the Community.

Value added services and networks tend not to be subject to tariffs. VANs are a reflection of an underlying dichotomy in telecommunications: the split between voice and data services (discussed earlier in this chapter). With increasing digitization of the voice network in switching and transmission, this distinction between voice and data is no longer clear. Data can be routed through PBXs; data facilities can be switched at the circuit level; packet-switching technology makes it possible to move data over both private and public packet networks.

The convergence of voice and data resulted in an uneasy regulatory and tariff situation. Regulators have difficulty determining where the line between voice and data services should be drawn.[36] At the base of the argument is the theory that voice communication is the proper realm of the regulator. Voice communication is the simple transmission of a message from one point to another. Nothing is done to the message; it isn't changed or affected in any way. It is merely transmitted. This compromises traditional regulated phone service. With the introduction of computerization to the voice network, the network can do more than just transmit a message. However, anything beyond pure transmission is more than traditional voice communication.

Customers who want more than just pure transmission (such as leased telephone circuits) can obtain those additional services from VAN providers. VAN providers often resell basic transmission facilities. In other words, they lease lines or purchase services from the PSN provider and then offer their enhanced services over those PSN facilities. VAN providers aren't usually regulated, and their services aren't subject to tariff. They do, however, purchase services that are subject to tariffs.

Because of technological advances in their networks, PSN providers can transmit data and voice messages. They can provide circuit-switched data facilities and shared packet-switching networks. Whether these facilities carry voice, data, or images isn't relevant. The PSN provider merely supplies transmission and switching services, and—of significance to this discussion—these PSN services are subject to tariffs.

Message transmission services offered by PSN operators can take a variety of forms subject to tariff; both price structures and price levels can differ significantly. A PSN operator's approach to pricing in its tariffs says a great deal about the operator's strategy for attracting and keeping customers on the public network. Pricing strategy also offers a variety of incentives for customers.

For all practical purposes, switched voice services come in three varieties: local, national long distance, and international long distance. The underlying characteristics of each are the same. A connection is made between an originating party and a terminating party for the duration of the call. The originating party dials the number of the terminating party; transmission facilities are provided by the PSN between the caller and the switching facility that serves the caller; transmission facilities are provided by the PSN between the called party and the switch serving that party. For local calls, the same switching facility serves both parties. For national long distance calls, routing and transmission services are provided by the PSN or by the long distance company between the two switching facilities. For international calls, significant routing and transmission facilities are required to carry the message across national borders and, perhaps, continents. Figure 2.4 illustrates how the telephone network handles each of these call types.

Local Service

Local service comprises the use of a line between a customer premise and a switching facility, the use of a telephone number that represents the customer's address on the telephone network, and the use of switching facilities that route the customer's calls to other customers within the local calling area. The local calling area might encompass only those customers served by the same switch. In communities large enough to contain several switching facilities, the local calling area might contain customers served by any of the switches in the community.

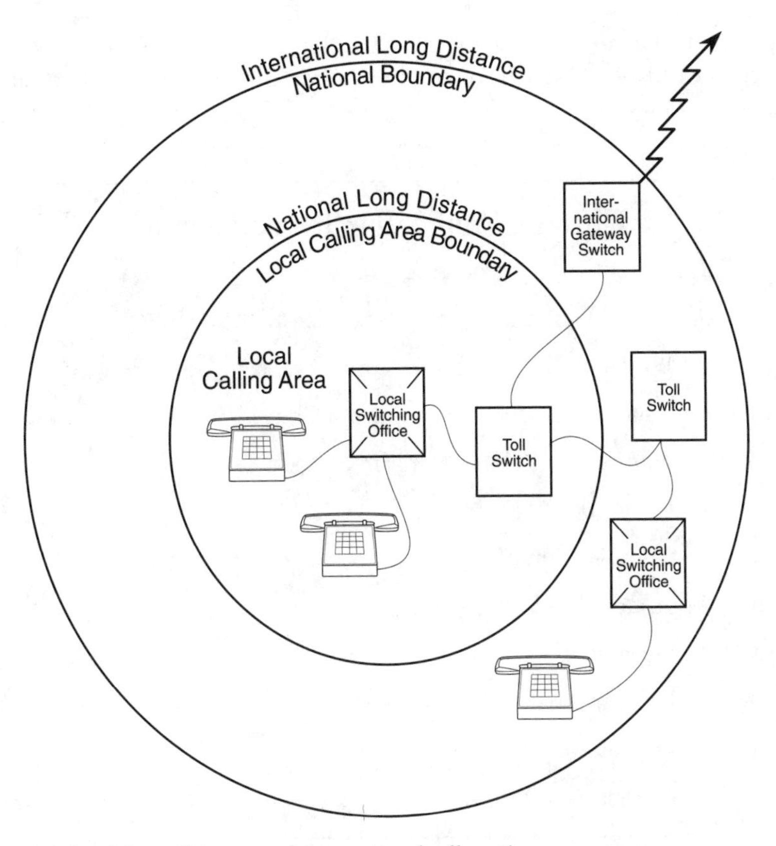

Figure 2.4. *Local, long distance, and international call routing.*

Local service charges include nonrecurring installation charges and recurring monthly charges. Installation charges can have an important effect on the number of subscribers that can connect to the PSN. High installation charges may present a barrier to entry and discourage connection. For nations with limited resources and limited capacity, high installation charges might not pose a problem because they deter demand for additional capacity. If a country has a strong commitment to universal service, it must have low installation charges.

Local service charges also include recurring monthly charges that can take several forms. They can entail a flat monthly charge and unlimited local calling. This billing structure encourages calling and thus the need to increase

capacity. However, because the basis for charging isn't use, the flat monthly charge could be raised in order to finance the additional costs of new facilities.

Differing Approaches to Local Service

I. Treatment of Nonrecurring Charges

A. High installation charge—barrier to entry

B. Low installation charge—universal servicestrategy

II. Treatment of Recurring (monthly) Charges

A. Flat monthly charge, unlimited local calling

B. Usage-based charges

C. Combination of flat rate plus usage component

III. Approaches to Usage Charging

A. Per message

B. Per minute

1. Time-of-day charging

2. Distance component

Another approach is to charge per unit of traffic. This is often referred to as *measured service*. Charging can be on a per-call or a per-minute basis. Per-call charging encourages calls of long duration. Per-minute charging tends to ration the duration of calls. Per-minute charges can also entail distance-sensitive and time-of-day components. Time-of-day charging is designed to encourage off-peak calling. Because facilities are sized to accommodate peak usage with minimum blockage, these facilities can be relatively idle at off-peak times. Prices sensitive to the time of day are intended to spread calling to less busy times, avoid blockage during peak times, and encourage the use of resources that would otherwise be idle. Figure 2.5 illustrates the typical daily variation of traffic in a network.

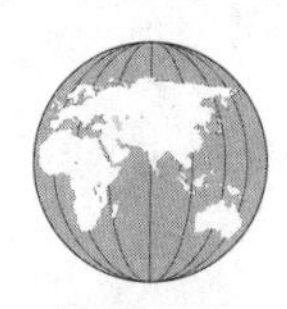

a

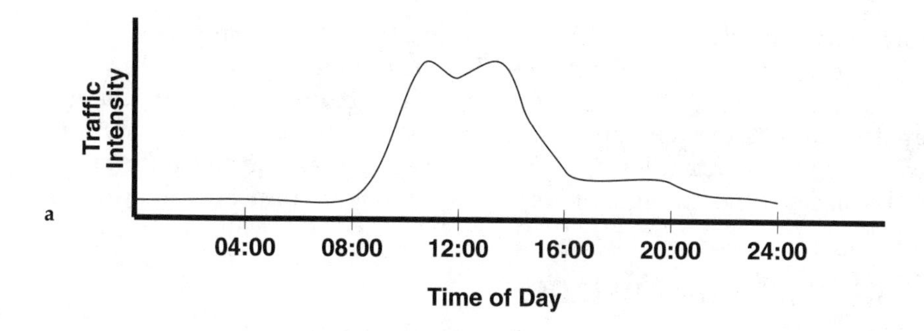

b

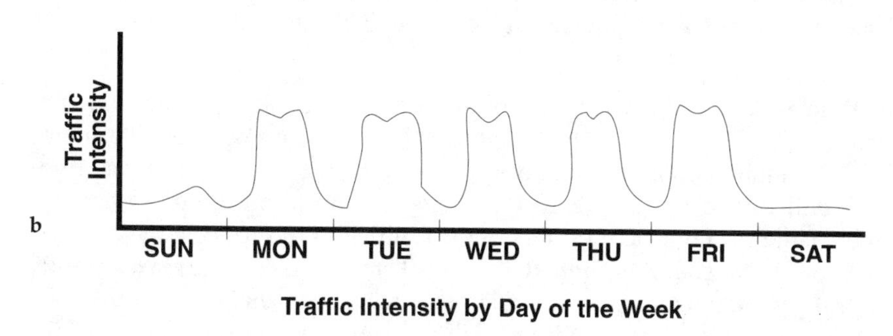

Figure 2.5. *Traffic density by time of day (weekday)(a). Traffic density by day of the week(b).*

Any type of usage-sensitive pricing requires metering equipment and more sophisticated billing capabilities. Usage-sensitive pricing is more reflective of true cost. Some economists argue that such pricing is optimal.[37] Interestingly, usage-based local service pricing isn't the norm in the United States, although measured service is prevalent in many other developed nations.

Local service represents the building block in providing access to the PSN. Without a local connection, customers can't call someone in their local serving area. Without a local connection, a customer can't access long distance switches and can't place a long distance call. The local connection represents access to telecommunications services.

Customers who don't have a local service connection, either because they don't want one or because the cost of connection is beyond their means, can access the PSN through a public telephone. In most nations, the network provider is required to provide public telephone service. In the United States,

for example, part of the franchise given to local telephone companies by state commissions contains a requirement that the local telephone company provide at least one public telephone in each local service area. Usage charges are levied for using the public telephone so that the network provider can cover the cost of providing one. These usage charges are levied at the time a call is placed, either through the depositing of coins or the use of calling cards.

National Long Distance

National long distance calls are usually charged on a usage-sensitive basis. Distance, measured in zones or bands, is usually a component of charging. The size and number of these zones differs by size of nation. In 1990, U.S. long distance companies employed twelve mileage bands, while Denmark had two.[38]

The time of day is also a component of long distance charging in many nations. Of the Organisation for Economic Cooperation and Development (OECD) members, only Luxembourg doesn't include a time-of-day component.[39] Both distance and time-of-day pricing involve implications for the consumer. Time-of-day pricing offers opportunities to reduce cost by strategic placement of calls. In this context, the use of store-and-forward faxes results in significant savings. Distance-sensitive pricing may make use of the PSN more costly for companies with widely dispersed locations because of increased long distance charges to reach farther points.

Some long distance rate tables reflect a difference in charges for the initial minute, or minutes, of a call and the call's additional minutes. Initial-minute charges are higher than rates for additional minutes. This difference in charging is meant to capture the cost of call setup. *Call setup* involves the time and resources required from the time a call is dialed until the connection is made with the called party.

Another component of long distance charging is the method used to establish the call. Rate distinctions are made between calls that are dialed directly by the calling customer and those that are established with the assistance of a telephone operator. The use of an operator increases the cost of the long distance call.

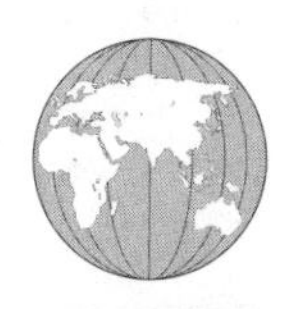

> **Elements of Long Distance Pricing**
>
> I. Usage-based charging
>
> II. Time-of-Day discounts: encourage off-peak usage
>
> III. Distance bands or zones
>
> IV. Differential between charges for initial minute and for additional minutes
>
> V. Operator service charging component

International Long Distance

International long distance calling presents a whole range of pricing alternatives. Whereas national long distance pricing is based on the cost structure of only one nation, international calling entails the use of facilities in at least two countries. Issues of pricing among countries involves more than simple time-of-day and distance factors.

International calling rates are based on call-pairs. Revenue sharing agreements exist between individual nations. The results of these sharing arrangements are reflected in long distance calling charges. These revenue sharing arrangements have many implications for the telecommunications manager. For a more detailed discussion, see the International Settlements section in Chapter 3, "Economic Issues."

In a discussion of tariffs and their effect on telecommunications managers' choices, one implication of real significance is the variation in cost for calling between nations. For high-cost routes, means other than the PSN, such as a private network or a VAN, might be better. The details of international pricing patterns often encourage the use of complex routes between countries to reduce costs. The cost-effective approach in some cases is to use a third country as an intermediate routing point because of favorable international tariffs. In Europe, the United Kingdom has played this role.

Packet Switching

In many countries, PSN providers can offer public packet-switching services. Packet-switching technology provides a means for using one facility to accommodate several customers. Rather than dedicating one circuit, or path, to a specific customer, packet switching allows the disassembling of

messages into packets that can be dispersed over a circuit and then reassembled. Because packets can be interspersed in time, several messages can share the same facility. Both ends of the message require a packet assembler/disassembler (PAD).

Packet switching presents an efficient means of moving data without the expense of leasing a dedicated line or acquiring private facilities. Packet-switching tariffs include installation and rental charges for the connection to the packet network and for other necessary interface equipment. Usage charges are usually based on volume—or number of packets sent—and duration of the connection.

The level of pricing for packet-switching tariffs indicates whether a nation encourages the use of the public network or the growth of private networks. High packet-switching prices, especially monthly lease charges for connection to the network, encourage the formation of private facilities.

Mobile Telephone

Mobile telephone services include fixed charges for equipment and distance-sensitive usage. Distance is usually a component and time of day can also be a factor. In the United States, mobile telephone subscribers must pay a charge, not only for calls they originate, but also for calls they receive. Thus, billing arrangements for cellular systems can vary by country and can differ from rates for regular telephone service.

Modern mobile telephone systems are called *cellular systems* because the service area is divided into distinct cells. Each cell has a base station, and mobile units within a cell communicate with it. As a mobile unit crosses a cell boundary, the call is handed off to the base station in the new cell. The cellular system is preferred to the more traditional single base station for the entire service area because frequencies can be reused inside the cellular system, allowing for a much higher cellular telephone density.

In some countries, such as the United States, different cellular operators can serve different cities. For cellular service to be considered universal, subscribers must be able to access the network regardless of their location. Subscribers not assigned to the local system are referred to as *roamers*. In Europe, efforts are being made to allow roaming between countries even though each country's system is administered separately. A single organization can provide cellular service nationwide.

As with other services, several provision modes for cellular service are possible. These can be privately or publicly owned, the provider can be a monopoly, or several providers in a service area can offer the service competitively. To encourage competitive price and service structures, the FCC in the United States has granted cellular licenses to two operators in each service area.

Telex

Though the use of facsimile (fax) machines is undoubtedly causing some decline in the use of telex, many nations still provide public telex services, which are subject to tariff. In developing nations where the PSN doesn't yet offer the type of infrastructure that can support fax, dial-up data, and other such services, telex is still an important service. Charging for telex usually involves an installation fee, monthly charges to cover connection to the network, monthly charges to cover equipment (unless equipment has been deregulated in that nation), and usage charges based on both duration and distance.

ISDN

Many countries, the developed countries in particular, are now offering ISDN services to their customers. For all countries, ISDN is a new service; many countries regard the deployment of ISDN as strategic. While the CCITT has defined many types of ISDN service, most countries offer only Basic Rate ISDN (BRI) and Primary Rate ISDN (PRI). BRI is equivalent to two voice channels and a data channel. PRI can offer 23 voice channels or a combination of higher speed channels (faster than a 64 kbps voice channel).

While ISDN pricing varies widely, a consensus seems to be emerging that BRI should be offered at prices no higher than 1.8 times the price of a voice channel. This gives users an incentive to adopt ISDN, even in the face of expensive terminal equipment. The pricing of PRI varies more than BRI. Some countries offer PRI as the only method for high-speed connectivity, eschewing high-speed digital leased lines.

Leased Lines

Tariffs for leased lines is a very strategic issue for most nations. Many customers construct dedicated networks for their own use through leased lines. Over these dedicated networks, customers can move voice, data, and video

without paying the usage-based charges that such services would incur on the PSN. Indeed, some of the services accommodated by these dedicated networks might not be available through the PSN. For example, video requires such great amounts of bandwidth that switched video services are, for all practical purposes, unavailable at present. Customers order leased lines by technology (analog or digital), grade of service (voice or data), speed, and bandwidth. They pay installation and fixed monthly fees but no usage-based charges.

The amount of freedom accorded customers—in the use of these leased facilities, in how easily these facilities are made available, and under what conditions these facilities can connect to the PSN—is a measure of how strongly that country wishes to keep traffic on the PSN. Indeed, some nations, Germany for example, have added volume charges for leased lines; and others, such as Switzerland, have included compensation payments for lost traffic to make leased lines a less attractive alternative to the PSN.

Types of Leased Lines and Their Features

Facilities dedicated to specific customer, nonswitched service

I. Analog Lines

A. Low Speed (up to 1,200 bits per second): telegraph grade, used for telex or other such message services

B. Medium Speed (2,400 to 9,600 bits per second): voice grade, used for voice; can be conditioned for data

C. High Speed (19.2 kilobits per second): wideband, used for videoconferencing, teleconferencing, digital data links

II. Digital Lines

A. Data channels (2,400 bits per second to 56 kilobits): used for digital data service

B. DS-1, U.S. and Japan (1.544 megabits per second; equivalent of 24 voice channels at 64 kilobits each): used for data, compressed video

C. E-1, European standard (2.048 megabits; equivalent of 30 voice grade channels)

D. DS-3, U.S., (equivalent of 672 voice channels): used for full motion video and movement of large amounts of data

III. Point-to-point Lines: ordered and provided between two specific points

IV. Multipoint Lines: ordered and provided between several locations

Value-of-Service Versus Cost-Based Pricing

Nations display little uniformity in tariffs of the services discussed here. As a recent study noted, "The actual tariff structure in any particular country is often quite arbitrary, and there are considerable differences between countries in the means by which charges are levied."[40] The study lists some reasons for such great disparity, citing the "imprint" of history rather than current technology; political factions that resist radical tariff changes; and the public service obligations of PSN providers. All these elements help explain why, instead of uniform charges, countries have ranges of charges, most of which bear little resemblance to underlying costs.

As this study points out, and as the CCITT makes very clear in its explanation of the basic objective behind tariff pricing, PSN operators don't take a cost-based approach to pricing. PSN prices were, and for the most part still are, based on a value-of-service concept.

As the CCITT statement (cited earlier in this chapter) indicates, PSN providers set prices to recover the cost of providing service, including a return on investment. However, the costs of providing services aren't necessarily easy to identify. The PSN provides a classic example of the problem of setting prices for jointly provided services. Most PSN services use the same equipment. Local service is offered over the same lines and through the same switches used for national and international long distance services. Allocating those facilities to each service is difficult, so accurately allocating costs is also difficult.

Moreover, PSN providers incur overhead costs, including those of administration, billing, and customer service that must be allocated to the services on the same basis. PSN providers have great flexibility in deciding how much overhead to allocate to each service. Indeed, when the state owned the PSN and when no competitors existed for PSN services, the issue of underlying

cost seemed unimportant. The important issues were to generate revenues (which were often used to subsidize other areas of the government) and to maintain the viability of the PSN.

In that context, value-of-service pricing was reasonable. Under that regimen, prices were set with an eye toward keeping certain services affordable and certain customers on the network at the expense of other services and customers. In short, pricing was built on a system of cross-subsidization, with prices determined by the ability to pay or the perceived value of the service, rather than underlying costs.

For this reason, though no technical difference exists between a regular line that terminates in a business and one that terminates in a residence, the charge for the business line is usually greater. Businesses are perceived not only as being better able to afford higher rates but also as finding greater value in the line because of its necessity for their use. As a corollary to this point, residential rates can be more affordable so the PSN can connect more customers to it.

Because greater numbers of subscribers enhance the value of the PSN (that is, the value of using the network increases if that network can reach more parties), this subsidization of residential customers by business users has some merit. Long distance calls, the majority of which are for business purposes, also provide subsidies to local service rates.

Another important subsidy stream flows from urban users to rural ones. The underlying costs are higher when providing service to more remote areas. Rural areas require more miles of facilities per user, and fewer users must bear switching costs than users in more densely populated areas. Rather than making rural customers bear the full costs of their service, PSN providers tend to geographically average rates, a practice the CCITT finds a "commonly recommended solution."[41]

The benefits of value-of-service pricing are clear for its beneficiaries; the drawbacks for those paying the subsidies are also clear. Large business customers, seeking ways to reduce their telecommunications expenses, pursue alternative services. Because of the high level of long distance charges and business rates, enough of a margin between price and cost exists to encourage competitive providers. In response to these competitive pressures, PSN operators began to phase out some subsidization and move closer to cost-based pricing. Here, as elsewhere, accurately allocating costs is difficult. In many cases, the cost of providing a circuit to the first customer in a region is higher than providing the same service to subsequent customers in that region because subsequent subscribers take advantage of systems already in

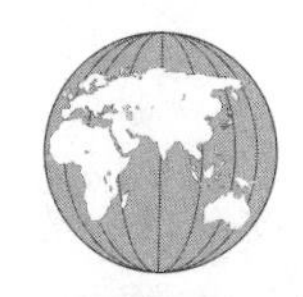

place. Although this is not a problem when rates are averaged, it can be when subscribers must pay the cost of the equipment needed to provide service, as is the case in Chile.[42]

This process of rate rebalancing suggests that local charges will increase, or at least not decrease relative to long distance rates. Evidence indicates that the trend has begun in many nations.[43]

Strategic Implications of Pricing Relationships

People who read national tariffs might find little uniformity, but they should look for certain patterns when determining the best approach for obtaining telecommunications services in a specific country. One important pattern is the treatment of leased lines. In some countries the use of leased lines is severely restricted. For example, until recently, German VAN providers weren't permitted to use leased lines in the provision of their services.[44] In the Netherlands, leased lines were provided only if the network had excess capacity.[45]

Beyond the availability and allowed uses of leased lines is the question of the break-even point between using the PSN and obtaining a leased line. If the PSN charges for long distance and usage-based local services are low and leased lines are high, the PSN provider's strategy is obviously to keep traffic on the PSN and discourage the creation of private networks. If the converse is true, leased lines are encouraged. A recent study found that the break-even points were lowest in the United Kingdom, Ireland, Australia, and the Scandinavian countries; and highest in Germany, Spain, and the Netherlands.[46]

The pricing relationship among types of leased lines can also be significant. In some countries, the pricing of digital lines can be so high in relation to analog facilities that it slows the deployment of digital facilities. Other important break-even points may involve line speed: the relative cost of a high-speed line when compared to lower speed facilities.

In addition to the relationship between business and residential rates, the relationship between long distance and local charges is also important. In rate rebalancing efforts designed to drive prices closer to cost, long distance charges should decline in relation to local rates. Because of technological changes in transmission media, distance doesn't appreciably affect cost. Cost-based pricing should, then, decrease distance sensitivity of long distance pricing and the disparity between usage-sensitive local rates and long distance charges.

The movement in pricing of international long distance is also significant. International rates are relatively high compared to national long distance charges. For those seeking to establish international networks, such pricing disparities provide incentives to seek alternative services such as private networks or leased lines.

Another element of interest should be the willingness of the PSN provider to offer services on an individual basis. Rates and services subject to tariffs are in many ways lowest-common-denominator services because they might lack desired features and because they can lag behind technological developments. Some PSN providers might be willing to accommodate the special needs of specific customers.

Factors to Look for in Reading Tariffs

I. Existence of pricing cross-subsidies
   - A. Pricing relationship between business rates and residential rates
   - B. Pricing of local charges versus level of national long distance rates
   - C. Relationship of international long distance rates to national long distance charges and local rates

II. Flexibility for meeting specific needs
   - A. Treatment of leased lines
     1. Availability
     2. Pricing in relation to switched services
     3. Pricing among types of leased lines
   - B. Availability of new services subject to tariffs (VPN, ISDN)
   - C. Possibility for Individual Case Basis offerings

New services also require new tariffs. Many countries are beginning to offer fully digital services, such as ISDN and other switched digital services. As mentioned earlier, the pricing of these services suggests how a particular nation views that service. If the service is priced competitively, the country is effectively stimulating subscribers to use that technology. If the new service carries a high price, most subscribers have little incentive to switch to it unless they have a specific application that requires the new service.

Another new service being offered by PSNs and other common carriers is virtual private network (VPN). With a VPN, a firm purchases a collection of switched services designed to appear to the subscriber as a private network. For example, a firm with locations throughout a country might use a single dialing plan for all its locations, simplifying the use of the network and its administration. Usually, to achieve a feature of this type, the firm must construct a private network from PBXs and leased lines; with VPN, the service is actually provided over the public network, and the numbers are appropriately translated by the PSN so that the VPN appears to operate as a private network. This service requires a sophisticated network infrastructure and isn't available in all nations. Again, the pricing of VPN indicates how the PSN views this service from a strategic perspective.

In summary, the important factors to look for in tariffs are the relationships among prices for various services and the amount of flexibility available. The challenge in reading national tariffs is to compare offerings and prices among countries and determine exactly what can and cannot be done in each nation. The lack of uniformity does present a challenge to telecommunications managers.

Service Quality

As organizations become increasingly dependent upon telecommunications, telecommunications mangers place a higher premium on the reliability and consistency of telecommunications services, including those provided through the PSN. Managers of telecommunications networks must know what levels of performance they can expect. They also need some assurance that those levels are acceptable. For managers of global networks, the mechanisms for providing such assurances aren't always available because formal service quality measurement and monitoring processes aren't in place in most countries.

No matter how many services are declared competitive and no matter how many providers can offer those services, a major concern for a telecommunications manager operating in any nation must be the performance of the PSN. Even if the manager oversees a private network, that private network might, at least partially, consist of leased lines, and that private network undoubtedly must be linked with the PSN at some point. Organizations that use fax machines, dial-up data services, and narrow-band ISDN must use those

services over the PSN. Managers of telecommunications networks need assurance that calls over the PSN are completed, that the PSN is operational and easily accessible, and that problems are solved in a timely manner.

The lack of formal service quality monitoring procedures in many countries is attributable to former, and in some cases current, industry structure. If the PSN is publicly owned and operated by a governmental department, no mechanism may exist for independent monitoring. The service provider and the service quality overseer are one and the same. For effective monitoring to take place, a regulatory entity must exist that is separate from the service provider, and some system must be in place for developing standards and enforcing them.

Therefore, not surprisingly, an active program of service quality monitoring exists in the United States. The PSN in this country is privately owned and operated. These private companies, however, must be designated as common carriers by a regulatory body in order to provide service. In return for this designation, or franchise, the regulators hold the common carrier to specific service quality standards. Indeed, because of the U.S. system of dual regulation, where state public service commissions oversee traffic within state boundaries and the FCC regulates traffic that crosses state boundaries, PSN operators often must comply with two sets of service quality standards.

Although pricing regulation of the PSN in the United States is being eased and competitors are now allowed to offer services that once were provided by a sole monopoly provider, service quality reporting isn't decreasing. Because the assumption is that the force of competition serves as a sufficient policing tool, those services which U.S. regulators have deemed significantly competitive have no substantial reporting requirements. For this reason, although the FCC granted AT&T significant pricing flexibility when it put a price cap regulatory regime in place, it didn't levy any additional service quality reporting requirements on AT&T.[47] In the FCC's view, competition from MCI, Sprint, and a host of other long distance companies was enough incentive for AT&T to maintain quality service.

The FCC, however, when easing pricing regulation of the seven Regional Bell Operating Companies (RBOCs) and the larger of the non-Bell local operating companies, didn't believe that sufficient competition existed to guarantee that the local telephone companies wouldn't take advantage of relaxed regulation to drive up profits at the expense of quality. The relaxed pricing scheme that the FCC crafted for the local companies was very similar to the AT&T price cap plan. Under the price cap plan, companies could keep a substantial portion of profits generated by decreased costs and increased efficiencies.

Whereas the FCC believed that competition would keep AT&T from cost cutting efforts that could harm quality,[48] it held no such conviction about the RBOCs and the other local companies. One result of relaxed regulation for these companies has been an expansion of reporting.

Local telephone companies governed by the price cap plan must now report a wide range of service quality data to the FCC. Reports include quarterly data regarding installation and repair intervals, trunk blockage, switch downtime, service quality complaints, and call set-up time. Twice each year, these same companies report on customer satisfaction, dial tone response, and transmission quality. Each year they provide information regarding miles of fiber-optic cable installed, digital switches deployed, and other information about their infrastructure.[49] These reports ensure that companies aren't trading quality for cost containment and increased profits.

Service Quality Reporting Elements Required of RBOCs in the United States

I. Quarterly Reports

A. Installation and Repair Intervals for Interexchange Access (separated into switched access and special access)

1. Installation

a. percentage of commitments met

b. average missed commitments

2. Repair

a. total trouble reports

b. average repair interval

B. Installation and Repair Intervals for Local Service (separated into business and residence; urban and rural)

1. Installation

a. percentage of commitments met

b. average missed commitments

2. Repair

a. total access lines

b. number of initial trouble reports

c. number of repeated trouble reports

d. instances of no trouble found

C. Trunk Blockage by month

1. Number of Trunk Groups exceeding Threshold

D. Total Switch Downtime

1. Listed by rural and urban
2. Listed by size of switch
3. Specific occurrences of two minutes or more of downtime
4. Service Quality Complaints
5. Separated into Business and Residence

II. Semiannual Reports

A. Dial Tone Response

1. Percentage of offices at which dial tone provided within 3 seconds of lifting telephone instrument

B. Transmission Quality

1. Percentage of offices meeting loss, balance, distortion, and noise objectives
2. Percentage of trunks meeting loss, balance, distortion, and noise objectives

III. Annual Reports

A. Infrastructure Report for Switching Equipment (urban and rural areas)

1. Type of switches
2. Number of switches equipped with ISDN, SS7, Equal Access capability

B. Infrastructure Report for Transmission Facilities (in sheath kilometers) for trunk and loop plant

1. Digital versus analog
2. Copper, radio, fiber

Source: FCC, Memorandum Opinion and Order; CC Docket 87-313; DA 91-619; May 17, 1991.

Service quality monitoring hasn't been lessened at the state level either, despite the easing of regulation. Even in Nebraska, which has moved the farthest of any U.S. state toward deregulating the provision of virtually all services, the state regulatory agency retains service quality oversight of telecommunications companies.[50] The Nebraska deregulation law completely deregulated mobile and cellular services along with the pricing of long distance service and restricted the Nebraska Commission's latitude in setting rates for basic local service. For all practical purposes, the state regulatory agency has neither a role in pricing telecommunications services in the state nor a substantial role in the introduction of new services.

The Nebraska Commission, however, has the authority to specify the time frame in which service outages must be repaired, how quickly calls to the operator must be answered, the percentage of calls (98 percent) that should receive a dial tone within three seconds, and acceptable levels of transmission loss.[51] Whatever the price of services or the range of services offered, they must be provided at specific quality levels.

The U.S. experience is worth consideration for many reasons. The development of U.S. telecommunications during the past decade has almost become synonymous with competition and deregulation. However, despite the push of competition, U.S. regulators continue to monitor service quality. Indeed, in some instances they have increased reporting requirements. This continued insistence on service reporting reflects a commitment to the continued viability of a quality public network. Although U.S. regulators eased rate regulation and allowed competitive entry into various services, they haven't eased service regulation for PSN providers.

The U.S. approach is different from that taken in most countries. A report issued by the OECD comments that, of 14 nations surveyed, only Televerket in Sweden was required to provide statistics and meet service quality standards. None of the other 13 nations listed had regulatory requirements for service reporting, although British Telecom (BT) does publish semiannual statistics in response to pressure from OFTEL in the United Kingdom.[52] The 14 European nations surveyed in the OECD report have fairly well-developed telecommunications infrastructures with sophisticated users. If reporting requirements don't exist in these countries, they are even less likely to exist in developing nations with fewer resources and much less advanced networks.

The lack of reporting requirements in nations such as the 14 surveyed by OECD might be a temporary situation. As nations privatize their telecommunications resources and establish regulatory bodies to oversee these newly

private companies, the mechanism for establishing service quality standards and monitoring compliance with those standards is being established. Those standards will be in demand if for no other reason than the large number of U.S. companies that are internationalizing their operations and their telecommunications networks. Pressure is inevitable from the managers of these global networks for the type of service quality monitoring they have come to expect in the United States.

The introduction of competition makes such standards inevitable as well. As competitors emerge for various telecommunications services, service providers have an incentive to neglect noncompetitive services and concentrate on maintaining their market share in competitive areas. Because competitive services tend to be aimed at large business customers, usually in urban areas, residential services and rural areas might suffer. Service standards and monitoring mechanisms are a method for assuring that all parts of the PSN, including those for which no strong rivals are imminent, provide high-quality service.

For global telecommunications managers, uniformity in service standards and enforcement would be a major benefit. They could expect, and provide to their users, a uniform level of service network wide. However, developing global service quality standards might be an impossible task. Service providers and regulators in all countries would need to agree on a common set of standards. Various international standards bodies developed recommendations for what those standards should include. For example, the European Conference of Postal and Telecommunications Administrations (CEPT), proposed the following quality of service categories:

- Waiting time
- Fault reports
- Call set-up time
- Call failure rates
- Transmission quality and privacy
- Account queries[53]

The CCITT also drafted recommendations. Those recommendations are more technically oriented and detailed.

The CCITT has prepared some recommendations on the technical aspects of service levels. Table 2.1 lists the recommendations pertaining specifically to service quality. In addition, the G.100 series for analog circuits and the G.800 series for digital circuits cover technical aspects of transmission quality.

Table 2.1. CCITT recommendations pertaining to service quality.

Recommendation	Title
E.420	Checks on the quality of the international telephone service—General considerations
E.421	Service quality observations on a statistical basis
E.422	Observations on international outgoing calls for quality of service
E.423	Observations on traffic setup by operators
E.424	Test calls
E.425	Internal automatic observations
E.426	General guide to the percentage of effective attempts that should be observed for international telephone calls
E.427	Collection and statistical analysis of special quality of service observation data for measurements of customer difficulties in the international automatic service
E.428	Connection retention
E.800	Quality of service and dependability vocabulary
E.810	Model for the performance of service on a basic call in the telephone network
E.830	Models for the allocation of international telephone connection retention, accessibility, and integrity
E.845	Connection accessibility objective for the international telephone services
E.850	Connection retention objective for the international telephone service
E.855	Connection integrity objective for international telephone service

continues

Table 2.1. continued

Recommendation	Title
E.862	Dependability planning of telecommunication networks
E.880	Field data collection and evaluation on the performance of equipment, networks, and services

In addition to these technical parameters, service quality is also related to the overall design of the network. The manner in which telephone operators determine how many circuits to locate between facilities is called *traffic engineering*. In traffic engineering, network designers must estimate the telephone traffic that must be carried between two facilities, establish a target grade of service,[54] and calculate the number of circuits needed to meet the grade of service objective given the expected traffic load. The CCITT recommendations in the E.500, E.600, and E.700 series are designed to support this process.

At the basis of any discussion regarding the establishment of service standards lies the question of what should appropriately be measured. The OECD study (mentioned earlier) observes that most recommendations might be too focused on technical standards and not focused enough on customer satisfaction issues. A good case, of course, can be made for the importance of technical standards, but to disregard customer service standards would be a mistake. For the PSN to serve the needs of telecommunications managers and keep them committed to using the PSN rather than bypassing it with private facilities or leased lines, issues such as billing accuracy and detail should definitely be addressed. The issue that all telecommunications managers face, including those who manage global networks, is efficient, cost-effective service. Without billing detail, telecommunications managers have difficulty assessing the true costs of using the PSN.

Information about infrastructure and its planning might be useful to address as well. In the United States, local telephone companies must report on the level of digitization and fiber-optic miles deployed. Such information can be useful to a telecommunications manager in deciding between the PSN and the features offered by competitive providers.

Some Suggested Categories of Quality Standards

- Waiting Time—for service installation and for service delivery
- Fault Reports—frequency and clearance rate
- Call Set-up Time
- Call Failure Rates—caused by trunk blockage or other technical problems
- Transmission Quality—amount of loss or noise
- Privacy
- Account Queries—dealing with billing questions

When standards are adopted, appropriate indicators for determining compliance with those standards must also be identified. To be useful, indicators should be simple to gather and not unduly burdensome to monitor. A recent study suggests that indicators should consist of "a small number of basic quality indicators which can be collected easily, [and] can be compared on an international basis…."[55] The suggested indicators include

- The elapsed time in seconds from the time a call is dialed until the called party responds—to measure response time for operator services
- The percentage of attempted calls that meet CCITT recommendations for transmission quality—to measure transmission quality
- The percentage of technical faults that are cleared by the end of the next working day—to measure fault clearance
- The percentage of local and national long distance calls that fail to reach their destinations because of technical failure—to measure call failure rates
- The number of working pay telephones per 1,000 inhabitants —to measure effective pay telephone density
- The number of accounts questioned by customers—to measure billing accuracy

Even if standards and indicators are determined, other questions remain. Performance measurements can be gathered and published at the national or local level. They can be gathered randomly from all subscribers or by type

of subscriber (business versus residential, for example). Monitoring can apply to various services: local, national long distance, or international calling. What level of reporting is optimal? What services should be monitored? These questions will need to be answered if a system of service quality reporting is established.

Assumptions about what constitutes quality service lie beneath these recommendations for standards and indicators. The emphasis on failure rates and fault clearances makes it clear that reliability is an all-important feature of a quality PSN. Calls are expected to reach their destinations without technical problems. Any problems that develop should be solved quickly. Quick response time is also a desired feature, as evidenced by measurements of operator response time, dial tone delay, and call set-up delay. Pay telephone density measures availability of access to the PSN. In short, quality service assures that customers can rely on the network, can access it quickly, and are billed accurately for its use.

Service providers collect statistics on many of these indicators. France Telecom, for example, developed an elaborate monitoring system for a number of service quality variables and can analyze the resulting statistics by region. Statistics are gathered and reported for such items as the number of subscriber trouble reports per 100 main lines, the percentage of those reports that are rectified on the same day, and the percentage cleared up in seven days. In addition, France Telecom reports on dial tone delay, call set-up delay, and completion rates for local, national long distance, and international calls.[56]

Some service providers voluntarily publish these statistics in annual reports. PSN providers in Ireland, Italy, and Spain, for example, publish annual figures on such issues as waiting time until installation of service and traffic. Other PSN operators use such information only for internal purposes. If regulators require the collection and publication of such statistics, the next question is whether providers who don't meet service standards will suffer repercussions. When that step is taken, service quality regulation, not just service quality reporting, becomes a reality.

Suggested Categories of Service Quality Indicators

- Elapsed time from dialing to called party pick-up (to measure response time)
- Percentage of attempted calls meeting CCITT recommendations for transmission quality (to measure transmission quality)

- Percentage of technical faults cleared by end of next working day (to measure fault clearance)
- Percentage of calls failing to reach destination because of technical failure (to measure failure rates)
- Number of pay telephones per 1,000 population
- Number of accounts questioned (to measure billing accuracy)

Whatever the level of required reporting, service quality monitoring can help ensure that the PSN operates at a high level despite competitive inroads into telecommunications markets. Moreover, monitoring reports can provide telecommunications managers with valuable information as they decide between staying on the PSN or seeking alternative services.

The possibility of establishing globally uniform service standards might not have strong support. The establishment of service quality standards depends on political issues, the status of network development, the amount of pressure from sophisticated users, and the resources available to gather and analyze statistics in each nation.

Information Policy and Transborder Data Flow

Information policy concerns itself with a country's approach to managing and regulating information flow. The nature of a telephone network can depend on the information policy adopted by a country.

A common viewpoint in the research community holds that innovation flourishes with the least information restrictions. Innovation often depends on easy and ready access to information and ideas so that innovators can combine and extend the existing state of the art in their field. Innovation is vital to all economies—developing and developed—so incentives exist to remove as many restrictions on information flow as possible. The risks of such a policy, however, include the potential for asymmetrically opportunistic behavior and transborder data flow (TDF) issues. Asymmetrically opportunistic behavior results in exploitation by those engaged in this behavior.[57] Developing countries that are historically information poor have raised TDF issues.

Asymmetrically opportunistic behavior is essentially an issue that a firm frequently deals with as a matter of policy and is also a matter with some domestic policy implications in the domain of patent, copyright, and trade secret law. These issues have implications for economic development, but their discussion is beyond the scope of this book.

TDF issues, on the other hand, are fairly new. In their survey of information policy issues, O'Brien and Helleiner frame the tension between the information rich and the information poor nations in terms of "North-South" political tensions.[58] Many developing (information poor) countries are in the southern hemisphere of the planet, and many of the developed (information rich) economies are in the northern hemisphere. This asymmetry puts developing countries at a serious disadvantage as marketplace participants and negotiating partners. This occurs because the information rich nations can readily find out about resources, weather, technology, and market conditions, whereas the information poor nations cannot.

Smith carries these arguments further by raising the issue of remote sensing as such a case.[59] In *remote sensing*, satellites in orbit photograph the earth, sensing climactic conditions, vegetative growth, and so forth. Although these satellites are a boon to scientists who study the earth, one can also imagine the owners of this information, for instance, manipulating the prices of commodities so they can increase their profits. The satellite data is often provided at low cost to developing countries, but they often don't possess the expertise and infrastructure to process the data into useful information. Therefore, many developing countries view this as a violation of their sovereignty.

More appropriate to the topic of this book is when a firm provides services to firms or consumers in another country. For example, consider banking between the United States and Canada. The databases and information necessary to operate a bank are often stored at centralized computer sites, in this case, in the United States. The information itself pertains to Canada and Canadian citizens. This fact concerns the Canadian government because information pertaining to its citizens is stored in another country. After all, a sovereign nation possesses this information. Data movement is commonplace in the operations of large international firms, which must routinely transmit, process, and disseminate data on production, sales, and operations to their other divisions, which are often in other countries.

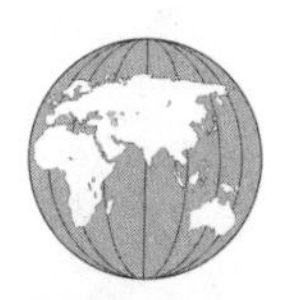

The OECD published guidelines for TDF in 1982 that were designed to protect national TDF concerns without unduly limiting trade. The recommendations themselves are concerned with the implementation of the Guidelines. The Guidelines outline a series of principles that countries should consider in formulating domestic legislation:

- **Collection Limitation Principle**—Limits should exist for the collection of personal data; such data should be collected legally, and, wherever possible, with the consent of the data's subject.
- **Data Quality Principle**—Data should be accurate, relevant, complete, and up-to-date.
- **Purpose Specification Principle**—The purpose of the data collection should be specified by the time of collection, not after, and the use of the data should be limited to that purpose. If the data is to be used for other purposes, these other purposes should be specified in advance and be consistent with the original purpose.
- **Use Limitation Principle**—Personal data should not be disclosed except with the consent of the data subject or the authority of law.
- **Security Safeguards Principle**—Personal data should be protected by reasonable security safeguards.
- **Openness Principle**—A policy of openness about practices and policies should exist with respect to the collection, storage, and use of personal data. Individuals should be able to determine the existence, nature, extent, location, and purpose of the personal data.
- **Individual Participation Principle**—Individuals should have the right to obtain confirmation from a data controller whether personal data on the indvidual exists; view the individual's personal data in a reasonable time, at a reasonable cost and in intelligible form; be given reasons for denied information requests, with the opportunity to challenge the denial; and challenge the data relating to the individual and, if successful, force the erasure, modification, or ammendment of the data.
- **Accountability Principle**—Controllers of data should be accountable for complying with the previous principles.[60]

OECD Recommendations Concerning Guidelines Governing the Protection of Privacy and Transborder Data Flows of Personal Data

- That Member countries take into account in their domestic legislation the principles concerning the protection of privacy and individual liberties set forth in the Guidelines...;
- That Member countries endeavor to remove or avoid creating, in the name of privacy protection, unjustified obstacles to transborder flows of personal data
- That Member countries co-operate in the implementation of the Guidelines...;
- That Member countries agree as soon as possible on specific procedures of consultation and co-operation for the application of these Guidelines."[61]

Issues such as these have led to TDF concerns in international information policy. An information policy that is too restrictive can result in lower levels of direct foreign investments because firms might not be able to export the information necessary to operate efficiently. On the other hand, a policy that is too permissive can intensify the information asymmetries present in the relationship between developing and developed countries.

The nature of the telecommunications system depends on the information and TDF policies as much as it does on the provision policy. A restrictive information policy must have tightly controlled public gateways to international networks because the information flow must be controllable. A permissive policy, on the other hand, allows free connection to international networks from a number of sources, both public and private.

Telecommunications managers of international firms must acquaint themselves with the country-specific regulations. Most countries—developing countries in particular—have such laws and regulations in place. These restrictions can have significant implications on the design of an information system to support operations in such a country. Interestingly, TDF regulations typically focus on data communications because large quantities of data can be moved efficiently; the same information, read over a telephone, might not violate the laws.

Special Arrangements

In response to the entry of new players and new services into the telecommunications arena, governmental and regulatory entities have both amended existing policies and created new ones. Because the United States stands at the forefront of fostering competitive entry and new services, examining what U.S. policymakers do in this regard is instructive. In particular, an interesting exercise is to see what changes have been made to accommodate noncarrier (private) transoceanic cables and enhanced service providers.

As other countries introduce competition into their telecommunications markets, their national authorities must deal with these issues. Indeed, many nations have a policy regarding landing rights, particularly as this concept is expanded to include satellite-based communications facilities.

Landing Rights

Any entity seeking to land, or terminate, a transoceanic cable in any part of the United States must first receive approval under the Cable Landing License Act. Prior to passage of this act in 1921, the President of the United States granted landing licenses to foreign governments, which reciprocated by giving landing privileges to U.S. companies.[62] The 1921 act gave the President the authority, exercised through the Department of State, to grant, deny, or revoke submarine cable landing licenses. Presidential Executive Order Number 10530, dated May 10, 1954, delegated this authority to the Federal Communications Commission. However, the FCC must receive approval from the Secretary of State before granting a cable landing license.

The Cable Landing License Act is very specific about what can be done without a landing license:

> *No person shall land or operate in the United States any submarine cable directly or indirectly connecting the United States with any foreign country, or connecting one portion of the United States with any other portion thereof, unless a written license to land or operate such cable has been issued by the President of the United States.... The conditions of...this title shall not apply to cables, all of which, including both terminals, lie wholly within the continental United States.*[63]

The act, moreover, makes it very clear that U.S. security interests, the interests of U.S. citizens, issues of reciprocity, and pricing issues all play a role in the granting of landing licenses:

> *The President may withhold or revoke such license when he shall be satisfied after due notice and hearing that such action will assist in securing rights for the landing or operation of cables in foreign countries, or in maintaining the rights or interests of the United States or of its citizens in foreign countries, or will promote the security of the United States or of its citizens in foreign countries, or may grant such license upon such terms as shall be necessary to assure just and reasonable rates and service in the operation and use of cables so licensed.*[64]

U.S. common carriers traditionally apply for landing licenses before constructing submarine cables. Before beginning construction, they also apply to the FCC for what is called Section 214 authorization.

Section 214 of Title II of the Communication act (the act that created the FCC and that outlines the parameters of the FCC's duties and authority) requires that carriers receive FCC approval before constructing, extending, or terminating a service or facility, including a service or facility that crosses international boundaries. The onus is on the carrier to show that the service or facility is needed, cost effective, and technically sound. The common carrier falls under the regulatory aegis of the FCC and is, therefore, bound by the Section 214 requirements.

The responsibilities of the FCC also include consideration of the need for submarine cables. This consideration includes

- Assessment of need
- Examination of demand flexibility
- Cost
- Service reliability
- Foreign correspondent acceptance
- Digital connectivity
- Furtherance of pro-competitive policies
- Defense needs[65]

The Section 214 approval process reinforces the FCC's role in submarine cable considerations.

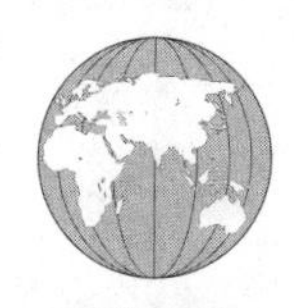

In 1985, however, the FCC's commitment to competition led to the waiver of the Section 214 process for private international submarine cables. In order to enhance competition in international traffic, the FCC determined that organizations seeking to build private submarine cables need not first prove to the FCC that the cables were necessary, cost effective, or technically sound. Common carriers seeking to purchase capacity on these private cables, however, must still provide a Section 214 proving necessity, cost effectiveness, and technical soundness in order to do so.

Although not obligated to acquire Section 214 approval, private cable providers must apply to the FCC for landing licenses. The goal of easing competitive entry is secondary to the issues of reciprocity, security, and U.S. interests raised by the Cable Landing License Act.

Indefensible Rights of Use

In a telecommunications environment characterized by diversified ownership and competition, the rights of the interested parties must be specified. Interested parties are the owners and users of the transmission facility. With multinational partners owning the cable (and perhaps contracting their operation to a separate entity) these issues become significantly more complex.

Indefensible Rights of Use (IRUs) began, for all practical purposes, as a form of capital lease for the telegraph industry. Record carriers (telegraph companies) owned no interest in the first voice-grade submarine cable, TAT-1. Instead, they leased capacity on that facility: they held IRUs on the submarine cable. The FCC allowed the record carriers to treat their IRUs as an investment upon which they could earn a return.[66]

This treatment was necessary because of the necessity to compute a rate base so the carriers could earn a specified rate of return through their prices. A smaller rate base called for smaller total revenues so that the permitted rate of return wouldn't be exceeded. Only a reduction in prices could achieve a reduction in revenues.

Today the concept of an IRU is still in existence, although now the policy questions regarding IRUs revolve around private users, not record carriers. The situation compounds itself, in a sense, because submarine cables are jointly owned by U.S. companies at one end and by other entities, often PTTs, at the other.

The building of private cables has caused the FCC to revisit the treatment of IRUs. IRUs are permanent rights to use a circuit on a cable. IRUs infer no right to participate in the management of the cable operation, to salvage value, or to control the cable facility. With an IRU, an entity, "in return for a one-time capital outlay, plus monthly maintenance and operational contributions," can reap the benefit of being assured "the availability of leased-channel-type service on a long-term basis at known and relatively fixed costs."[67]

Although in the late 1800s and early 1900s telegraph cables were owned by private companies, both American and foreign, and operated on a whole circuit basis, the advent of voice-grade submarine cables like the first TransAtlantic Telephone cable (TAT-1) changed that dynamic.[68] Prior to laying TAT-1, high-frequency radio circuits provided transoceanic voice service. PTTs deemed that ownership of these circuits be shared by carriers at each end, with ownership extending to a mythical midpoint. This same "legal fiction" was adopted for TAT-1 and subsequent cables, with each owner's share terminating at a theoretical midpoint in the ocean.

When AT&T owned most of the U.S. share in submarine cables and PTTs owned most of the shares at the other end of the cables, the relationship between owners was fairly symmetrical and clear. The inclusion of IRUs owned by record carriers didn't muddy the waters greatly. PTTs recognized the status of record carriers. Record carriers operated under the oversight of the FCC, which exercised an interest in their participation in TAT-1 and other cables.

With the introduction of private users, however, the relationship between U.S. owners and those holding ownership of the half-circuits at the other end of the cable is less clear. Again in the interests of competition, the FCC considered the feasibility of requiring U.S. common carriers who own the U.S. portion of a submarine cable to sell IRUs on that cable to private users who were interested in such a purchase. The FCC decided not to make such exchanges mandatory but provided for the voluntary purchase of IRUs by private entities. In making this decision, the FCC had to tread lightly in one area:

> *Finally, our policy will not infringe upon the rights of the PTT co-owners. Our policy recognizes that cable circuits are owned on an undivided-half-interest basis by U.S. entities and the PTTs. Our policy does not require PTTs to transfer their interests and makes the transfer of the U.S. half interest contingent upon the agreement of the PTTs. Our policy also requires users to negotiate all necessary agreements with the PTTs.*[69]

In its policies regarding submarine cable, the FCC must deal with issues of private use. On one hand, the FCC determined that cables owned by private

entities could provide capacity to common carriers; on the other hand, the FCC also determined that common carriers could provide ownership-type interest to private users. In both instances, however, the international aspect of the service was very evident. When getting permission to land a cable in the United States, private cable owners still must meet U.S. foreign policy interests. Private users seeking capacity on common carrier cables must recognize the interests and wishes of the telecommunications administrations owning the non-U.S. end of those cables.

Recognized Private Operating Agencies

The issue of recognition by PTTs and other foreign telecommunications administrations is one the FCC must deal with in regards to enhanced service providers. The basic problem is a lack of uniformity across national borders. The United States draws a distinction between basic and enhanced services. The International Telecommunications Convention of the International Telecommunications Union (ITU), however, draws a distinction between telecommunications and data processing. The boundaries between those two sets of definitions don't coincide. This lack of congruence creates some confusion regarding what in the United States are referred to as enhanced service providers. These entities aren't regulated in the United States, but because they provide services that the ITU deems as telecommunications, they appear as subject to ITU guidelines and requirements. (Examples of these services are discussed in the next few paragraphs.) In an effort to foster competition, the FCC presented a tightly argued case for according these enhanced service providers with recognized private operating agency (RPOA) status as defined by the ITU. The purpose of designating these service providers as RPOAs is to "assure PTTs that such providers will obey the ITU Convention and CCITT regulations" and so perhaps "help the providers obtain operating agreements."[70]

Responsibility for conferring the RPOA designation falls on the signatory to the ITU Convention. In the United States, that signatory is the Department of State. The FCC's role in the process is to accept and screen applications and make recommendations to the State Department. Before finalizing an application procedure, the FCC must determine how enhanced service providers fit into the ITU framework and whether the Department of State, upon FCC recommendation, has the authority to decide that the applicants do indeed fit.

The ITU Convention recognized entities other than PTTs or telecommunications administrations. It also recognized:

> *...any private operating agency...which operates a public correspondence or broadcasting service and upon which the obligations provided for in Article 44 of the Convention are imposed by the Member in whose territory the head office of the agency is situated, or by the Member which has authorized this operating agency to establish and operate a telecommunications service in its territory.*[71]

In other words, as long as the signatory ensures that the private operating agency abides by Article 44 of the Convention, the ITU will recognize that agency. Article 44 of the Convention specifies that:

> *The members are bound to abide by the provisions of this Convention and the Administrative Regulations in all telecommunications offices and stations established or operated by them which engage in international services or which are capable of causing harmful interference to radio service of other countries....*[72]

The ITU further defined *private operating agency* as:

> *...any individual or company or corporation, other than a governmental establishment or agency, which operates a telecommunications installation intended for an international [telecommunications] service or capable of causing harmful interference in such a service.*[73]

The FCC's task then was to determine whether U.S. enhanced service providers provided a "public correspondence" service. If they did, they could be eligible for RPOA status. Further, the FCC needed to determine whether the enhanced service providers' lack of "common carrier" status made them ineligible. Finally, the FCC needed to determine what sort of procedure should be adopted for entities seeking RPOA designation.

The FCC determined that enhanced service providers who offered message services rather than pure data processing were eligible for RPOA status. Entities offering only data processing services didn't fall under the ITU definition of telecommunications and so weren't eligible. The distinction between message service and data processing was based on the treatment of the communication message. If the message was transmitted essentially unchanged, the service clearly fell under the ITU definition of telecommunications. If the message was altered, manipulated, or changed in any way, the service fell under the definition of data processing.

Definitional problems could exist, however, with services like packet switching and code and protocol conversion. The ITU still defined these services as

telecommunications. The FCC, however, determined that they were enhanced services and subsequently not subject to the FCC's regulatory control. These services were offered by a host of competitive providers in the United States, providers who had no status with PTTs when they sought overseas markets. These providers weren't common carriers like AT&T and MCI, which fell under the regulatory mantle of the FCC and enjoyed recognized status overseas.

Even though the United States regarded them as enhanced services, the FCC determined that services such as packet switching and protocol conversion still constituted public correspondence rather than data processing. Providers of these services could then be RPOAs. The FCC further decided that these providers' lack of common carrier status should not affect their ability to be designated as RPOAs. The FCC determined that nothing in the definition of an RPOA encompassed the notion of common carriage. Further, the FCC noted that the term *common carrier* was a creation of U.S. law and had no legal significance outside U.S. borders.[74]

The FCC's motive in examining this issue was to facilitate the ability of enhanced service providers to move into the international telecommunications market. As the FCC states:

> *We are thus in this proceeding seeking to reassure overseas communications entities that neither our basic/enhanced dichotomy nor our decision to forbear from regulating enhanced-service providers will prejudice any of their rights under the Convention and regulations and to develop policies which will facilitate the introduction of competitive, international data services between the United States and overseas points.*[75]

Due to the lack of across-the-board uniformity in industry structure and in service definitions, the FCC needed to clarify a situation that could impede the U.S. goal of encouraging competition and the growth of data communications in the international arena. In this proceeding, the FCC addressed the differences in terminology between the ITU (an international body) and U.S. domestic regulatory provisions. It also attempted to define the boundary between an individual nation's autonomy and that nation's obligations under international treaties.

Basically, the FCC determined that the United States had an obligation to meet its international commitments. The FCC further decided that the United States had a certain amount of latitude in meeting those obligations. The burden lay on the U.S. signatory to interpret ITU provisions and determine which U.S. entities met those provisions.

Because the FCC wanted to facilitate, rather than hamper, the efforts of enhanced service providers, it determined not to mandate application for RPOA status. PTTs were free to formulate operating agreements with U.S. enhanced-service providers not holding an RPOA designation. To expedite matters even more, the FCC developed a fairly simple application procedure.

Part 63 of the FCC's Rules and Regulations (Sections 63.701 and 63.702) outlines the procedures for RPOA application. Applicants need only do the following:

- File an original and two copies of an application with the FCC stating the nature of the service to be provided and a statement making clear that the applicant is aware of ITU obligations.
- Provide the name and address of the applicant entity.
- Provide information regarding ownership of the entity.
- Provide a copy of the entity's articles of incorporation and its bylaws.
- Provide a statement regarding the entity's status as a common carrier or as an enhanced service provider.
- Provide information regarding the points between which the service will be provided.
- Provide information about whether the facilities used will be owned by the applicant, leased, or subject to another type of arrangement.

No formal denial process follows the submission of applications. The FCC provides notice of the filings. Those with objections can voice them through an informal letter. After making its determination, the FCC forwards its recommendation to the Department of State.

By virtue of their status as U.S. companies, even those service providers not designated as RPOAs are subject to ITU obligations. All U.S. entities involved in telecommunications are bound by the agreement signed by the U.S. signatory. Despite this situation, however, the FCC felt compelled to formulate a mechanism that provides further evidence of compliance. Because of regulatory and policy decisions made for the domestic market—decisions that resulted in the creation of entities with few international counterparts—the FCC had to adjust its policy and procedures to facilitate international developments.

Conclusions

The amount of time and attention that the FCC devoted in recent years to the issues of landing rights, IRUs, and RPOAs is indicative of the complications presented by new services and new players in the international arena. As new entities emerge, the old way of doing business must be examined and adjusted to accommodate new procedures and relationships. As other nations begin to emulate the U.S. example by allowing competitive entry and the introduction of new services, they too must reexamine traditional arrangements and agreements.

Endnotes

1. G. John Ikenberry, "The International Spread of Privatization Polices: Inducement, Learning and 'Policy Bandwagoning,'" in *The Political Economy of Public Sector Reform and Privatization*, ed. Ezra N. Suleiman and John Waterbury (Boulder, Colo.: Westview Press, 1990), pp. 99-106.

2. Ikenberry.

3. N.C. Lerner, "Telecommunications Privatization and Liberalization in Developing Countries," *Telecommunication Journal*, vol. 58 V/1991, pp. 281-282.

4. Bjorn Wellenius, "Beginnings of Sector Reform in the Developing World," in *Restructuring and Managing the Telecommunications Sector*, ed. Bjorn Wellenius, Peter A. Stern, Timothy E. Nulty, and Richard D. Stern (Washington, D.C.: World Bank, 1989), pp. 89-98.

5. Jeffrey Herbst, "The Politics of Privatization in Africa," in *The Political Economy of Public Sector Reform and Privatization*, ed. Ezra N. Suleiman and John Waterbury (Boulder, Colo.: Westview Press, 1990), pp. 88-110.

6. Lerner, p. 281.

7. Jean Paul Vercruysse, "Telecommunications in India: 'Deregulation' vs. Self-Reliance," *Telecommunications and Informantics*, vol. 7, no. 2, p. 112.

8. For further explanation of these points see John Redwood, "A Consultant's Perspective," in *Privatization and Deregulation in Global Perspective*, ed. Dennis J. Gayle, and Jonathan Goodrich (New York and Westport, Conn.: Quorum Books, 1990), pp. 55-56.

9. Redwood.

10. *Telecommunications Reports International*, February 7, 1992, pp. 12-13.

11. Mark A. Hukill and Meheroo Jussawalla, *Trends in Policies for Telecommunication Infrastructure Development and Investment in ASEAN Countries* (Honolulu, Hawaii: East-West Center, 1991), p. 23.

12. Eli Noam, *Telecommunications in Europe*, (New York: Oxford Unversity Press, 1992), p. 80.

13. See Bjorn Wellenius, "Beginnings of Sector Reform in the Developing World," in *Restructuring and Managing the Telecommunications Sector*, pp. 89-98, for some examples of different approaches.

14. Ezra Suleiman, "The Politics of Privatization in Britain and France," in *The Political Economy of Public Sector Reform and Privatization*, ed. Ezra N. Suleiman and John Waterbury (Boulder, Colo.: Westview Press, 1990), p. 115.

15. For a discussion of the privatization process and some assessment of its results in Britain, see John A.C. King, "The Privatization of Telecommunications in the United Kingdom," in *Restructuring and Managing the Telecommunications Sector*, ed. by Bjorn Wellenius, Peter A. Stern, Timothy E. Nulty, and Richard D. Stern (Washington, D.C.: World Bank, 1989), pp. 55-60.

16. *Telecommunications Reports International*, December 13, 1991, pp. 9-10.

17. Hukill and Meheroo, p. 25.

18. Hukill and Meheroo, p. 21.

19. See Vernon L.B. Mendis, "Phased Privatization with Proposed Foreign Participation: The Sri Lanka Experience," in *Restructuring and Managing the Telecommunications Sector*, pp. 99-106.

20. *Telecommunications Reports International*, June 14, 1991, pp. 3-4.

21. Peter R. Scherer, "Perspectives on World Telecom Reform" in *Transnational Data and Communications Report*, May/June 1992, pp. 27-33

22. Pablo T. Spiller and Cezley I. Sampson, "Regulation, Institutions and Commitment: The Jamaican Telecommunications Sector" (Delivered at the Twentieth Annual Telecommunications Policy Research Conference, Solomons Md., September 1992).

23. Ted Johnson, "Caribbean Basin Becomes Large Telecommunications Market," *Telematics and Informatics*, vol. 7, no. 1, 1990, pp. 1-7.

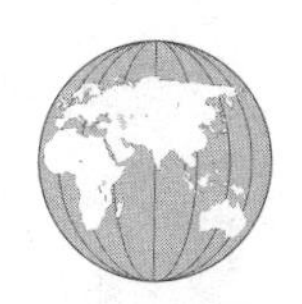

24. Economic systems such as a telephone network possess what economists refer to as network externalities. An *externality* is a factor normally external to a consumer's cost-based decision making process. In a network externality, the magnitude of the externality is a function of the size of the network.

25. Carol Weinhaus and Anthony Oettinger, *Behind the Telephone Debates*, (Norwood, N.J.: Ablex, 1988), p. 5.

26. The courts decided that customers could attach any equipment to the network, as long as it wasn't "publicly detrimental," in *Hush-a-Phone Corp. v. United States*, 238 F.2d 266 (D.C. Cir. 1956); *Hush-a-Phone Corp. v. AT&T*, 22 FCC 112 (1957).

27. Whether IXC service is part of universal service is not at all clear. AT&T is still the carrier of last resort for long distance, but that definition may apply to locations that other IXCs might not want to serve rather than to individual customers. Interestingly, AT&T can deny a customer service for nonpayment of charges, but the local telephone company can't disconnect a customer who has paid his local telephone bill but not his long distance bill. This presents some interesting questions about what actually constitutes universal service in the United States today.

28. According to some critics, not all nations see the U.S. experience as positive. According to Kenneth Dyson and Peter Humphreys, "the normative code underpinning West European telecommunications policies could remain remarkably resistant to American-style deregulation in certain key countries," in *The Political Economy of Communications: International and European Dimensions* (London: Routledge, 1990), p. 8.

29. Commission of the European Communities, "Towards a Competitive Community-Wide Telecommunications Market in 1992: Implementing the Green Paper on the Development of the Common Market for Telecommunications Services and Equipment," COM (88), Brussels, 1988.

30. See, for example, *Telecommunications Reports International*, October 16, 1992, pp. 8-9.

31. Kenneth Baker, Minister for Information Technology, in a speech before the House of Commons Standing Committee on the Telecommunications Bill, delivered November 17, 1983, as quoted by William B. Garrison, Jr., in *Four Case Studies of Structural Alterations of the Telecommunications Industry*, The Annenberg Washington Program in Communications Policy Studies, January 1988, p. 17.

32. Ingo Vogelsang, *Welfare Consequences of Selling Public Enterprises: Case Studies from Chile, Malaysia, Mexico, and the U.K.*, The United Kingdom, Volume 1: Background; British Telecom, World Bank Conference, (Washington, D.C., 11-12 June 1992), p. 7.

33. Mark A. Hukill and Meheroo Jussawalla, *Trends in Policies for Telecommunications Infrastructure Development and Investment in the ASEAN Countries* (Honolulu, Hawaii: East-West Center, 1991), p. 13.

34. For example, some competitive IXCs in the U.S. file informational tariffs on their common services, even though they do not need to abide by them in all cases. In contrast to tariffs filed by the regulated carrier, such tariffs need not represent the complete set of available services.

35. CCITT Blue Book, *General Tariff Principles: Charging and Accounting in International Telecommunications Services*, Series D Recommendations, Volume II: Fascicle II.1, 5.1.1.1 (Melbourne 1988), p. 354.

36. Indeed, in the United States, the FCC's Computer Inquiries dealt with this very difficult question, with the FCC trying to define the services and draw boundaries to ensure no cross-subsidization between the two.

37. See John T. Wenders, *The Economics of Telecommunications: Theory and Practice* (Cambridge, Mass.: Ballinger Publishing Company, 1987), for examples.

38. *Performance Indicators for Public Telecommunications Operators*, OECD ICCP Series Report No. 22 (Paris: OECD, 1990), p. 37.

39. *Performance Indicators*, p. 37.

40. *Performance Indicators*, pp. 25.

41. *General Tariff Principles*, 5.1.2.3 iv, p. 356.

42. Again, this is an example of inadvertent price discrimination, where the earlier subscribers pay more because they value the service more.

43. *Universal Service and Rate Restructuring in Telecommunications*, ICCP Series Report No. 23 (Paris: OECD, 1991), pp. 149-153.

44. Garrison, p. 70-71.

45. Garrison, p. 41.

46. *Performance Indicators*, p. 71.

47. A price cap regulatory regime is one in which the movement of prices is regulated, not the level of underlying costs and investment used in providing services. Price caps are explained more fully in Chapter 3.

48. This has recently been called into question in the United States, however. Several recent outages affecting a large number of subscribers have been attributed to cost-cutting measures.

49. See *Policy and Rules Concerning Rates for Dominant Carriers*, CC Docket No. 87-313, Memorandum Opinion and Order, DA 91-619, released May 17, 1991.

50. Legislative Bill 835, amendment to Revised Statutes of Nebraska, sections 75-109 and 75-604, and Revised Statutes Supplement, section 75-609.

51. *Telecommunication Rules and Regulations*, Nebraska Public Service Commission.

52. *Universal Service and Rate Restructuring*, pp. 70-82.

53. *Performance Indicators*, p. 105.

54. Grade of service is frequently synonymous with blocking probability. It establishes the probability with which a user is likely to experience an unavailability of resources on any call carried over those facilities. CCITT recommendation E.540 specifies that the grade of service for the international part of an international connection should be 1 percent (except for six or fewer circuits in a group, when the grade of service can be 3 percent): 99 out of 100 international calls by a user should be successful in the sense that transmission facilities are available to handle the call.

55. *Performance Indicators*, pp. 135-137.

56. *Universal Service and Rate Restructuring*, pp. 72-73.

57. The following situation exemplifies asymmetrical opportunism. Suppose Firm A utilizes information developed by Firm B for Firm A's advantage. Although this is not necessarily problematic in itself (because Firm B can possibly take advantage of Firm A's initiative), it becomes so when Firm A doesn't freely share information it developed with Firm B. This behavior is described as asymmetrically opportunistic.

58. Rita Cruise O'Brien and G.K. Helleiner, "The Political Economy of Information in a Changing International Economic Order," in *Communications Economics and Development*, eds. Meheroo Jussawalla and D.M. Lamberton, (New York: Permagon Press, 1982), pp. 100-132.

59. Anthony Smith, *The Geopolitics of Information*, (New York: Oxford University Press, 1980).

60. OECD, *Guidelines on the Protection of Privacy and Transborder Flows of Personal Data*, (OECD, Paris, 1981).

61. OECD, *Guidelines on Protection.*

62. Henry Goldberg discusses the history of the Cable Landing License Act in "One-Hundred and Twenty Years of International Communications," in *Federal Communications Law Journal*, vol. 27, no. 1, January 1985, pp. 131-154.

63. U.S.C. 47 Section 34 (1988).

64. U.S.C. 47 Section 35 (1988).

65. Michael K. Kellogg, John Thorne, and Peter W. Huber, *Federal Telecommunications Law* (Boston: Little, Brown and Company, 1992), p. 731.

66. Goldberg, p. 139.

67. Federal Communications Commission, *Report and Order*, "In the Matter of International Communications Policies Governing Designation of Recognized Private Operating Agencies, Grants of IRUs in International Facilities and Assignment of Data Network Identification Codes," CC Docket 83-1230, 104 F.C.C. 2d (1986), pp. 256-257.

68. Goldberg, pp. 133, 139-140.

69. *Report and Order*, p. 258.

70. *Report and Order*, p. 214.

71. *Report and Order*, p. 214, footnote 20.

72. *Report and Order*, p. 216, quoted in footnote 25.

73. *Report and Order*, p. 214, quoted in footnote 20.

74. *Report and Order*, p. 246.

75. *Report and Order*, p. 214.

Economic Issues

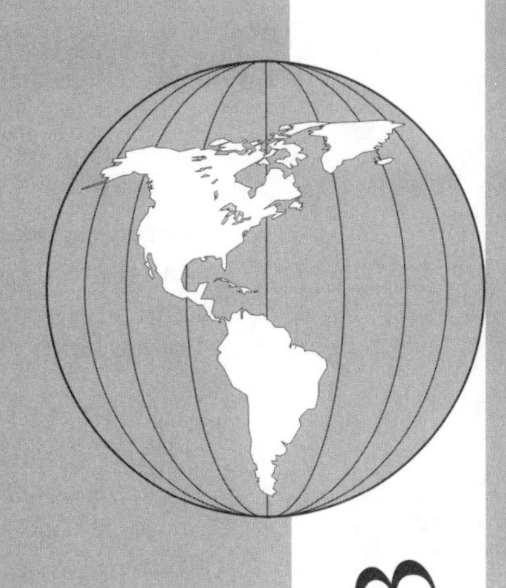

CHAPTER 3

In addition to regulatory and organizational issues, a variety of economic issues influence international telecommunications. Drawing a sharp line between these areas is difficult because regulation has economic and organizational consequences, and organizational structures arise from economic need. In this chapter, we will consider pragmatic economic issues (such as settlements and pricing), more theoretical issues (such as trade in services), economic consequences of standards, and economic development issues. The more theoretical issues are important because they form the foundation upon which much of the services and arrangement of international telecommunications are built. By understanding the role and motivation of standards and economic development, you can understand many of the phenomena that can be observed in today's marketplace.

International Traffic: The Problem of Accounting and Settlement Rates

When a small business owner in Paris, Texas, places a call to Paris, France, technical arrangements must be made to ensure that the call

is routed accurately and delivered to its destination. Many nontechnical arrangements also must be made to establish routes between the United States and France. Issues of pricing, cost recovery, and revenue sharing must be addressed.

At least two entities are involved in handling most international calls, the carrier or telecommunications administration where the call originates and the carrier or telecommunications administration where the call terminates. Each expects compensation for providing its part of the service. The carrier or PTT at the originating end bills the customer for placing the call, collects the money, and compensates the carrier or PTT at the termination point. Compensation for the terminating carrier or PTT, the amount charged for the call by the originating carrier or PTT, and whether the entities involved are carriers or government entities (PTTs) are becoming issues of great importance, as you will see in the following discussion.

During the 1980s, the volume of international messages (both voice and data) increased significantly. However, increased traffic flow among nations hasn't been uniform.

Calls that originate from some countries, notably the United States, have increased more rapidly than calls originating from other nations. Because the carrier or PTT at the originating end of a call compensates the carrier or PTT at the terminating point of a call, nations with increased originating traffic have experienced increased outlays of money. If these nations have not experienced a corresponding increase in terminating traffic and so have not received compensation payments, a payments deficit occurs. Conversely, nations terminating a larger portion of traffic have experienced greater income of compensation than outlays (a surplus). Some nations use these surpluses as development tools; others nations use the corresponding deficits as significant balance of trade issues.

Moreover, as some nations open their telecommunications markets to more players, a once simple process has become complex, with the usual rules of the game coming under increasing scrutiny. In the area of settlements, as in other aspects of telecommunications, competitive forces, changes in industry structure, and the process of regulatory reform put pressure on traditional institutions.

Accounting and Settlements: The Way It Was

Compensation arrangements between countries (referred to as accounting and settlement issues) were fairly simple before privately owned carriers began to replace telecommunications administrations and before competitive providers were allowed into former monopoly services. An example from that simpler era is a useful introduction to the topic of accounting and settlements.

Traditionally, international long distance calls (including telex, packet-switched calls, and other switched services) were treated as services jointly provided by PTTs and monopoly carriers. A customer in Country A placed a call to Country B. The telecommunications administration, or PTT, in Country A charged the customer for the call. This charge is known as the *collection rate*. The PTT in Country A collected the money and then addressed the issue of compensating the PTT in Country B.

Before any calls were placed between Countries A and B, the PTTs in both nations had negotiated an accounting and settlement agreement, which specified the *accounting rate* for a route between the two nations. In other words, a determination was made that the service providers' costs involved in the joint provision of service between Countries A and B would be a certain amount. This accounting rate is more a matter of negotiation than an accurate reflection of underlying costs. As we will discuss later, this lack of correlation between the accounting rate and underlying costs has become a matter of great controversy. As a further point of negotiation, a determination was made on how the two countries split the accounting rate. This split was usually 50/50 and was called the *settlement rate*.

The carrier or PTT in Country A (assume the country is Peru) collected monies from the originator of the call then compensated the carrier or PTT in Country B (assume Country B is Venezuela) for terminating the call by paying out half the accounting rate. The call went through; the caller paid the bill; the Peruvian carrier or PTT was compensated for originating and billing the call; the Venezuelan carrier or PTT was compensated for terminating the call. The Peruvian carrier or PTT paid the Venezuelan carrier or PTT 50 percent (the settlement rate) of the agreed amount for that route (the accounting rate). Figure 3.1 illustrates this arrangement.

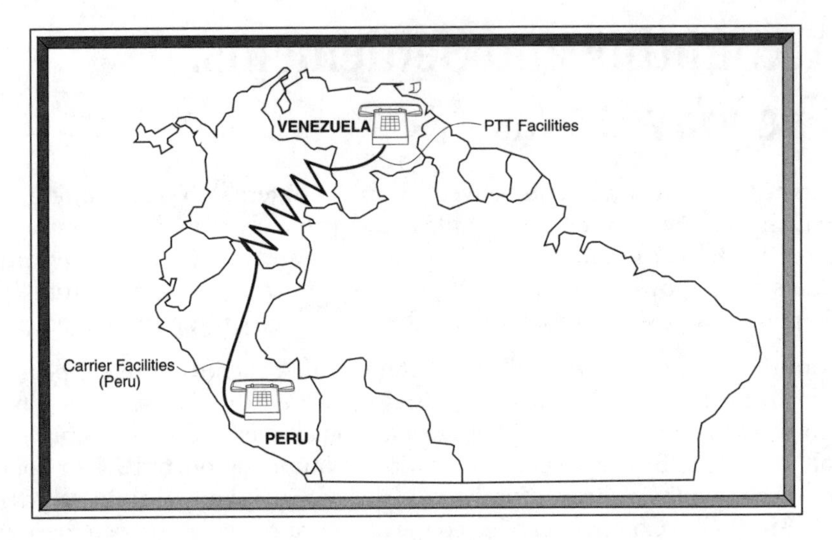

Figure 3.1. Collection, accounting, and settlement rates for an international call.

Emerging Problems and Complications

If this fairly simple arrangement pleased everyone involved, certain circumstances had to exist. First, the amount of traffic from Country A to Country B had to equal the amount travelling in the opposite direction. This assumption of symmetrical traffic flows was the basis for a 50/50 settlement rate. Compensation payments between two countries cancelled themselves out. In other words, assuming that Country A originated 1,000 minutes of calling time to Country B and that the accounting rate negotiated for those calls was $1.00 per minute (to be split 50/50), Country A owed Country B $500. Assuming symmetrical traffic flows, Country B also handled about 1,000 minutes of calling time to Country A and therefore owed Country A $500. The net result was a zero balance of payments for the international settlement.

For this arrangement to work smoothly, some relationship should exist between the collection rate charged the customer and the accounting rate split between the two PTTs. The PTT in Country A, after paying half the accounting rate to Country B, still needed to cover its own cost for originating the call, billing and collecting the call, and paying for its share of the cable or

satellite link between the two countries. The only payment the PTT in Country A received was the collection rate. If the collection rate wasn't adequate to cover those costs plus half the accounting rate, Country A lost. If the accounting rate was a great deal higher than the collection rate, Country A was in a difficult position while Country B received a goodly compensation.

Another factor that facilitated this simple arrangement was the assumption that only two entities were involved in handling the call and negotiating arrangements, the two PTTs. With only two entities involved, there was an equal balance in the negotiating process.

In the last decade, the simpler features of the accounting and settlement arrangements were no longer the rule. Traffic flows between nations are far from symmetrical, particularly traffic originating or terminating in the United States. As a result, the accounting and settlement process created large surpluses and corresponding deficits in trade for nations originating proportionally more calls than they terminated. The relationship between collection rates and accounting rates is nebulous at best. While collection rates are declining in response to competitive pressures and increased network efficiencies, accounting rates aren't declining at the same rate. Indeed, accounting rates have been determined arbitrarily rather than based on underlying costs. As more competitive providers enter the market in some nations—but not in others—the negotiating process for accounting and settlement rates becomes unbalanced and much more complex.

Uneven Traffic Flows

Anyone doubting that international traffic flows are uneven need only read the comments of the U.S. FCC Chair, Alfred Sikes. According to Sikes, "The United States generates far more international calls than it receives."[1] Because it originates more calls and pays out high accounting rates, the United States is facing a deficit in net settlements that increased from $40 million in 1980 to more than $2 billion in 1989. At current growth rates this deficit could reach more than $7 billion by 1998.[2]

The U.S. situation is central to a discussion of accounting and settlement issues because the United States is involved in such a large percentage of the total international market. The United States either terminates or originates over half the traffic carried by the Intelsat satellite system. (See Chapter 4, "Organizational Issues.") That dominance of international traffic is also true for transoceanic cables, because so many of them connect the United States

and other continents. The U.S. deficit is an international issue, if for no other reason than the growing insistence by the United States that something be done to decrease it.

Information offered as part of the recent FCC proceedings regarding international accounting rates presents a compelling argument as shown in Table 3.1.[3] Between 1985 and 1989, the annual growth rate in settlement payments from the United States to the United Kingdom was 52.8 percent; U.S. payments to China increased by 64.5 percent; and payments to Belgium grew by 61 percent. Of 50 nations with whom the United States had a net deficit in settlement payments, only one , India, showed a decline in the growth rate of settlement payments. That decline was only 3.1 percent. In total, the United States paid out $2.4 billion in 1989, with $2.2 billion going to 50 nations. Of those 50 nations, Mexico received the largest amount ($533.9 million) and Jordan received the smallest ($7.9 million).

Table 3.1. U.S. deficit in settlements with 50 nations: 1985-1991.

Country	Settlement Payments 1989 (millions)	Annual Growth in Settlements (1985-1989)	Percentage of Total 1989 Deficit ($2.4 billion)
Mexico	$533.9	24.5%	22.25%
West Germany	167.2	31.8	6.97
Philippines	114.8	16.1	4.78
South Korea	111.6	11.6	4.65
Japan	78.5	10.4	3.27
Dominican Republic	75.2	33.1	3.13
Colombia	70.9	17.4	2.95
Italy	69.5	23.3	2.90
Israel	57.4	9.5	2.39
United Kingdom	46.2	52.8	1.93
Taiwan	45.0	11.9	1.88
Brazil	43.7	19.5	1.82

Country	Settlement Payments 1989 (millions)	Annual Growth in Settlements (1985-1989)	Percentage of Total 1989 Deficit ($2.4 billion)
El Salvador	42.6	24.6	1.78
Pakistan	40.1	32.8	1.67
Peru	40.0	18.9	1.67
Poland	36.2	23.0	1.51
Ecuador	34.8	22.2	1.45
France	32.9	19.5	1.37
Jamaica	32.6	19.5	1.36
Guatemala	32.4	34.7	1.35
Greece	30.2	17.9	1.26
Thailand	30.2	17.9	1.26
China	26.4	64.5	1.10
Trinidad and Tobago	23.9	30.8	1.00
Haiti	23.1	8.1	.96
Spain	22.9	18.5	.95
Egypt	21.3	10.0	.89
Iran	20.8	13.4	.87
Argentina	19.7	18.8	.82
Portugal	17.2	23.5	.72
Panama	16.6	33.8	.69
Ireland	16.5	12.5	.69
Indonesia	14.7	19.0	.61
Turkey	14.3	21.5	.60
Hong Kong	14.1	5.2	.59

continues

Table 3.1. continued

Country	Settlement Payments 1989 (millions)	Annual Growth in Settlements (1985-1989)	Percentage of Total 1989 Deficit ($2.4 billion)
India	13.8	–3.1	.58
Cuba	13.7	2.2	.57
Nicaragua	12.9	31.0	.54
Yugoslavia	12.5	1.2	.52
Honduras	12.1	33.8	.50
Malaysia	12.1	6.7	.50
Belgium	11.9	61.0	.50
Costa Rica	11.3	18.5	.47
Saudi Arabia	11.2	11.5	.47
Switzerland	10.8	12.6	.45
Chile	10.3	9.4	.43
Barbados	9.8	14.3	.41
Bolivia	8.5	12.4	.35
Romania	8.4	3.4	.35
Jordan	7.9	20.6	.33

Source: Federal Communications Commission, Further Notice of Proposed Rulemaking, CC Docket No. 90-337, Phase II (FCC 91-158), Appendix C, (6 FCC Rcd No.12, 3441-3442, May 1991).

The settlement situation during 1990 was no better, with the U.S. deficit reaching $2.9 billion, an increase of 20.7 percent over the previous year.[4] The net payout to Europe was $687 million; to Asia it was $545 million. Mexico again received the largest amount (almost $583 million). Amounts paid to the United Kingdom, Jamaica, Spain, Germany, France, Peru, and Brazil increased significantly from those paid in 1989. Payments to Japan declined over 26 percent, but still totalled more than $55 million. Table 3.1 has a more complete listing of 1985 through 1991 U.S. deficits.

Many reasons have been suggested to explain the large disparity in traffic flows between the United States and other nations. The explanation most often offered is the relationship between the collection rate for calls originating in the United States and the collection rate for calls originating in other nations. Since the United States introduced competition into its domestic and international long distance markets, AT&T and its competitors have had an impetus to drive collection rates lower. The result is a significant increase in calls originating in the United States. Because competition frequently doesn't exist in many other nations, their collection rates are still high in comparison to U.S. rates, and traffic stimulation of the same magnitude hasn't emerged.[5]

Traffic imbalances between the United States and other nations also can be fostered by differences in rate structure. In the United States, time-of-day discounts are built into the rate schedule. Networks are engineered to handle peak traffic loads, so in off-peak times, unused excess capacity can exist. In an effort to stimulate traffic in off-peak times and encourage use of excess capacity, AT&T and its competitors offer off-peak discounts for calling during evening and weekend time periods. This off-peak pricing situation hasn't normally been the case for other nations.[6] As a result, traffic originating in the United States has increased in off-peak periods, but other nations haven't experienced a similar off-peak increase in traffic.

This system of off-peak or time-of-day discounts, in addition to furthering asymmetrical traffic flows, can actually create a net loss for U.S. carriers because of the relationship between collection and accounting rates. The U.S. carriers collect discounted collection rates from their customers; however, accounting rates paid to the carriers or PTTs that terminate the calls aren't usually discounted for time-of-day or off-peak usage, or at least not discounted as significantly. As a result, in some cases the amount paid to the terminating PTT in accounting and settlement dollars can actually exceed the amount the U.S. carrier collects from its customer.

Collection rates are coming down in nations other than the United States. However, an examination of some examples (as shown in Table 3.2) suggests that U.S. collection rates are still lower, on balance, than most other collection rates, especially on longer duration calls. Although U.S. schedules tend to charge more for the initial minute of a call but significantly less for additional minutes, many other nations charge the same for all minutes.

Even if collection rates are lowered in all nations, the imbalance with the United States might not be solved. As we've pointed out, the "size of the imbalances" suggests that the deficit in payments may be "very much a function of the underlying determinants of traffic flows"; and that the "prime cause of the growing deficit has been an across-the-board increase in U.S.-originated

demand for telephone calls relative to the rest of the world."[7] The underlying determinants of traffic flows cited include differences in per capita income, economic and cultural differences, tourism, immigration, and the location of international businesses.[8]

Table 3.2. Examples of International Message Toll Service (IMTS) dial service rates.

Country	Initial Period Standard	Discount	Economy	Additional Minutes Standard	Discount	Economy
U.S. to Mexico*	$1.77	—	$1.19	$1.41	—	$0.95
Mexico to U.S.*	$1.56	—	$0.88	$1.20	—	$0.64
U.S. to Germany	$1.77	$1.42	$1.15	$1.09	$0.82	$0.65
Germany to U.S.	$1.60	—	—	$1.60	—	—
U.S. to Korea	$3.68	$2.80	$2.38	$1.37	$1.04	$0.89
Korea to U.S.	$2.94	—	—	$2.94	—	—
U.S. to Philippines	$3.25	$2.75	$2.30	$1.40	$1.10	$0.85
Philippines to U.S.	$3.30	$2.45	—	$2.45	$1.80	—
U.S. to Japan	$3.05	$2.55	$2.16	$1.24	$0.99	$0.81
Japan to U.S.	$2.57	$2.08	$1.53	$1.25	$0.97	$0.76
U.S. to Colombia	$2.39	$1.91	$1.55	$1.11	$0.84	$0.67
Colombia to U.S.	$2.33	$1.08	—	$2.33	$1.08	—
U.S. to Dom. Rep.	$1.45	$1.16	$0.94	$1.06	$0.80	$0.64
Dom. Rep. to U.S.	$2.45	$2.02	—	$2.45	$2.02	—

Country	Initial Period Standard	Discount	Economy	Additional Minutes Standard	Discount	Economy
U.S. to Italy	$1.88	$1.40	$1.16	$1.04	$0.77	$0.65
Italy to U.S.	$2.87	$2.16	—	$2.87	$2.16	—
U.S. to Israel	$2.98	$2.47	$2.08	$1.20	$0.90	$0.75
Israel to U.S.	$2.56	$1.92	$1.30	$2.56	$1.92	$1.30
U.S. to UK	$1.44	$1.15	$0.98	$0.94	$0.71	$0.60
UK to U.S.	$1.04	$0.94	$0.79	$1.04	$0.94	$0.79

* Rates shown for Mexico are between Mexico City and New York City because rates vary with mileage within the United States and Mexico. Rates from Mexico don't include government tax.

Source: Federal Communications Commission, Notice of Proposed Rulemaking, CC Docket No. 90-337, FCC 90-265, Appendix C (5 FCC Rcd No.16, 4959-4961, 1990).

If these determinants are indeed the reasons for the traffic flows, decreases in collection rates might not be enough to change the balance. The United States will continue applying pressure to all nations with which it has a deficit for lower collection rates. Such pressures might result in greater traffic stimulation and will undoubtedly continue to encourage a process of price decreases attractive to telecommunications managers.

Accounting Rates

The CCITT, in its *General Tariff Principles*, discusses four approaches to remunerate the terminating entity of a call:

- Flat rate payment per circuit used
- Payment based on a price per traffic unit
- Payment based on an accounting/revenue division method (the accounting/settlement method discussed earlier in this chapter)
- No payment to the terminating carrier (sender keep all)[9]

The accounting/revenue division method is used most frequently and is computed on net traffic between two nations. An accounting rate (split as explained earlier) is assessed to each minute of international traffic. This accounting rate is usually expressed in terms of special drawing rights (SDRs).

SDRs are based on International Monetary Fund exchange rates for U.S., German, British, French, and Japanese currencies.

The complaint about accounting rates (voiced most strongly by countries facing a settlement deficit) is that they aren't reflective of underlying costs. The FCC in the United States recently claimed that U.S. carriers might be overpaying Asia and Europe alone as much as $500 million per year because of high accounting rates. According to the FCC, U.S. carriers should be paying half of current accounting rates to terminate traffic, or at most $.39 to $.60 in Asia and $.23 to $.39 in Europe.[10] Of even more concern to the FCC is the fear that U.S. carriers are also subject to discriminatory treatment in this regard and in Europe, for example, are paying from $.50 to $1.40 more per minute to terminate U.S.-originated traffic than do other nations.

As part of its campaign to solve settlement imbalance problems, the FCC, in its proceeding regarding international rates, has directed U.S. carriers to negotiate lower, more cost-reflective accounting rates. The FCC further determined to encourage changes in CCITT recommendations that invite cost-based rates. Some of these efforts have begun to bear fruit. CCITT Study Group III, in March 1992, adopted a draft Recommendation on cost-oriented accounting rates for consideration by the CCITT membership. The Recommendation allowed for some discretion in the amount of time permitted for introducing significant reductions. The suggested time period was one to five years, but the actual time period chosen could be tied to the individual country's state of development.[11]

The inclusion in the draft language of a phase-down period for substantial reductions in rates recognizes the importance of accounting and settlement surpluses to many nations, including developing nations. These nations use revenues from settlement surpluses to subsidize other areas of the government. Developing nations have a strong incentive not to change current accounting rates, or if they must change rates, they have a phase-down period of adjustment.

The actual level of accounting rates is not only cost-reflective, but the usual method of splitting those rates can greatly misrepresent underlying costs. The costs involved in call origination are more extensive than those involved in call termination. Costs are involved in call setup, billing, and collection. Also, noncollection of payments is absorbed at the originating end. None of this is reflected in a 50/50 settlement split.

Issues other than underlying cost are involved in the settlement amount. The International Telecommunications Union's (ITU) 1984 *Maitland Commission Report* suggested that funding for developing nations could be provided by setting aside a small portion of account revenues or that developed nations

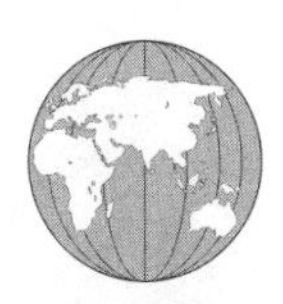

could provide assistance to their lesser developed counterparts by agreeing to other than a 50/50 split.[12] In other words, the developing nations would keep more than 50 percent. Beneath the assumption justifying such an uneven split was the idea that the cost for terminating calls was higher in developing nations. This idea was reinforced by an ITU study, which found that the total cost per minute for telephone calls in developing nations was 2.08 times the cost in industrialized nations.[13]

The United States remains adamant in its position that no U.S. carrier accept less than 50 percent of the accounting rate for traffic. One reason the United States maintains this position is fear that the settlement split could become a tool that foreign PTTs use in the negotiating process to play competing U.S. carriers against one another. This process is known as *whipsawing* and is discussed in the next section.

Those dissatisfied with the current accounting and settlement approach have suggested a range of alternatives:

- The creation of an access charge arrangement by which tariffs are levied for termination of a message
- Cost-based access fees to terminate international calls at rates that are no higher in price than rates for placing a domestic call of equivalent distance
- Off-peak accounting rates
- Two-tiered accounting rates where a lower rate kicks in after a certain level of traffic imbalance is reached
- Growth-based accounting rates where lower rates apply after a certain volume of traffic is reached[14]

Other suggestions are related more to changes in industry structure than to changes in rates. These include the introduction of more competition, the creation of end-to-end ownership arrangements (so that only one entity is involved in handling the call), and the creation of joint ventures by carriers. Any changes made in the accounting and settlement scheme will affect who handles international calls and what is ultimately charged for those calls.

Whipsawing

If competitive carriers sit at one end of the bargaining table and a PTT at the other end, accounting and settlement negotiations can result in a phenomenon called *whipsawing*. The PTT can have a superior bargaining position,

especially because negotiations with the carriers are bilateral. The PTT can use its position to play one carrier against another. As a further bargaining point, the PTT doesn't need to do business with all the carriers.

The PTT might have an existing 50/50 settlement agreement with carriers A and B and then make a 60/40 settlement split a condition for carrier C's entry into the market. The PTT could then refuse to do business with carriers A and B unless they adopted that same split.

To avoid such a situation, the FCC requires all U.S. carriers to adhere to a 50/50 split in their negotiations. Indeed, the avoidance of whipsawing caused the FCC to adopt a Uniform Settlements policy until 1987. That policy forced adherence by all carriers to uniform accounting rates for all parallel routes, split 50/50 and expressed in the same financial terms, such as SDRs or dollars. In 1987, the FCC allowed carriers to agree to nonuniform accounting rates, but only after the FCC was notified and other carriers had an opportunity to file protests. Recently, the FCC liberalized its approach a bit further. To give U.S. carriers greater flexibility in negotiating accounting rates, the FCC decided to allow a carrier to adopt a lower accounting rate upon simple notification of the FCC. However, carriers seeking to adopt differently structured accounting rates (growth-based, for example) must seek a waiver from the FCC before proceeding. The FCC did comment that waivers involving noncost-based surcharges or departure from 50/50 settlement splits would be difficult to obtain.[15]

In its proceeding, the FCC sought other means of preventing the whipsawing phenomenon, in addition to the 50/50 requirement. While not requiring that every carrier immediately implement a reduced accounting rate negotiated by one carrier, the FCC did require that carriers file with their notification letters, notarized statements attesting the following:

> *...that the filing carrier (1) has not bargained for, nor has knowledge of, exclusive availability of the new accounting rate; and (2) has not bargained for, nor has any indication that it will receive, more than its proportionate share of return traffic.*[16]

By not allowing carriers to negotiate exclusive rates, these provisions reduce the chance of whipsawing. They also reduce the chance of whipsawing by requiring adherence to the rule of proportionate return.

If three carriers all terminate traffic to one PTT, and the PTT originates traffic back to the country where the carriers operate, the question arises: To which carrier should the PTT terminate its traffic? This is an important question because the PTT pays that carrier for terminating those calls. Under

proportionate return, the PTT allocates its terminating traffic to the three carriers in the same proportion it received traffic from them. Deviation from this rule could lead to whipsawing. The PTT might get concessions from one of the carriers by offering to route all its terminating traffic to that carrier.

Transit Routes

Carriers seeking to terminate traffic in a specific country have some options. The carrier can negotiate a direct agreement with the PTT or telecommunications administration in that country. Alternatively, if the PTT, perhaps thinking that the carrier won't generate enough traffic to justify the administrative burdens associated with a direct arrangement, doesn't wish to negotiate such an agreement, the carrier can utilize a transit route.

For instance, a carrier in Peru (Figure 3.2) that can't arrange a direct agreement with a carrier in the United States (let's say AT&T) might have an agreement to terminate traffic with the PTT in Brazil. Meanwhile, Brazil's PTT has an agreement to terminate traffic in the United States. Brazil might charge the carrier in Peru a transit fee for using its facilities to reach the United States. The carrier would pay Brazil a transit fee and pay for the ultimate termination of the calls to the United States.[17] We realize that Peru does have a direct agreement with the United States. We are just using this as a hypothetical scenario for illustrative purposes only.

In this example, the Peruvian carrier's termination costs increase because it uses a transit route. The use of transit routes can also decrease termination costs. If the best accounting rate the carrier can negotiate to terminate traffic in Venezuela is higher than the total of the transit fee and the accounting rate to terminate traffic from Brazil to Venezuela, the carrier could actually lower its settlement payments with a transit route. Such arrangements can develop specific countries as hub locations. A country serves as a point to which carriers initially route their traffic, and the telecommunications administration in that country routes the calls to their ultimate locations.

Forces for Change

Many forces exist that will change the current accounting and settlements process. The entry of more competitors into the international market will make

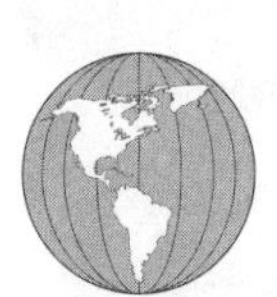

the current negotiating process inadequate. Negotiations might begin to take place between carriers, rather than between carriers on one end and PTTs on the other. In other words, AT&T in the United States now negotiates with Mercury in the United Kingdom rather than with the former British PTT in the traditional model. Such balanced negotiations will place more downward pressure on accounting rates. Lower accounting rates will lead to lower collection rates and so to greater stimulation of traffic. Instead of negotiations, joint ventures by carriers could be created. For example, carriers such as BT in the United Kingdom, France Telecom, and KDD in Japan could form a joint venture to carry traffic between these countries and divide the profits according to the joint venture agreement. Competition for traffic among such joint ventures might also lead to lower prices for international calls.

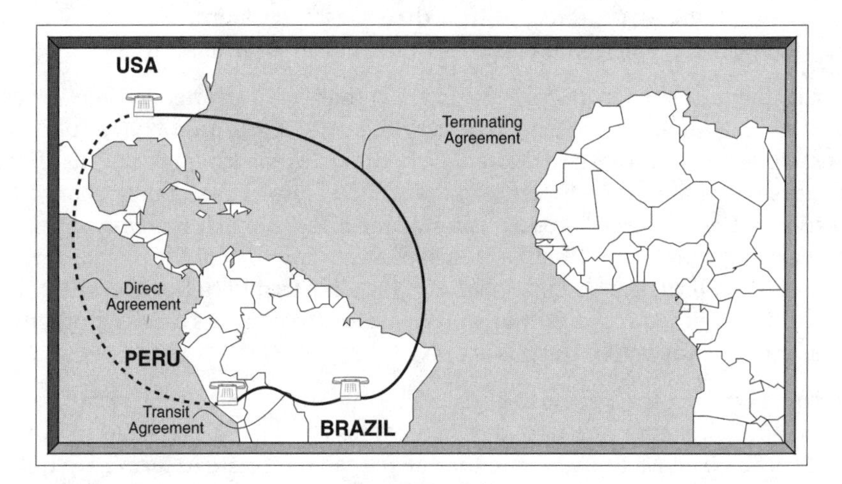

Figure 3.2. Example of a transit route.

If nations allow foreign carriers to operate inside their borders, end-to-end provision of service might become another alternative to the current arrangement. If KDD is allowed to provide service in the United Kingdom, for example, one carrier (KDD) could handle traffic between Japan and the United Kingdom; no settlement arrangements would be needed. Issues of foreign ownership are matters of trade and domestic policy. The emergence of transit hub locations (as described previously) might provide yet another alternative.

One major development that could affect the accounting and settlement arena would be the institution of widespread shared use and resale of international circuits. In its proceeding on international rates, the FCC targeted this issue as one avenue for decreasing its net settlements deficit and determined that U.S. carriers should permit resale of their international private lines for provision of basic telecommunications services (long distance) where equivalent resale opportunities are allowed by the other nation.[18] The CCITT addressed the issue of international shared use and resale by submitting liberalized language for its *Recommendation D.1* regarding resale of private lines. The CCITT used its new accelerated mail ballot approval process in seeking approval of the liberalized recommendation. The entrance of new players (resellers) would present an incentive for the lowering of accounting and collection rates.

Pricing

Telecommunications pricing has been a frequent subject for economic study. Because economists have traditionally regarded telecommunications as a utility most efficiently supplied by a monopoly provider, competitive market forces have not governed pricing. Instead, the monopolist sets prices, often under the scrutiny of a regulatory body. While most economists assume that a competitive market disciplines pricing and holds prices closer to cost, monopoly pricing has no such inherent discipline.

Pricing for services offered by a monopoly provider is subject to inefficiencies and abuses:

- Prices may be substantially higher than a monopolist's underlying costs.
- Prices—for which demand remains high regardless of price level (usually essential services)—may be kept high, despite relatively low underlying costs, and the resulting profits may be used to subsidize less popular services by keeping prices artificially low.
- A monopoly has little incentive to pass along the benefits of increased efficiencies (lower costs) to consumers. Indeed, little incentive may exist for seeking efficiencies.

Because of these inefficiencies and abuses, customers may be overpaying for services, service providers may operate at less than optimal efficiency, and the monopolists' ability to keep prices for some services artificially low may discourage potential competitors. The possibility of such negative results is one reason for economists' interest in telecommunications pricing issues.

From an economist's perspective, telecommunications services have the following attributes:

- Network components and facilities are geographically located in relation to the final consumer.
- Production is capital intensive.
- The product can't be stored in inventory for later use.
- Production facilities have well-defined capacities.
- The costs of operation are largely independent of use of the production capacity (networks), so variable costs are small.
- The incremental (marginal) cost of servicing a call is near zero when the network is underutilized, and very high when the network is utilized at capacity, because additional capacity must be provided, or the congestion must be managed by nonprice mechanisms, such as waiting lines or blocked calls.[19]

In economics, prices ration scarce resources among consumers and cover the costs of producing the resources that are rationed. One objective of telecommunications pricing, therefore, is to ration telecommunications services equitably among users while covering the operating expense of the telecommunications provider. In practice, this normally results in charging different prices when the network load varies, so that consumers who value services more will pay higher prices at peak times, and consumers who value services less will postpone their consumption until prices are lower.

Ideally, prices should vary with the network load (*dynamic pricing*). However, this is inconvenient and impractical for users. Fortunately, network loads vary in a regular way in the course of a day and in the course of a week, allowing prices to vary consistently when higher network loads are expected. Thus, we have the common practice of charging more for services during normal business hours when loads are higher and less when loads are lower. This is referred to as *peak load pricing* and is practiced widely in the telecommunications and electric power industries.

Pricing Challenges

Telecommunications services aren't always priced according to economic theories or underlying costs. No single universally accepted approach has been identified as the best for balancing the needs of the general consumers,

owners of telecommunications resources, and specific telecommunications users. Also, no unanimous agreement has been reached regarding the optimal pricing mechanism for setting rates. Underlying the controversy regarding pricing are the questions of who should bear the cost of the service, what those costs actually are, and what barriers should be built between services offered on a monopoly or franchise basis and competitive services.

One major pricing controversy concerns the issue of fully allocated versus incremental pricing. Underneath this controversy lies the question of who should pay for which network costs. This is an issue because telecommunications network equipment is jointly used to provide a variety of services. For example, a subscriber uses the same loop plant—connection between the premise and the first switching office—when placing local calls, domestic long distance calls, and international calls. The question is how should these various services share in recovering the cost of the loop plant?

Those who take an incremental pricing approach make the assumption that the loop is necessary for local calls, even if no long distance services are used. In this view, the basic component of telecommunications service is the local connection to the exchange office. Long distance calls are merely additions to that basic local service. Those arguing in this manner suggest that local rates should completely bear the cost of the loop plant. Allocating any of that cost to long distance rates, in effect, creates a subsidy from long distance services to local services. Indeed, in the United States, the cost of loop plant was allocated more heavily toward long distance rates because of the public policy goal of keeping local rates low to aid in the achievement of universal service provision. However, as competition for long distance services pushes long distance rates lower, this practice is being phased out.

Those who don't espouse the incremental approach argue that a fully allocated or fully distributed approach should be taken. Some overhead costs associated with the provision of joint services can't be neatly allocated to each individual service offering. In a fully allocated approach, these indirect costs, which can't be allocated directly, are allotted to each service on an agreed upon basis. The fully allocated approach is one in which the policymaker wants each service to bear its full costs. This goal is especially desirable for those seeking to avoid the problems of cross-subsidization between regulated and competitive services. Cross-subsidization of services results when the profits from one service are used to subsidize, or keep artificially low, the prices of another service.

Table 3.3. Differences in pricing/cost approaches.

	Purely Incremental Approach*	Fully Allocated Approach*
Local Rates	100%	XX%**
Long Distance Rates	0%	XX%**

* Local loop costs (equipment, installation, maintenance, etc.)
** Allocated based on usage or some other basis acceptable to the regulator

The fear of cross-subsidy in this regard is especially evident in mixed mode approaches to service regulation. Network providers can offer both regulated services (often on a monopoly basis) and competitive services using the same network and facilities. The allocation of costs among those services is especially important if the network provider is to be kept from monopoly abuses. The network provider can use revenues derived from monopoly services (rates) to keep the prices of competitive services down. This creates an unfair competitive advantage for the network provider and creates a situation in which the rate payer is subsidizing the carrier's competitive offerings. In the United Kingdom, BT's operating license requires the company to guard against cross-subsidization of regulated and competitive services. In the United States, the FCC mandates a complex series of accounting rules to guard against cross-subsidization. At times, the FCC has required that monopoly and competitive services be offered through separate subsidiaries to avoid cross-subsidization.

Telecommunications pricing isn't a simple matter. Policy issues, underlying economics, and regulatory structures come into play when considering a pricing approach. Many of the same concerns involved in pricing issues are involved in discussions about regulatory approaches and appropriate industry structure.

Alternatives for Industry Pricing Structure and Their Implications

As we discussed earlier, a monopoly has traditionally provided telecommunications services. If the monopoly is a publicly owned administration, pricing is generally assumed to be set in the public interest. Although the

administration can be supervised by other branches of government or commissions appointed to oversee the operations of the monopoly, having government regulate itself doesn't make sense. Corrections to a misbehaving public monopoly must come through the political system and the press.[20]

An alternative to public ownership of the monopoly provider is to allow the monopolist to be private, as has been the case in the United States. In this case, regulatory supervision is necessary because otherwise the monopoly provider would probably operate in its own interest, which might not coincide with the public interest. In the absence of regulation, a private monopoly would abuse its market position to charge high prices and engage in predatory pricing to prevent new entrants from emerging as significant marketplace forces. From the standpoint of the social arrangement between the provider and the government, the monopoly is granted a franchise to operate as a monopoly in exchange for agreeing to be regulated.

The social objectives of regulation are often codified in law, although they might also exist in a less formal document or agreement. Frequently, the objective of regulators is to establish prices so that the service provider earns a "reasonable" return on investment, so that social goals such as universal service are carried out, and so prices don't rise at too rapid a pace (faster than general inflation). Thus, regulators effectively direct the monopoly in terms of how much it may invest in its infrastructure and, in some cases, where the investment must be made. If a regulatory agency considers an investment frivolous or unnecessary, it might not allow the carrier to earn a return on that investment from subscribers. As a result, user-oriented observers of the industry consider that the subscribers pay for investments because the provider was permitted to charge users for an investment in the infrastructure.

If true competition exists, no regulation is necessary for the firms, because market competition prevents the exercise of monopoly abuses. Competition is desirable in this framework, because regulators might not make socially optimal decisions. Therefore, if regulators are removed from the pricing equation, the markets operate more efficiently.

Approaches to Regulation

Many approaches to regulation of telecommunications carriers exist. Of these, the most common are *rate of return regulation*, *price cap regulation*, and *social contract regulation*. Each form has different implications for regulators, users, and telephone providers. In addition to applying these regulatory plans,

regulators frequently require regulated service providers to average their prices geographically and distribute their allowable revenue on a value of service basis so as to meet social goals such as universal service. (See Chapter 2, "Regulatory Issues.")

In rate of return (ROR) regulation, the regulatory commission fixes tariffs so that the telecommunications provider is permitted to earn—in addition to covering its variable costs—a specified return on its rate base. The rate base consists of those capital costs that the regulatory agency deems necessary for the provision of telecommunications service. To adequately construct a rate base, the regulatory agency must specify a rate at which these capital investments could be depreciated.

Setting low depreciation rates decreases the rate base, resulting in lower prices for users. Therefore, regulatory agencies endeavor to achieve this goal. On the other hand, telecommunications providers, because they can't maximize profits, seek to maximize revenues. They do this by attempting to increase the rate base and the variable costs. In this way, providers can charge higher prices for services. Because regulators aren't involved in the decision-making process of the telecommunications provider, they operate under imperfect information about actual costs.

So, substantial opportunity and incentive exists for the telecommunications provider to inflate the actual investments and costs necessary to operate the network. This results in incentives to "gold plate"—overinvest in capital equipment—the telecommunications in addition to introducing other pricing distortions.[21]

Because of these difficulties with ROR regulation, economists have sought regulatory approaches that are more efficient and that don't cause the price distortions of ROR regulation. These alternative approaches are sometimes referred to as *incentive regulation* because they are intended to provide incentives to invest efficiently in technology instead of overinvesting.

One such approach is to use *price cap regulation*. In price cap regulation, services are separated into "baskets" that correspond to general service types, such as residential services and business services. The prices for these baskets of services are set at a predetermined level. The carrier is free to redistribute prices within the basket of services, as long as they don't exceed the overall price cap for the basket. The level of each basket is readjusted periodically to account for inflation and costs outside the carrier's control (such as energy prices and taxes). The price level can also be decreased to account for improvements in productivity due to the deployment of new technologies. Price bands might also be specified so that the price of the basket of

services can't be lower than a floor, thereby eliminating predatory pricing. Finally, services can be periodically rebalanced across baskets to accommodate new services and changes in the demand for existing services.

The price cap scheme regulates prices without regard to costs. So, if the service provider reduces its operating costs, prices are unaffected, and the provider increases its profits. Because of this, providers have an incentive to invest efficiently, because cost saving measures don't produce decreased revenues.

Finally, a variety of *social contract tariffs* exist. In general, these tariffs provide strict regulation for "basic" services and no regulation for "enhanced" services. Typically, prices for basic services, such as basic analog residential and business local and long distance service, aren't allowed to rise faster than the rate of inflation. Enhanced services aren't regulated at all. While these services require little regulatory oversight, no controls exist for the cross-subsidization between basic and enhanced services. In many cases, the basic service rates are indexed to the inflation rate, with no provision for productivity improvements. This can give service providers an incentive to use some of the profits from basic services to subsidize unprofitable or underpriced enhanced services.

Telecommunications and Trade in Services

International telecommunications only recently has been recognized as an important topic in trade negotiations. Traditionally, international telecommunications was regarded as a utility, usually provided by a government entity or by a privately owned monopoly. Because facilities between countries must be connected in order for communication to occur, negotiations were held between the network providers. These negotiations were bilateral and their results were usually confidential.

This situation wasn't regarded as a matter of trade. It was more a matter of governments (and usually, in the case of the United States, AT&T) getting together and making arrangements to provide a fairly limited set of services. No competitive providers existed. The users of the network weren't involved. All entities had a clear understanding that the services in question were jointly provided, with the network operator in each country providing its portion of the service.

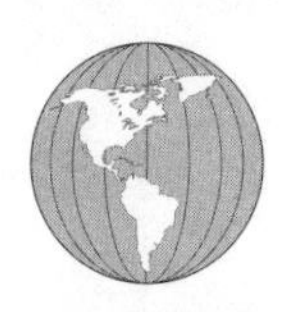

Telecommunications has changed drastically in the last decade. Providers compete to provide many services in many nations. International corporations have built private networks for their own use and have a loud voice in requesting more services and interconnection to the public network. In many nations, the network is no longer owned and operated by the government.

The status of telecommunications itself has changed. No longer perceived merely as a convenient utility, telecommunications has become essential to industry. Banking, insurance, transportation, and many other industries now depend on telecommunications networks. Telecommunications is now a commodity, and a very important one internationally. The manner in which entities deal with the sale and purchase of this commodity has become a matter of international trade. What this means for the future of international telecommunications and the effect trade agreements will have on existing telecommunication standards, regulatory structures, and service provision are matters that merit the close scrutiny of telecommunication managers.

The General Agreement on Tariffs and Trade (GATT)

A central ingredient in world trade is the General Agreement on Tariffs and Trade (GATT). Twenty-two nations negotiated and signed the GATT in 1947. Though it began as a multilateral trade agreement, it has become, according to Leah Haus,

> *...[the] main institution responsible for guiding the conduct of world trade. The norms and rules of the GATT aim to ensure the maintenance of an open, nondiscriminatory market in which government intervention is minimized and tariffs and prices guide the decisions of private firms.*[22]

The GATT comprises rounds of multilateral negotiations that take place every ten years and last three to five years. The eighth round of these negotiations began in Punta del Este, Uruguay, in 1986 and is called the Uruguay Round. Currently, 103 parties contract to the GATT.

The traditional subject of GATT negotiations is goods, such as fuels, textiles, iron and steel, and agricultural products. A framework exists for these negotiations, which are usually held to reduce trade barriers like tariffs and import quotas.

GATT negotiations center on some basic principles:

- **The market principle**—The belief that the free market should govern transactions. In this context, any domestic monopolies or regulations should not infringe on market rules.
- **The reciprocity principle**—A reduction in barriers by one nation should be followed by a reciprocal benefit on the part of the other nations.
- **The most-favored-nations principle (or nondiscrimination)**—A benefit accorded one nation should be accorded all nations party to the multilateral agreement.
- **The legitimacy of domestic regulation principle**—Nations reserve the right to formulate domestic regulations necessary to fulfill needed policy goals.
- **The national treatment principle**—Foreign entities are treated in the same manner as domestic concerns. No preferential treatment is accorded domestic companies.
- **The orderly adjustment principle**—Nations can implement a reduction in trade barriers in such a manner that mitigates negative effects on their social, political, and economic situation.
- **The transparency principle**—Details regarding tariffs and prices should be open to all importers and exporters.
- **The consultation principle**—Open discussions should be the mechanism for dealing with conflicts between nations.
- **The dispute settlement principle**—A means of arbitration should exist for the orderly settlement of disagreements.[23]

These principles have served the contracting parties to the GATT well when crafting trade agreements regarding goods. Efforts are now underway to apply such principles to trade in services as well.

The General Agreement on Trade in Services (GATS)

Because of the growing importance of services, the focus of international trade negotiations has now expanded to include services as well as goods. The importance of services is increasingly apparent as nations' economies become

more dependent on services. Financial services such as banking, insurance, and real estate; professional services such as engineering, accounting, and law; computer and data processing services; and telecommunications services are ever greater sources for employment, revenue generation, and international activity. Given the increasing importance of services, not surprisingly, the usual framework for handling trade issues is being extended to encompass services as well.

At the Punta del Este meeting in 1986, the issue of trade in services was placed on the agenda. This marked the culmination of much study and discussion by groups in the United States and by the OECD about the need for such a step.[24] The decision was made to form a Group of Negotiations on Services (GNS) to formulate an overall framework of principles and rules for international trade in services.

The GNS worked to develop a framework that applies to all service sectors and identifies those sectors for which specialized rules (or annexes) are needed. The GNS had made some progress by its midterm review in Montreal in 1988. A few basic principles and rules agreed to in Montreal are greatly reminiscent of the basic principles underlying the GATT:

- Provisions should exist "to promote transparency of domestic laws, regulations, and administrative practices related to services."
- A commitment should exist to liberalize—encourage competition and open markets in—some service sectors, with the recognition that such liberalization could present difficulties in some developing countries.
- National treatment of service exports and exporters should exist with no preference given to domestic services and service providers.
- Most-favored-nation treatment should apply.
- Entities should make efforts to encourage the increasing participation of developing countries.
- Entities should recognize that exceptions might be necessary in order for countries to deal with "adverse temporary circumstances" or more permanent problems.
- Entities should recognize each nation's right to regulate its service sector in an appropriate manner. However, new regulations should be consistent with each nation's commitment to liberalizing the market.[25]

Although the previous basic notions were agreed to in Montreal, the GNS continued for four more years to work through disagreements on the actual wording of a final document. Conflicts between developed and developing

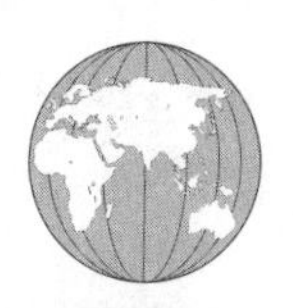

nations hampered negotiations. For example, in the field of telecommunications, although developed nations sought to move fairly aggressively on issues of market access, national treatment, and the easing of domestic regulation, many developing nations sought to contain the force of liberalization and deregulation and to urge greater attention in helping the developing world develop infrastructure and export opportunities.[26]

Even among the developed nations, some strife was apparent. In its proposed draft for a GATS framework, the United States sought to allow "special agreements" in some service sectors. Such special agreements could exempt some services from coverage by the proposed GATS. This provision met with resistance. Disputes between the United States and the European Community (EC) further complicated the proceedings. Their "ambivalent relationship with each other across a number of key issues" was discussed by the Canadian political science professor R. Brian Woodrow as the "fulcrum on which the Uruguay Round and the services negotiations would be balanced."[27]

One focal point of conflict between the United States and the EC is indicative of negotiating problems caused by underlying differences in industry structure. In the area of telecommunications equipment, the EC feared that most of the benefits of EC market liberalization would accrue to non-EC firms. To minimize this possibility, EC negotiators sought to gain access to the U.S. telecommunications equipment market, which is dominated by privately owned Bell Operating Companies, GTE, and AT&T. Because the telecommunications administrations in the EC have quasi-governmental status, they would be subject to provisions regarding liberalized government procurement procedures. The Bell Operating Companies, GTE, and AT&T, because of their private status, wouldn't be subject to such provisions. EC negotiators, claiming that these U.S. companies had an unfair advantage, sought to make the procurement policies of the Bell Operating Companies, GTE, and AT&T subject to GATT provisions.[28] The United States protested such an action.

In early 1992, progress was being made in discussions regarding a draft framework for a GATS, according to Arthur Dunkel, Director General of the GATT.[29] The elements of this draft agreement are, for the most part, those found in the GATT. Included in the draft agreement are

- A most-favored-nation clause.
- A provision regarding transparency, or the requirement to make public "all relevant laws, regulations, administrative guidelines and all other decisions, rulings, or measures of general application...which pertain to or affect the operation of this Agreement."

- Provisions for encouraging increased participation by developing nations.
- Stipulations regarding appropriate actions of monopoly providers to ensure that their actions are in keeping with the trade agreement provisions.
- Provisions for liberalized market access.
- Provisions regarding national treatment so that foreign entities are treated in the same way as domestic companies.
- Plans for progressive liberalization through ongoing negotiations.
- Creation of a dispute settlement procedure.[30]

According to Director General Dunkel, application of these fundamental principles to trade in services will result in "predictable and transparent trading conditions as well as progressively higher levels of services trade liberalization."[31]

Those crafting the GATS framework are aware that specific service sectors should be covered under special rules or annexes. Telecommunications is identified as one service area meriting such treatment. Indeed, telecommunications is singled out in the GATS process. The United States specified the need for such an annex when it submitted its proposed draft for a GATS framework. The GNS selected telecommunications as the first service sector to be studied.

The Treatment of Telecommunications

Not surprisingly, the GNS singled out telecommunications as not only a service sector in its own right but a service upon which other services increasingly depend. According to Geza Feketekuty, who has served as counselor to the U.S. trade representative since 1976, telecommunications is unique because both user and supplier issues are important to the trade arena.[32] For users, important trade issues arise because services are purchased from monopoly providers of communications services. For suppliers, trade issues surround the competitive provision of services.

Instead of being a utility provided by a monopoly, telecommunications (because of regulatory, economic, and technical developments) is very much a sector of traded and tradable services. Specifically:

- Because new technology leads to the creation of new services—such as data processing, electronic databases, banking, insurance, and transportation, that are dependent upon communications facilities—any disputes regarding communications charges or restrictions placed on the use of communications facilities are issues of trade.
- Because private entities have the economic and technological capability to build and maintain their own networks, issues regarding the liberty of entities to build private networks and connect them to the public network are now issues of market access and trade.
- Because technology makes it feasible for competitive entities to supply communication services, issues regarding the amount of competition allowed in various segments of the international telecommunications market are issues of concern.[33]

Recognizing all these concerns, the GNS conducted a series of unproductive meetings in 1989 and convened a very active Working Group on Telecommunications Services in 1990.

Some very basic areas of disagreement emerged from the working sessions. As mentioned earlier, a difference in viewpoint emerged between developed and developing nations. Many developing nations subsidize other parts of their economy with revenues from their government-owned monopoly communication enterprises. Competitive inroads into those enterprises could have far-reaching effects on the whole economy, causing the developing nations to move slowly in liberalization efforts. Moreover, many developing nations feared that, without some assistance from their trading partners in the more developed world, the gap would widen between the state of their infrastructure and services and the state of the developed world. For developing nations, a link between infrastructure development assistance and trade concessions made some sense. Developed nations seeking market access could reciprocate with development assistance.

Two important areas of dispute discussed in the working sessions merit further discussion here. One issue is treatment of the mode of delivery of telecommunications versus matters of network access and use. Mode of delivery encompasses the question of who is allowed to provide the network services. Matters of access and use deal with methods for connecting to the public network and allowed uses for network services. The United States sought to cover mode of delivery in the framework principles of the GATS, rather than in a separate sectorial annex. This would have pushed the delivery of telecommunications into a far more liberal venue than is presently the case. In other words, such a move would encourage competition in the

actual provision of the network. This approach wasn't followed. Rather, issues involved in mode of delivery were separated from matters of network access and use; both mode of delivery and matters of access were covered under a sectorial annex. This allowed nations greater leeway in negotiating the conditions under which market access occurs. The net effect is to slow the process of liberalizing telecommunications in those nations seeing a need to move slowly in opening up that market.

Another important issue of discussion was defining the services that the GATS process should cover. Most nations distinguish between what they term "basic services" and "enhanced services." *Basic services* tend to be those which, for policy and economic reasons, a monopoly provider still regulates and usually offers. Enhanced services are those which are deemed competitive in nature and therefore open to provision by a variety of providers. Because the whole GATS process involves the lessening of barriers and the encouragement of open markets, enhanced services seem viable areas for trade negotiations. Basic services, on the other hand, present more difficult policy considerations. Many nations want to move very slowly in opening basic services to competition.

A fundamental problem arose during the working group discussions. No real consensus emerged on where the line should be drawn between basic and enhanced services; no real consensus emerged on whether the GATS should cover basic services. An example of this confusion appears in the various approaches taken in comments filed by working group participants in early 1991. While Japan thought that all Type II services (those provided by a facilities-based carrier) should come under the agreement, Switzerland cited only value-added services. Meanwhile, Hong Kong listed a host of local services, including local videotext, local public mobile data, local electronic data interchange (EDI), and local telex; and Indonesia listed database access service and computer time sharing.[34] Table 3.4 lists services suggested for GATS coverage and the working group members that suggested them.

Table 3.4. Services listed by working groups for GATS coverage.

Country	Sector/subsector
Canada	Enhanced telecom services
EC	Value-added services* Packet- and circuit-switched data services* Other types of satellite mobile service*

Country	Sector/subsector
Hong Kong	Local public mobile data Local public community repeaters Local electronic mailbox Local value-added data Local electronic data interchange Local videotext Local teletex
Indonesia	Database access service Computer time sharing Videotext service
Japan	Type II telecommunications business
Korea	Online databases and remote computer services* Computer and communications services* Data transmission services*
New Zealand	Value-added services Public telecommunications transport services (including leased lines)
Switzerland	Value-added services*
United States	Telecommunications network-based enhanced services

* With conditions on scope or effective date

Source: "Telecom Important to Restarted Services Negotiations," *Transnational Data and Communications Report* (March/April 1991), p. 5.

Surprisingly, the United States, which is an avid proponent of liberalization, now seeks to exempt basic services from the fundamental most-favored-nation clause of any GATS agreement.[35] The U.S. action seeks to "permit the [United States] to accord different treatment among basic telecommunications services providers of foreign countries in its objective to obtain a more liberal regime in basic telecommunications services."[36]

The motive for this U.S. action is to deny those nations that don't reciprocate by liberalizing their own basic services markets the benefits of liberalized access to the U.S. telecommunications market. The most-favored-nation clause allows "free riders" to take advantage of agreements between the United States and those nations that are actively liberalizing telecommunications

services. The U.S. exemption focuses on basic long distance services. The most-favored-nation clause specifies that if the United States allows liberalized access to the EC, for example, it must do the same for Singapore, which still maintains strong monopolies over basic services. In other words, if EC nations enter the U.S. basic long distance market, Singapore can do the same, even though the GATS provides no mechanism for assuring that the United States can enter the Singapore market.

The United States proposes to rescind this exemption for those nations that make commitments to do all of the following:

- Place no limits on the number of competitors permitted to participate in the basic long distance services market.
- Permit foreign entities to provide basic long distance services through facilities-based competition and resale.
- Permit foreign investment in basic long distance services.
- Give new providers transparent, nondiscriminatory, and cost-based access to basic telecommunications service.
- Create an independent regulatory body to administer fair and transparent regulatory procedures.[37]

By its actions, the United States seeks to make commitment to liberalization the price of open access to the U.S. basic long distance market. This action responds to what may be a shortcoming in the most-favored-nation clause as it has commonly been applied. The clause intends to provide the benefits that arise from bilateral negotiations to all parties contracting to the trade agreement. In the telecommunications arena, where the approach to regulation and market structure is so divergent, a most-favored-nation approach, with no further requirements from the contracting parties, might actually retard the liberalization process rather than enhance it.

The Telecommunications Annex to the GATS

Despite disputes and exemptions, a draft telecommunications annex to the *Draft Agreement on Trade in Services* was crafted. The annex reflects many of the disputes that arose during the work sessions. Although it moves the process of telecommunications forward, it also provides mechanisms for nations to use in retarding that process.

The annex is a significant document and merits close analysis. In defining its scope, the document exempts cable and broadcast distribution of radio and television programming and specifies that the provisions of the annex apply to "all measures of a Party that affect access to and use of public telecommunications transport and network services."[38] Section 4 of the annex requires that contracting parties

> *...ensure that relevant information on conditions affecting access to and use of public telecommunications transport networks and services is publicly available, including: tariffs and other terms and conditions of service; specifications of technical interfaces...information on bodies responsible for the preparation and adoption of standards...conditions applying to attachment of terminal equipment...notifications, registration or licensing requirements.*[39]

The GATT concept of transparency will then apply to telecommunications services under the GATS. This provision is important to telecommunications managers of private networks and those engaged in service provision because it requires that all information about interconnection, pricing, regulations governing provision of service, and standards be made available to all.

Section 5 of the annex deals with the important issue of access and use of public telecommunications networks and services. Reasonable and nondiscriminatory access is specified at cost-oriented prices. Access is accorded to private leased lines and public providers. Provision is made for the connection of private leased or owned circuits to the public network and for the purchase and attachment of terminal equipment. Use of the public transport network for the movement of information within and across borders is allowed, including intracorporate communications and access to databases. Parties are allowed to take necessary measures to safeguard the public telecommunications network (Section 5.6). Such measures include, for example, a requirement that equipment be certified as not harmful to the network before it can be connected to the public network.

This section appears to afford protection for very liberal access to the network. However, the provision allowing for measures necessary to protect the integrity of the network is fairly broad. Providing that any resulting conditions imposed on access are deemed necessary for security and integrity purposes, the following conditions for network access can be imposed:

- Restrictions on resale or shared use
- Requirements to use specified technical interfaces
- Requirements for the interoperability of such services

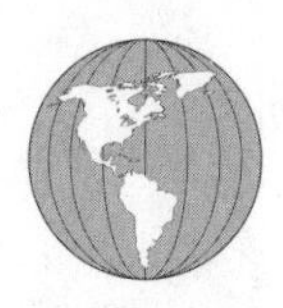

- Type approval of terminal equipment that interfaces with the network
- Restrictions on the connection of private leased or owned circuits
- Notification, registration, and licensing (Sections 5.7.1-5.7.6)

For developing nations, a further caveat is offered that can

> *...place reasonable conditions on access to and use of public telecommunications transport networks and services necessary to strengthen its domestic telecommunications infrastructure and service capacity and to increase its participation in international trade in telecommunications services.* (Section 5.8)[40]

Whereas the tone of the annex is very open regarding network use and access, measures available to protect the network provide a fairly wide loophole for nations wanting to impose restrictions on resale, connection of private circuits, and use of terminal equipment. All are issues of concern in the dispute regarding the face of telecommunication services.

Section 6 of the annex deals with the assistance for and technical cooperation with developing nations; it specifically endorses the efforts of the ITU, the United Nations Development Program, and the International Bank for Reconstruction and Development. Section 7 of the annex recognizes the importance of global standards for compatibility and interoperability of the network and specifies further cooperation with standards bodies like the ITU and the International Standards Organization (ISO).

General Agreement on Trade in Services
Annex on Telecommunications

(Draft)

1. *Objectives*

1.1 Recognizing the specificities of the telecommunications services sector and, in particular, its dual role as a distinct sector of economic activity and as the underlying transport means for other economic activities, the Parties have agreed to the following Annex with the objective of elaborating upon the provisions of the Agreement with respect to measures affecting access to and use of public telecommunications transport networks and services. Accordingly, this Annex provides notes and supplementary provisions to the Agreement.

2. *Scope*

2.1 This Annex shall apply to all measures of a Party that affect access to and use of public telecommunications transport networks and services.*

2.2 This Annex shall not apply to measures affecting the cable or broadcast distribution of radio or television programming.

2.3 Nothing in this Annex shall be construed:

2.3.1 to require a Party to authorize a service supplier of another Party to establish, construct, acquire, lease, operate, or supply telecommunications transport networks or services, other than as provided for in its schedule; or

2.3.2 to require a Party (or to require a Party to oblige service suppliers under its jurisdiction) to establish, construct, acquire, lease, operate or supply telecommunications transport networks or services not offered to the public generally.

3. *Definitions*

For the purposes of this Annex:

3.1 *Telecommunications* means the transmission and reception of signals by any electromagnetic means.

*Interpretative notes relating to provisions marked with asterisks are provided following the text of this annex.

3.2 *Public telecommunications transport service* means any telecommunications transport service required, explicitly or in effect, by a Party to be offered to the public generally. Such services may include, *inter alia*, telegraph, telephone, telex, and data transmission typically involving the real-time transmission of customer-supplied information between two or more points without any end-to-end change in the form or content of the customer's information.

3.3 *Public telecommunications transport network* means the public telecommunications infrastructure which permits telecommunications between and among defined network termination points.

3.4 *Intra-corporate communications* means telecommunications through which a company communicates within the company or with or among its subsidiaries, branches and, subject to a Party's domestic laws and regulations, affiliates. For these purposes, "subsidiaries," "branches," and, where applicable, "affiliates" shall be as defined by each Party. "Intra-corporate communications" in this Annex excludes commercial or non-commercial services that are supplied to companies that are not related subsidiaries, branches or affiliates, or that are offered to customers or potential customers.

3.5 Any reference to a paragraph or subparagraph of this Annex includes all subdivisions thereof.

4. *Transparency*

4.1 In the application of Article III of the Agreement, each Party shall ensure that relevant information on conditions affecting access to and use of public telecommunication transport networks and services is publicly available, including: tariffs and other terms and conditions of service; specification of technical interfaces with such networks and services; information on bodies responsible for the preparation and adoption of standards affecting such access and use; conditions applying to attachment of terminal equipment; and notifications, registration or licensing requirements, if any.

5. *Access to and use of Public Telecommunications Transport Networks and Services*

5.1 Each Party shall ensure that any service supplier of another Party is accorded access to and use of public telecommunications transport networks and services on reasonable and non-discriminatory terms and conditions, for the supply of a service included in its schedule. This obligation shall be applied, *inter alia,* through paragraphs 5.2 through 5.7 below.*

5.2 Each Party shall endeavour to ensure that pricing of public telecommunications transport networks and services is cost-oriented.

5.3 Each Party shall ensure that service suppliers of other Parties have access to and use of any public telecommunications transport network or service offered within or across the border of that Party, including private leased circuits, and to this end shall ensure, subject to paragraphs 5.6 and 5.7, that such suppliers are permitted:

5.3.1 to purchase or lease and attach terminal or other equipment which interfaces with the network and which is necessary to supply a supplier's services;

5.3.2 to interconnect private leased or owned circuits with public telecommunications transport networks and services or with circuits leased by another service supplier; and

5.3.3 to use operating protocols of the service supplier's choice in the supply of any service, other than as necessary to ensure the availability of telecommunications transport networks and services to the public generally.

5.4 Each Party shall ensure that service suppliers of other Parties may use public telecommunications transport networks and services for the movement of information within and across borders, including for intra-corporate communications of such service suppliers, and for access to information contained in data bases or otherwise stored in machine-readable form in the territory of any Party. Any new or amended measures of a Party significantly affecting such use shall be notified and shall be subject to consultation, in accordance with relevant provisions of the Agreement.

5.5 Notwithstanding the preceding paragraph, a Party may take such measures as are necessary to ensure the security and confidentiality of messages, subject to the requirement that such measures are not applied in a manner which would constitute a means of arbitrary or unjustifiable discrimination or a disguised restriction on international trade in services.

5.6 Each Party shall ensure that no condition is imposed on access to and use of public telecommunications transport networks and services other than as necessary:

5.6.1 to safeguard the public service responsibilities of suppliers of public telecommunications transport networks and services, in particular their ability to make their networks or services available to the public generally;

5.6.2 to protect the technical integrity of public telecommunications transport networks or services; or

5.6.3 to ensure that service suppliers of other Parties do not supply services unless permitted pursuant to commitments in a Party's schedule.

5.7 Provided that they satisfy the criteria set out in paragraph 5.6, conditions for access to and use of public telecommunications transport networks and services may include:

5.7.1 restrictions on resale or shared use of such services;

5.7.2 a requirement to use specified technical interfaces, including interface protocols, for inter-connection with such networks and services;

5.7.3 requirements, where necessary, for the inter-operability of such services and to encourage the achievement of the goals set out in paragraph 7.1;

5.7.4 type approval of terminal or other equipment which interfaces with the network and technical requirements relating to the attachment of such equipment to such networks;

5.7.5 restrictions on inter-connection of private leased or owned circuits with such networks or services or with circuits leased or owned by another service provider; or

5.7.6 notification, registration and licensing.

5.8 Notwithstanding the preceding paragraphs of this section, a developing country Party may, consistent with its level of development, place reasonable conditions on access to and use of public telecommunications transport networks and services necessary to strengthen its domestic telecommunications infrastructure and service capacity and to increase its participation in international trade in telecommunications services. Such conditions shall be specified in the Party's schedule.

6. *Technical Co-operation*

6.1 Parties recognize that an efficient, advanced telecommunications infrastructure in countries, particularly developing countries, is essential to the expansion of their trade in services. To this end, Parties endorse and encourage the participation, to the fullest extent practicable, of developed and developing countries and their suppliers of public telecommunications transport networks and services and other entities in the development programmes of international and regional organizations, including the International Telecommunication Union, the United Nations Development Programme, and the International Bank for Reconstruction and Development.

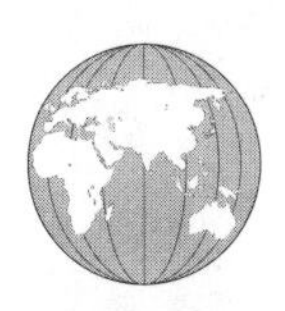

6.2 Parties shall encourage and support telecommunications co-operation among developing countries at the international, regional and sub-regional levels.

6.3 In co-operation with relevant international organizations, Parties shall make available, where practicable, to developing countries information with respect to international telecommunications services and developments in telecommunications and information technology to assist in strengthening their domestic telecommunications services sector.

6.4 Parties shall give special consideration to opportunities for the least developed countries to encourage foreign suppliers of telecommunications services to assist in the transfer of technology, training and other activities that support the development of their telecommunications infrastructure and expansion of their telecommunications services trade.

7. *Relation to International Organizations and Agreements*

7.1 Parties recognize the importance of international standards for global compatibility and inter-operability of telecommunication networks and services and undertake to promote such standards through the work of relevant international bodies, including the International Telecommunication Union and the International Organization for Standardization.

7.2 Parties recognize the role played by intergovernmental and non-governmental organizations and agreements in ensuring the efficient operation of domestic and global telecommunications services, in particular the International Telecommunication Union. Parties shall make appropriate arrangements, where relevant, for consultation with such organizations on matters arising from the implementation of this Annex.

Interpretative Notes

Note to paragraph 2.1

This paragraph is understood to mean that each Party shall ensure that the obligations of this Annex are applied with respect to suppliers of public telecommunications transport networks and services by whatever measures are necessary.

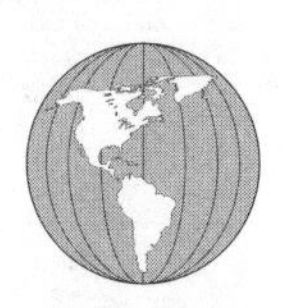

Note to paragraph 5.1

The term "non-discriminatory" is understood to refer to most-favoured-nation and national treatment as defined in the Agreement, as well as to reflect sector-specific usage of the term to mean "terms and conditions no less favourable than those accorded to any other use of like public telecommunications transport networks or services under like circumstances".

Challenges and Questions

The application of trade principles to services represents a challenge to those crafting a framework that covers trade in services. Though tracking the flow of goods across borders is fairly easy, the progress of services isn't so simple. Goods can be stopped at a border and import duties or quotas can be applied. Deciding how many cars are allowed to enter a nation or how much the import duty should be on a vehicle is difficult. Services aren't so clear. Identifying borders when an entity in one nation accesses a database in another is difficult. Moreover, services are often transaction-based. Keeping track of transactions for purposes of quotas or import duties is difficult and probably impractical.

If telecommunications is brought under a trade agreement, equipment in any nation would potentially be provided by a range of companies, domestic and foreign. A range of domestic and foreign providers, some of whom would build their own facilities, would also offer telecommunications services, whereas others would engage in resale. Access to the public network would be open to public and private entities, both foreign and domestic, on an equal basis.

A fully open telecommunications market is an interesting prospect, but it may not be easy to attain. Of all services, telecommunications is one of the more complex to bring under a trade umbrella. As mentioned earlier, telecommunications is both a service sector in itself and a mechanism for other services. User interests and supplier interests must be considered. The type of access users have to services and networks is important, as is the whole issue of the market structure underlying the provision of those services.

The issue of trade in telecommunications services is particularly difficult because telecommunications holds a unique position in many nations. Telecommunications began as a heavily controlled, often government-owned

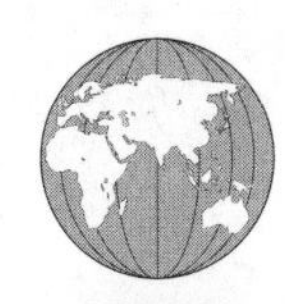

utility. In some nations, it is still regarded in this manner; in others, it isn't. In either case, it is subject to a range of policy considerations. For many nations, the attainment of universal service is important; if incursions from competitive forces seem to threaten the attainment of universal service, policymakers in those nations try to repel those forces. Yet, the one goal of the GATT and now the GATS is a liberalized market.

The application of trade principles to telecommunications provision is even more complicated due to the vast range of approaches taken by the various nations of the world. Nations don't agree on what is a regulated service and what isn't. Nations don't agree on the services that should be open to competition and those that should remain monopolies. In some nations, customers are free to purchase terminal equipment from any provider; in other nations, terminal equipment must be approved, and often provided, by the monopoly provider (often the government). Nations don't structure the market in the same manner. In some nations, the government owns and operates the network; in others, private entities own the network and a governmental entity acts as a regulatory body. Pricing policies aren't the same across all nations. Some nations adhere to cost-based pricing; others don't. The existence of private networks is encouraged in some nations and discouraged in others.

Those engaged in international telecommunications are accustomed to this diversity. Perhaps under the framework of a trade umbrella, some of that vast diversity will be lessened and some uniformity of treatment will emerge across nations.

Because networks must be able to connect to one another and terminal equipment must be compatible with networks in order to function, the whole issue of standards is important—and will become an important trade issue. Without some uniformity of standards, networks can't operate and services can't be delivered. However, insistence on specific types of standards might become a barrier to trade. If a network provider insists that a specific type of modem be used, for example, trade in modems is affected. Again, this is an issue of importance to telecommunications professionals.

For example, many felt that the United States' insistence on developing the ISDN "U" interface (see Figure 5.14) was such a barrier to trade. In its original conception, the service provider would own the NT1, so the "U" interface would have been unnecessary. A fully competitive CPE maker, however, requires the "U" interface. The development of the "U" interface added several years to the development of ISDN.

The existence of a trade umbrella for telecommunications services may substantially change the landscape. As has been noted by the editors of *Transnational Data and Communications Reports*, the GATS will

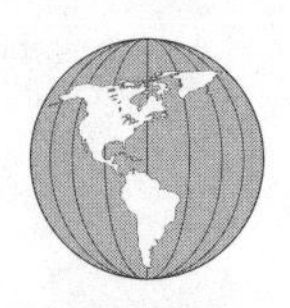

> *...for the first time establish that telecommunications networks and services are a commercially traded activity...telecommunications will be subject to multilateral trade rules, including a dispute settlement mechanism, which may result in some overlap with the exclusive jurisdiction [of] the ITU.... The traditional bilateral relationship between national operators for establishing pricing arrangements also may have to yield to a multilateral trade regime.*[41]

The reference to the ITU is telling here, largely because the trade umbrella would move telecommunications services toward liberalization at a faster pace than is the case with the ITU. The ITU has cooperated with the GATS process. The telecommunications annex mentions a need to adhere to ITU standards procedures. It will be interesting to observe which organization is more influential in effecting change in the telecommunications market.

Standards

In Chapter 4, we discuss some of the important standards-setting organizations in the international arena. Here, we'll consider some of the underlying motivations for standards. We will begin with a classification system that applies to standards of all kinds, proceed to a discussion of the emerging economic theory on standards and the motivation of firms that participate in standards committees, and, finally, examine the committee participants themselves.

Modes of Standards Development

In discussing categories of standards, we must consider two issues, emergence mechanisms (the manner in which standards come into the marketplace) and categories of standards of different types. Standards can emerge by three principal mechanisms. They can be the result of the free interplay of market forces (*de facto standards*); a governmental agency can establish them as law (*de jure standards*); or they can emerge from formal standards-setting bodies (*voluntary consensus standards*). Much of the focus in communications standards is on the development of voluntary consensus standards. Realize, however, that numerous de facto and de jure standards exist as well.

Some well-known examples of de facto standards are the IBM PC and the Bell 212 modem. Each of these products was produced by a market leader and

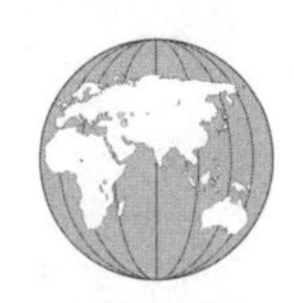

widely accepted by the public. As acceptance among users increased, other firms began manufacturing products to that standard. In the case of the 1,200 bits per second (bps) modem, the Vadic 3400 preceded the Bell 212 by several years, yet the marketplace chose the Bell 212 as the de facto standard.

De jure standards in the communications industry are normally those mandated by the FCC in the United States. In other countries, a state agency operates the telephone network, and as a result, choices made by that agency have the force of law and therefore are de jure standards. In contrast to de facto standards, the choice of users in the marketplace is largely irrelevant to the establishment of de jure standards; more likely, the political interests and strengths of the interested parties are relevant.

Finally, voluntary consensus standards have become prevalent in recent years. Manufacturers and users in accredited standards committees (such as ISO or CCITT) develop these standards. After the standard is approved, manufacturers build compatible products for sale in the marketplace. They must then successfully compete in the market with other manufacturer's products that conform to the same standard, and in some cases, with products conforming to other standards.

The type of standard, defined by emergence mechanism, is an important consideration in any discussion of standards. Users have no choice in adopting de jure standards, and de facto standards offer little utility to users if the marketplace changes. Because of the nature of the voluntary consensus process, those standards historically lag behind the leading edge of product capability.

In voluntary consensus standards, all interested parties periodically gather to collectively determine the details of the standard. If the preferences of the committee participants diverge, considerable negotiation and compromise must take place before they can reach a consensus on the details of the standard. This process can be very time consuming, particularly for complex standards. As a result, the technologies embodied in standards may not represent the state of the art.

Classification of Standards

In addition to considering the different emergence mechanisms for standards, examining different classifications of standards is useful. This enables us to focus more precisely on the standards that are generally applicable and most visible to end-users of communications standards.

Paul David has proposed a classification of standards that is fairly comprehensive.[42] David separates standards into *reference, minimum attribute,* and *compatibility* standards that one can apply to technical design as well as behavior. Table 3.5 illustrates this classification and presents some simple examples. For example, technical reference standards are weights and measures like those defined by the ISO and maintained by the National Institute for Standards and Technology (NIST) in the United States. An example of a technical minimum attribute standard is the minimum necessary strength of a material, such as steel. ISO, the American Society for Testing and Materials (ASTM), and the Institute of Electrical and Electronic Engineers (IEEE) frequently develop standards such as these. Although the IEEE has other types of standards, a large portion of the standards it develops are of the minimum attribute variety. In contrast, you can consider the accreditation of a college or institution a behavioral reference standard because it is compared to an absolute reference. In information systems, a key problem is the interoperability of systems, so the majority of industry standards pertain to technical compatibility.

Technical compatibility standards are further grouped by subcategory: *multivendor, multivintage,* and *product line* compatibility.[43] Much of the discussion of standards focuses on multivendor compatibility—compatibility among products produced by different vendors. However, other modes of compatibility are of considerable concern to users. If a product exhibits multivintage compatibility, it is compatible with products that preceded it;[44] product line compatibility refers to the capability of various products in an individual product line to interwork.[45]

Table 3.5. Classifications of standards.

	Technical	Behavioral
Reference	Weights and measures	Ettiquette Accreditation
Minimum Attribute	Stength of materials	Professional licensure
Compatibility	Nuts and bolts Modems Data Communication	Language and protocols

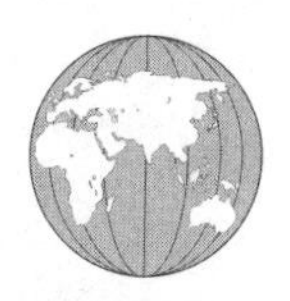

Motivation for Standards

Standards exist primarily because they foster economy. This economy is normally expressed in terms of lower product costs or lower operating costs. Standards promote lower product costs because manufacturers have a larger available market than they would in a market that lacked standards or had multiple competing standards. Therefore, with a unified standard, manufacturers can take advantage of production economies of scale to reduce product costs. The reduction of operating costs normally results from variety reduction. *Variety reduction* implies that a firm must stock fewer varieties of parts or supplies and that these are used for a greater range of applications. This can include parts for assembly, such as nuts and bolts, or other stock items, such as forms. If fewer items are needed, a firm can purchase them in larger quantities, resulting in increased economy. Furthermore, with a smaller variety of forms, personnel are more adept at completing them, which leads to further productivity gains.[46]

One phenomenon related to standards is referred to as *lock-in*. When a market has locked into a particular standard, consumers are unwilling to adopt a subsequent technology, even though it is better, because the costs involved with switching are high. Lock-in applies to compatibility standards in a significant way because a product that doesn't conform to the standard is incompatible with the rest. As a technology becomes more entrenched, encouraging users to adopt a superior one can be difficult.[47] Lock-in results from the formation of a bandwagon: as more people adopt a technology, even more are encouraged to follow suit. It has been demonstrated that establishing an early installed base is critical to the establishment of a bandwagon. This phenomenon can be used in predatory ways: a firm can announce a product before its release to begin the expectation of a bandwagon or prevent a bandwagon from forming around the product of a competitor.

Lock-in occurs because of an economic phenomenon called *network externalities*. A network externality is the hidden benefit that the consumer of a product conforming to a specific standard receives when another, unrelated, consumer purchases a product that also conforms to that standard. For example, a consumer of an IBM PC-compatible receives a benefit from all consumers who have purchased IBM PC-compatibles. This benefit comes in the form of an expanded variety of software from which to choose, such as hardware additions. As the market expands, the cost of such products tends to decrease while the variety tends to increase. In terms of industry structure, markets can exhibit excess inertia (too little incentive to standardize) or excess momentum (too much incentive to standardize). Excess inertia occurs

when the firms in a market desire a standard, but for fear that others won't follow, no one firm is willing to take the lead in establishing a standard. Excess momentum, on the other hand, occurs because the rush to adopt a new technology or standard (that would form a bandwagon) strands users of older technologies or standards. These users might not need to adopt the standard, but often must do so because of excess momentum.

Voluntary Consensus Standards

The market-based mechanism of de facto standards provides perhaps the simplest form of standards selection. It is, however, a costly, high stakes game that many vendors are reluctant to play because of the risks of market rivalry to establish a single standard under the competition model. In the market competition model, producers support their standard in hopes that it will be adopted. If it is adopted, the payoffs are great. The successful firm can collect profits from the sale of its products and licensing fees to other firms. If the technology isn't adopted, the firm loses the investment it made in supporting its technology. Furthermore, the customers of that firm are stranded if it ceases to manufacture its product. Hence, the risks of losing are great. Today, the risks of losing are perceived to be greater than the rewards of winning, so many manufacturers are supporting the voluntary consensus standards development process.

In fact, the current trend is toward standards preceding products into the marketplace. By developing the standards prior to market competition, firms reduce their risk of losing because the standard is in place prior to market competition. Unsuccessful firms in such a marketplace are those that cannot produce or market their products effectively, not those that lose a technological rivalry.

Firms interested in establishing a standard in the marketplace develop voluntary consensus standards. Representatives from these firms meet periodically to discuss technical details about the standard they are developing. A completed standard results from these discussions. Therefore, instead of incurring the costs and risks of technological rivalry, firms incur the costs and risks of committee participation. You can reasonably assume that the risks are fewer because the discussion centers on a succession of technological issues instead of requiring an all or nothing choice. However, it is unclear whether the costs are lower. Because of technological rivalry, determining the actual cost of committee participation is difficult to determine. The costs of participating actively and effectively in a standards committee are known to be quite high. The costs result from travel to committee meetings and the dedication of a professional's time for that purpose. Committees can meet as

frequently as monthly—though typically bimonthly or quarterly—and the deliberations can extend over several years. Many corporations have full-time professionals dedicated to standards development. Technical experts may also go to committee meetings as needed.

Firms and Sponsors

The firms that participate in the voluntary consensus process are generally either manufacturers wanting to build equipment that conforms to the standard, firms and organizations with a special need, or firms that purchase a significant volume of products that must work together. User participation is generally low because users are more aware of needs than solutions and because users are unable to make the cost and performance tradeoffs as efficiently as manufacturers can, so the committee debate is beyond their interest. However, in cases where users have special needs, they participate vigorously and effectively. A firm might choose to adopt a number of different strategies when participating in the standards committees:

- A firm can participate passively in the voluntary consensus standards development process. This means that it doesn't submit contributions or take part in the debate about the various technologies. Firms adopting this strategy usually participate in order to make use of the technology transfer feature of the standards process. Usually, these firms don't have a substantial stake in any particular technology.

- A firm can actively attempt to control the content of a standard. Firms that adopt this strategy might have a substantial stake (monetary or otherwise) in a particular technology.

- A firm can attempt to delay the adoption of a standard for its own strategic purposes. A firm adopting this posture in the standards process must make constructive contributions to the process in order to retain its credibility and control. It can't take the passive approach, although procedural delays also can be used.

The strategy chosen by a firm depends on its technical capabilities and its product development strategies. A firm that isn't knowledgeable about a technique or technology might adopt the first strategy for a portion of the debate so it can learn more about the technology. On the other hand, the technology leader might adopt the second strategy so that its product will be more competitive, either in cost, function, or market introduction. Finally, a firm that might not be able to bring a product to market as quickly as a firm adopting the second strategy might choose the third strategy to lessen the advantage of its competitor.

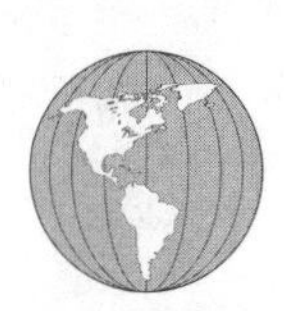

Committee Participants

Engineers, programmers, and other technically oriented people generally populate standards committees in the area of communications and computer standards. Because the content of the debate is critical, a firm can't afford to send a representative who isn't knowledgeable about the technical details of the debate. For this reason, some firms send their best product development engineers to standards meetings.

Engineers who participate in a standards committee might develop a loyalty to the objectives of the committee in addition to the loyalty they have to their company. So, they work to achieve the best possible technical solution with the given constraints, with less regard to their company's position than might be expected. This appears to be particularly true of the meetings that last five to ten days. Part of the responsibility of participants on a standards committee, then, is to convince their management that the committee's decisions are wise ones. The committee members, therefore, have two significant pressures on them. The first is to represent their firm to the best of their ability. This includes arguing for the particular technical position of a firm, and representing the general interests of the firm in this public forum. The second is to act as an emissary from the group to the firm. If a committee participant is consistently ineffective in doing this, that person's credibility and effectiveness in the committee is reduced significantly.

Just as technical competence is important, so too are interpersonal skills. A highly competent engineer who lacks interpersonal communication skills is generally not as effective as an individual with good interpersonal skills and some technical knowledge. Despite their highly technical content, standards are still developed by human beings. Communication skills are therefore quite important.

The Standards Development Process

A standards committee normally begins with a general project description that, in some respects, resembles a manufacturer's product proposal. The members of the committee study the project description and propose various alternatives for achieving the objective. Argument and debate in standards committees occurs by presenting and discussing these technical proposals. In some cases, technologies are designed in committee; in others, they are

brought to the committee by a sponsoring organization. When technologies are brought to a committee, they are sometimes part of a portfolio of technologies that represent the approach of the sponsor. In some cases, the technology is immediately acceptable to a majority of the participants and so is adopted without opposition. In other cases, alternative technologies are introduced by different participants, and a decision is required of the committee. Informal observation of standards committees indicates that most technologies are developed in committees or are adopted without opposition.

In practice, entities develop standards at multiple levels. International organizations, such as the ISO and the CCITT, actually issue most standards. In most developed countries, national or regional standards organizations exist to focus the standards development effort nationally or regionally. The various national organizations maintain liaisons with each other and with their international counterparts so that work is not duplicated. This reduces the cost to the committee participants because of the reduction in international travel and, in principle, presents unified national positions in the international standards committees. The increased globalization of firms reduces the impact of this somewhat, because a firm might participate in several national or regional bodies through its overseas subsidiaries.

Infrastructure/Economic Development

Governments are increasingly aware that a well-developed telecommunications system and economic activity are intricately related. Figure 3.3 shows this relationship. In this figure, the vertical axis is the average telephone density in the population, and the horizontal axis represents per capita Gross National Product (GNP). The relationship between these is strong and clear. What is not clear is the direction of causality: does a telecommunications infrastructure stimulate economic activity or does economic activity stimulate a telecommunications infrastructure?

The nature of the telecommunications investment varies substantially depending on the level of sophistication of the underlying system. Debate in the United States has focused on whether it is falling behind its economic competitors in investment in telecommunications infrastructure, with the potential consequence of producing a less hospitable business climate.

Developing countries face a different problem: the lack of basic telecommunications access for the majority of their population.

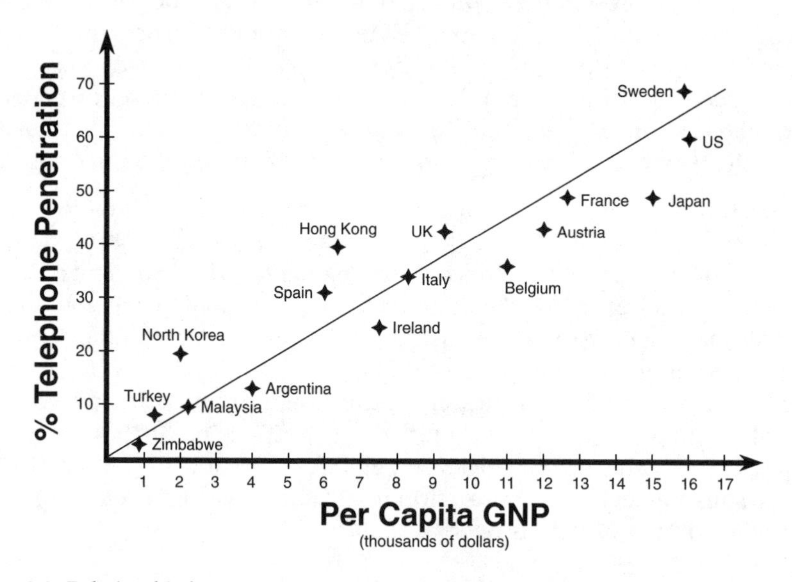

Figure 3.3. *Relationship between per capita GNP and telephone penetration (based on 1988 data).*

Although all telecommunications investment decisions must follow similar approaches, the basic questions are perhaps more easily articulated for developing countries. Telecommunications planning is defined as the process of creating, mobilizing, and allocating communications resources to achieve goals in a particular social, cultural, political, and economic context. This definition includes questions of resource allocation, policy goals, and the context in which the planning takes place. Consequently, an understanding of the planning process must include a discussion of these issues. Top down planning principles direct planners to begin with an understanding of the context and develop policies that are workable in that context.

The issues of investment planning for telecommunications in developing countries will probably soon receive increasing attention. Many of the Eastern European nations express interest in improving their telecommunications infrastructure in the aftermath of the political changes that occurred there.[48] Additionally, many of the Pacific Rim nations plan telecommunications investments as their economic power increases.

As illustrated in Figure 3.3, telecommunications and economic development are strongly correlated. Numerous attempts have been undertaken to establish causality in this relationship, but none is entirely satisfactory. Nonetheless, telecommunications generally serves as an enabling technology for economic development, particularly in developing countries. The supporting argument for this statement asserts that business and industry must have reliable telecommunications in order to function effectively in a competitive global marketplace. An inferior telecommunications infrastructure means that a firm trying to compete is at a disadvantage. Thus, competitive industry requires a sound telecommunications infrastructure. Other infrastructures, such as those that support electric power, roads and transportation, and clean water, serve a similar role. In addition to providing support for industrial enterprises, telecommunications simultaneously improves quality of life, particularly in rural areas, because customers are more closely linked with their peers, and resources, such as health agencies, libraries, and emergency communications, are more available.

In an attempt to further clarify the relationship between telecommunications and economic development, Heather Hudson, professor at the University of San Francisco, asserts that the following factors are necessary if a telecommunications infrastructure is to contribute to development:[49]

Organizational requirements—The telecommunications system must be designed to provide effective communications within an organization or an economic sector. Experience shows that an inappropriately designed system won't be used.

Organizational resources—The country must have sufficient resources to use and maintain the telecommunications system.

Social/institutional framework—Telephone use is gradually incorporated into a society based on existing social relationships. Thus, demand can take some time to materialize. No immediate use of communications devices can be expected among groups that normally don't communicate.

Complementary infrastructure—Other developmental aspects are necessary for the success of a telecommunications system.

Appropriate rates—The demand for telecommunications depends on the cost of the service. Jill Hills shows that the demand for telecommunications service in the United Kingdom differed significantly from U.S. demand early in the development of the telecommunications systems in those countries because of the rate structure.[50] The flat local rate used in the United States stimulated demand significantly.

The Role of Provision Policies

Most Western nations have universal service as their policy objective.[51] In the United States, universal service at a reasonable cost was the explicit policy expressed by the Communications Act of 1934. In its most general form, universal service promotes a high penetration of telephones into the population. In a market economy, a high number of network connections are obtained by keeping the basic connection charge as low as possible. In the United States, this connection (and the local telephone service) was priced below the cost of providing the service until the 1980s; subsidies from long distance telephone revenues made up the shortfall. This promoted connections to the network, thereby improving the economies of scale and scope inherent in the provision of telecommunications service, which further decreased cost and increased connections to the network. Also, urban areas effectively subsidized rural users, because the cost of service provision in rural areas is higher than in urban areas.

With approximately 90 percent of households connected to the network, U.S. policy effectively changed by moving from subsidized service connection to cost-based service provision.[52] The result is higher rates for local service (including connection to the network) and lower prices for long distance service. Another result is decreased telephone penetration, particularly among the poor and elderly, due to price increases.

Universal service isn't the only available policy objective. This policy emerged from principles of social equity and resulted in the economic regulation of firms providing service in the United States. Several alternative policies are available. For example, one could imagine the following:[53]

- If industrial development is the primary objective, efficient and cost-effective service to industry should be the principal objective (an "industrial development" policy).[54]
- If an economy has a strong need for foreign exchange, providing high-quality service to export-oriented firms is a priority. This is best described as an "export-oriented" policy.

In its early stages, a government investing under an industrial development policy directs the majority of its investments toward improving the telecommunications infrastructure serving existing or planned industries. The needs of industry are placed above the needs of the populace. Such service is often provided at lower cost because industrial facilities tend to be concentrated. If this approach is adopted, it is prudent to invest not just in an industry but also in the firms supplying materials and service to that industry.

Consistent with factors proposed by Hudson, the telecommunications infrastructure supports existing relationships among suppliers. As the industry grows, the national product sales and distribution network become more important and the supplier network increases so the need of industrial organizations for telecommunications penetration into the population increases, further driving the improvement of the infrastructure. Consequently, this policy might eventually result in a high penetration of telephone connections among the population, although its explicit objective is to provide service for industrial organizations. Of course, increased sales and profitability of the enterprise provide the motivation for these improvements.

The other policy objective is export-oriented. In this approach, firms and industries engaged primarily in the export of products are given priority in telecommunications investments over other individuals or organizations. This tends to stimulate the influx of foreign exchange because these export-oriented firms are as competitive as possible. Ideally, the telecommunications connections eventually trickle down to the nonexport-oriented industry and the populace at large (as in the industrial development policy), but the primary motivation is to support the development and growth of the export-oriented industry.

Each policy alternative has consequences for the tactical plan and the network architecture. Following are some of the consequences:

- **Universal Service**—To implement universal service, extensive infrastructure investments are made throughout the country in rural and urban areas, and both traffic sensitive and nontraffic sensitive tariffs are set to promote the network's use. Ultimately, the investment reaps a net benefit as economies of scale and scope drive down the cost of providing service. If extensive infrastructure improvements aren't possible in a relatively short time, the investment must be phased in over several years. In this case, two investment strategies are possible:

  Consistent with Hudson's factors, investments can be organized so they are consistent with existing organizational/social frameworks. This effectively results in islands of service that must eventually be connected. Care must be taken in this approach to ensure that social equity considerations are met (if those are policy objectives) in selecting the sequence of project implementation.

  Public service can be provided prior to private service. This results in facilities akin to the "telecottages" used throughout Scandinavia.[55] These facilities provide centralized public service to remote regions. In this case, access costs aren't directly borne by consumers,[56] so appropriate traffic sensitive tariffs must promote use.

- **Industrial Development**—Investments are directed primarily where industrial facilities are located. If a phased approach is necessary, the most profitable industry[57] and its supply and distribution networks receive the earliest investment, the next most profitable industry follows, and so on. Tariffs are set to maximize network use and access by industries. Tariffs for nonindustrial organizations are set higher to subsidize industrial connections. Eventually, the most populated regions have the highest penetration of telephones, with rural regions lagging, perhaps substantially. This approach tends to encourage regional centralization of industry because the services provided in industrialized areas are superior to those in nonindustrialized areas. Industries have little incentive to locate to rural areas.
- **Export Oriented**—Investments are directed to firms engaged in export business. Much of the population and domestic industry don't perceive the direct benefits of telecommunications investments. Those domestic consumers that do receive service must subsidize firms doing export business in order to continue financing infrastructure improvements.

Clearly, a critical element of a national telecommunications policy must be a provision policy, which guides subsequent investments over a long period of time. Such a policy can be altered, as it has been in the United States, after the initial objectives are met.

Planning

Other critical issues affect the allocation of investments and the design of the telecommunications network. Although included in the preceding general contextual analysis, several issues (such as complementary infrastructure, financing, and standards) require further elaboration.

Complementary Infrastructure

Complementary infrastructure is important to the success of a telecommunications network as a stimulant to economic development. Such an infrastructure contributes indirectly to the success of a telecommunications system, even though it doesn't perform telecommunications functions per se. Infrastructures included in this category are: electric power, transportation, supplies and spare parts, and a trained and educated technical staff.

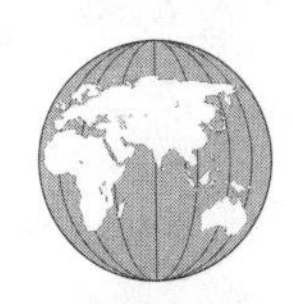

Because most telecommunications equipment is electronic, reliable power to all portions of the telephone network is critical. Similarly, a transportation infrastructure is necessary so that equipment can be installed and serviced in a timely fashion. In a social context, an adequate transportation system is necessary so that appropriate actions can be taken efficiently as the result of a telecommunications message. These infrastructures are clearly interdependent. For example, evidence exists that telecommunications actually stimulates the demand for transportation.[58]

In addition to these factors, a consistently effective telecommunications network is impossible without a trained technical staff to operate and maintain the network and the ready availability of parts and supplies for the network. Technical staffs must be trained on the general principles of telecommunications and the specific equipment they will be servicing. Western manufacturers of telecommunications equipment normally provide training programs for the latter purpose. Management level training in the principles of telecommunications is available at academic institutions.

Appropriate planning for a telecommunications infrastructure therefore goes far beyond the actual purchase and installation of equipment. Additional resources are critical and must be considered in the plan.

Financing

Financing a telecommunications infrastructure can be problematic, particularly when the network is in its early stages. At this point, the economies of scale and scope that characterize fully developed networks are not yet achieved. As a result, a disproportionate investment must be made early, adding to the difficulties of establishing a telecommunications infrastructure. As the network grows, it produces revenues that can be used to finance further expansion and enhancement. Ultimately, the revenues generated by the network can be used to subsidize other services.

Developed nations invest an average of 0.9 percent of their GNP in telecommunications, whereas developing nations invest an average of 0.35 percent of GNP.[59] Although these percentages aren't normative in any sense, they do provide an interesting contrast. Possible reasons for this underinvestment are:

- A lack of enumeration and quantification of the benefits of telecommunications with regard to the investments in other sectors
- A perception that telecommunications investment confers direct benefits on a relatively small portion (usually the elite or the privileged) of the population

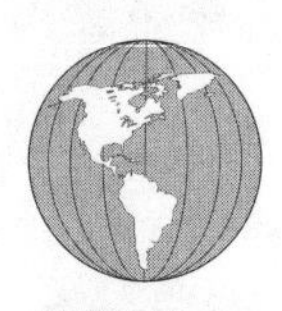

- Tariff policies that don't promote efficient allocation of telecommunications resources in the short run
- Institutional and organizational problems in the countries and/or within the telecommunications operating entities
- Lack of foreign exchange

These reasons don't apply to every situation; however, they do provide the basis for further analysis when projects don't meet with the success that was initially anticipated.

Development assistance traditionally has been available from a variety of international organizations such as the World Bank,[60] as well as in the form of credits from equipment manufacturers. Motions have been made in the ITU to provide direct monetary assistance as part of its general revenue as a regular budget item, but these efforts so far have resulted only in the provision of technical assistance.

Of course, an alternative to direct monetary and technical assistance lies in the development of a telecommunications manufacturing capability. This generally requires technology transfer to the emerging economy along with technical and management assistance from developed countries. Once this approach is fully established, the advantages are

- Foreign exchange issues are separated from infrastructure development issues.
- Provisioning of spare parts is fully under the control of domestic agents.
- Technical expertise for the telecommunications network is developed and maintained within the country.

A fully established infrastructure refers not just to the ability to manufacture and install switching equipment, transmission equipment, telephone instruments, and so forth but also the ability to manufacture the components needed to manufacture the equipment. Therefore, an independent manufacturing capability requires a substantial industrial base, including metal and plastic manufacture and semiconductor fabrication. With sufficient research, adequate development support, and good quality control in the equipment manufacturing facilities, a country could ultimately enter the export market for telecommunications equipment as a supplier.

Conclusions

In this chapter, we've surveyed a wide variety of economically oriented issues. Some of these issues were pragmatic (such as accounting and settlements and their implications) and others were more abstract (such as different telecommunications pricing strategies and the economic role of standards and standardization). As we stated in the beginning of this chapter and as should have become clear to you, separating these issues from regulatory and organizational issues is difficult. All are closely entwined. As we have indicated, some of the issues we have discussed, such as accounting and settlements and GATT/GATS, are still evolving. We hope we have provided you with enough information to understand future developments as they occur.

Endnotes

1. *Statement of FCC Chairman Alfred C. Sikes Regarding the Commission's Companion International Settlements & Accounting Rates Decisions*, CC Dockets 90-337, Phases I and II (Government Printing Office, 1991), pp. 3568-3570.

2. *Sikes.*

3. Appendix C to Further Notice of Proposed Rule Making, *In the Matter of Regulation of International Accounting Rates*, CC Docket No. 90-337, Phase II, 6 FCC Rcd No. 12 (1991), pp. 3441-3442.

4. *Telecommunications Reports* (September 16, 1991), p. 15.

5. This argument has been presented by Leland Johnson, *Competition, Pricing, and Regulatory Policy in the International Telephone Industry*, Rand Publication R-3790-NSF/MF (Santa Monica, Cal.: Rand Corporation, 1989).

6. Leland Johnson in *Competition, Pricing, and Regulatory Policy in the International Telephone Industry* claims that the reason for the unwillingness to provide off-peak or time-of-day discounts on the part of other nations' PTTs might stem from a fear of lost revenue and skepticism about resulting traffic stimulation, pp. 23-34.

7. Ken Cheong and Mark Mullins, "International Telephone Service Imbalances," *Telecommunications Policy*, vol. 15, no. 2 (April 1991), pp. 109-110.

8. Cheong and Mullins, p. 112.

9. CCITT *Recommendations* D150;151;155.

10. *Report and Order*, "In the Matter of the Regulation of International Accounting Rates," CC Docket No. 90-337, Phase I, 6 FCC Rcd No. 12 (1991), p. 3555.

11. *Telecommunications Reports* (March 23, 1992), p. 22.

12. Peter A. Stern, "The International Telecom Accounting and Settlements Debate," *Transnational Data and Communications Report* (July/August 1991), p. 31.

13. Stern, p. 31.

14. Stern, pp. 28-32.

15. *Report and Order*, p. 3554.

16. *Report and Order*, p. 3562.

17. See Johnson for further explanation of this arrangement, pp. 36-37.

18. *Telecommunications Reports* (December 16, 1991), p. 1-4; 37-38.

19. Bridger M. Mitchell and Ingo Vogelsang, *Telecommunications Pricing: Theory and Practice* (New York: Cambridge University Press, 1991). See Chapter 2, pp. 7-17.

20. The Japanese regulatory reforms were in part stimulated by a scandal in NTT in 1980, then a government-owned monopoly. This case illustrates how misuses can be corrected in public monopolists.

21. Mitchell and Vogelsang, p. 150.

22. Leah A. Haus, *Globalizing the Gatt, The Soviet Union's Successor States, Eastern Europe, and the International Trading System* (Washington, D.C.: Brookings Institution, 1992), p. 2.

23. This list of principles is explained in detail in Geza Feketekuty, *International Trade in Services: An Overview and Blueprint for Negotiations* (Cambridge, Mass.: Ballinger Publishing Company, 1988), pp. 152-156.

24. For a discussion of events leading up to the inclusion of trade in services issues see the Appendix to Feketekuty. See also *Trade in Information, Computer, and Communications Services*, ICCP#21, OECD 1990.

25. The following listing of the rules and principles agreed to at Montreal was presented by R. Brian Woodrow, "Tilting Towards a Trade Regime: The ITU and the Uruguay Round Services Negotiations," *Telecommunications Policy* (August 1991), p. 328.

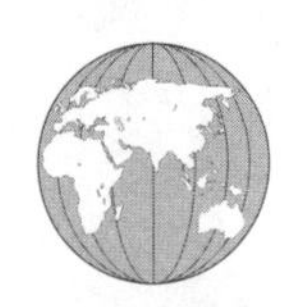

26. Christoph Doerrenbaecher and Oliver Fischer, "Telecommunications in the Uruguay Round," *Intereconomics*, vol. 25 (July/August 1990), pp. 185-192.

27. Woodrow, p. 337.

28. See for example a discussion in *Telecommunications Reports* (December 13, 1991), pp. 12-13.

29. These points are explained by Arthur Dunkel, "Telecom Services and the Uruguay Round" *Transnational Data and Communications Report* (January/February 1992), p. 18.

30. Trade Negotiations Committee, GATT Secretariat, *Draft Final Act Embodying the Results of the Uruguay Round of Multilateral Trade Negotiations, General Agreement on Trade in Services*, MTN.TNC/W/FA (December 20, 1991).

31. Dunkel, p. 18.

32. Feketekuty, p. 253.

33. Feketekuty, p. 253.

34. "Telecom Important to Restarted Services Negotiations," *Transnational Data and Communications Reports* (March/April 1991), p. 5.

35. "US Formally Exempts Basic Telecom from Uruguay Round," *Transnational Data and Communications Reports* (May/June 1992), p. 5.

36. Text submitted to the GATT Secretariat in March 1992, quoted in "US Telecom Formally Exempts Basic Telecom from Uruguay Round."

37. Text submitted to the GATT Secretariat in March 1992, quoted in "US Telecom Formally Exempts Basic Telecom from Uruguay Round."

38. "Telecommunications Annex to Draft General Agreement on Trade in Services," Section 2.1, reprinted in its entirety in *Transnational Data and Communications Reports* (January/February 1992), pp. 31-33.

39. "Telecommunications Annex to Draft General Agreement on Trade in Services," Section 2.1, reprinted in its entirety in *Transnational Data and Communications Reports* (January/February 1992), pp. 31-33.

40. "Telecommunications Annex to Draft," *Transnational Data and Communications Reports.*

41. "Multilateral Telecom Trade Regime Prepared" *Transnational Data and Communications Reports* (January/February 1992), p. 31.

42. Paul A. David, "Some New Standards for the Economics of Standardization in the Information Age," in The *Economic Theory of Technological Policy*, eds. P. Dasgulpa and P.L. Stoneman (London: Cambridge University Press, 1987).

43. H. Landis Gabel, "Open Standards in the European Computer Industry: the Case of X/Open," in *Product Standardization and Competitive Strategy*, ed. H. Landis Gabel (Amsterdam: North Holland, 1987).

44. For example, when IBM introduced its 360 computer, it was incompatible with previous models. This caused the standing of a significant installed base.

45. For example, no significant software conversions are required to run programs on an XT-type personal computer or an AT-type personal computer. In similar fashion, users of the VAX computers produced by Digital Equipment Corporation can retain the same software as they upgrade to more powerful models of this computer.

46. The discussion in this section is a distillation of work by several people including these principal contributors:

W. Brian Arthur, "Competing Technologies and Lock-In by Historical Small Events: The Dynamics of Allocation Under Increasing Returns," Technical Report 43, Center for Economic Policy Research (Stanford University, January 1985).

Sanford V. Berg, "Technical Standards as Public Goods: Demand Incentives for Cooperative Behavior," *Public Finance Quarterly*, vol. 17, no. 1 (January 1989), pp. 29-54.

Stan Besen and Leland Johnson, *Compatibility Standards, Competition, and Innovation in the Broadcasting Industry* (RAND Corporation, 1986).

Yale M. Braunstein and Lawrence J. White, "Setting Technical Compatibility Standards: An Economic Analysis," *The Antitrust Bulletin*, vol. XXX, no. 2 (Summer 1985).

Rhonda J. Crane, *The Politics of International Standards* (Norwood N.J.: Ablex, 1979).

Paul A. David, "Clio and the Economics of QWERTY," *American Economic Review*, vol. 75, no. 2 (May 1985), pp. 332-337.

Paul A. David, "Some New Standards for the Economics of Standardization in the Information Age," in *Economic Policy and Technological Performance*, eds. P. Dasgupta and P. Stoneman (Cambridge: Cambridge University Press, 1987).

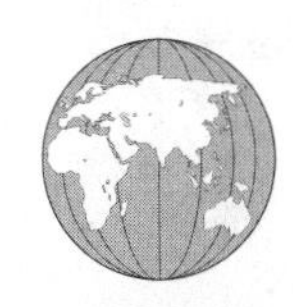

Paul A. David and Julie Ann Bunn, "The Economics of Gateway Technologies and Network Evolution: Lessons from the Electricity Supply History," *Information Economics and Policy*, vol. 3, no. 2 (1988), pp. 165-202.

Joseph Farrell and Garth Saloner, "Standardization, Compatibility, and Innovation," *RAND Journal of Economics*, vol. 16, no. 1 (Spring 1985), pp. 70-83.

Joseph Farrell and Garth Saloner, "Installed Base and Compatibility: Innovation, Product Preannouncements, and Predation," *American Economic Review*, vol. 76, no. 5 (December 1986), pp. 940-955.

Joseph Farrell and Garth Saloner, "Competition, Compatibility, and Standards: The Economics of Horses, Penguins, and Lemmings," in *Product Standardization and Competitive Strategy*, ed. H. Landis Gabel (New York: North Holland, 1987).

Joseph Farrell and Garth Saloner, "Coordination Through Committees and Markets," *RAND Journal of Economics*, vol. 19, no. 2 (1988), pp. 235-252.

Michael L. Katz and Carl Shapiro, "Network Externalities, Competition, and Compatibility," *American Economic Review*, vol. 75, no. 3 (June 1985), pp. 424-440.

Michael L. Katz and Carl Shapiro, "Technology Adoption in the Presence of Network Externalities," *Journal of Political Economy*, vol. 94, no. 4 (August 1986), pp. 822-841.

Donald J. Lecraw, "Some Effects of Standards," *Applied Economics*, vol. 16 (1984), pp. 507-522.

Albert Link, "Market Structure and Voluntary Product Standards," *Applied Economics*, vol. 15 (1983), pp. 393-401.

Marvin Sirbu and Kent Hughes, "Standardization of Local Area Networks," 14th Annual Telecommunications Policy Research Conference, Airlie, Va., April 1986.

Marvin Sirbu and Steven Stewart, *Market Structures and the Emergence of Standards: A Test in the Modem Market*, WP-8, MIT Research Program on Communications Policy (June 1986).

Marvin Sirbu and Laurence E. Zwimpfer, "Standards Setting for Computer Communication: The Case of X.25," *IEEE Communications Magazine*, vol. 23, no. 3 (March 1985), pp. 35-45.

Martin B.H. Weiss, "Compatibility Standards and Product Development Strategies: A Review of Data Modem Developments," *Computer Standards and Interfaces*, vol. 12, no. 2 (Sept. 1991), pp. 109-122.

Martin B.H. Weiss and Carl Cargill, "A Theory of Consortia in IT Standards Setting," *Journal of the American Society for Information Science*, vol. 43, no. 8 (Sept. 1992), pp. 559-565.

Martin B.H. Weiss and Marvin Sirbu, "Technological Choice in Voluntary Standards Committees: An Empirical Analysis," *Economics of Innovation and New Technology*, vol. 1 (1990), pp. 111-133.

47. A good example of this is the QWERTY typewriter keyboard. Even though the Dvorak keyboard has been shown to be more efficient, the QWERTY keyboard continues to dominate because of the embedded investment in typewriter training. See P.A. David, "Clio and the Economics of QWERTY," *American Economic Review*, vol. 75, no. 2 (May 1985), pp. 332-336.

48. For example, the USSR has undertaken a development plan to improve their voice and data telecommunications infrastructure. See G.G. Kudriavtzev and L.E. Varakin, "Economic Aspects of Telephone Network Development: the USSR Plan," *Telecommunications Policy* (February 1990), pp. 7-14.

49. Heather Hudson, "Toward a Model for Predicting Development Benefits from Telecommunications Investment," in *Communication and Economic Development*, eds. Meheroo Jussawalla and D.M. Lamberton (New York: Permagon Press, 1982).

50. Jill Hills, "Universal Service: Liberalization and Privatization of Telecommunications," *Telecommunications Policy* (June 1989), pp. 129-144.

51. Hardy's analysis shows that residential telephone penetration, not commercial telephone penetration, is a precursor to economic development. This lends support to the concept of universal service beyond social equity principles. See A.P. Hardy, "The Role of the Telephone in Economic Development," *Telecommunications Policy*, vol. 4, pp. 278-286.

52. The divestiture of AT&T was the expression of this implicit policy change. The Communications Act of 1934, with its language of universal service at a reasonable cost, remains in effect and is the explicit policy goal. The urban users continue to subsidize rural users under current telecommunications practice in the United States.

53. These are intended to represent the kinds of alternative provision policies that might be selected. Other policies might be selected that are not included here.

54. Garcia has described the Philippine telephone network. Based on this description, an argument can be that this system was based on an industrial development policy. See Renato B. Garcia, "Telecommunications and Investment Decisions in the Phillipines," *Telecommunications Policy* (March 1984), pp. 51-57.

55. Lars Quotrup, "Nordic Telecottages: Community Teleservice Centres for Rural Regions," *Telecommunications Policy* (March 1989), pp. 59-68.

56. Users must bear indirect costs in this approach, however. They bear the transportation cost of physically going to the facility and the coordination cost of arranging times and dates for telephone conversations with their peers.

57. Notice that profitability can be interpreted with some latitude. For example, the industry making the largest contribution to the foreign exchange could be targeted as the most profitable.

58. Hudson.

59. Garcia.

60. Saunders points out that the World Bank sees itself as a lender of last resort in telecommunications. Hudson notes that through 1980 telecommunications has accounted for approximately 3.6 percent of the World Bank's total lending. See Robert J. Saunders, "Telecommunications in Developing Countries: Constraints on Development," in *Communication and Economic Development*, eds. Meheroo Jussawalla and D.M. Lamberton (New York: Permagon Press, 1982) and Heather Hudson, *When Telephones Reach the Villiages* (Norwood, N.J.: Ablex Publishers, 1984).

Organizational Issues

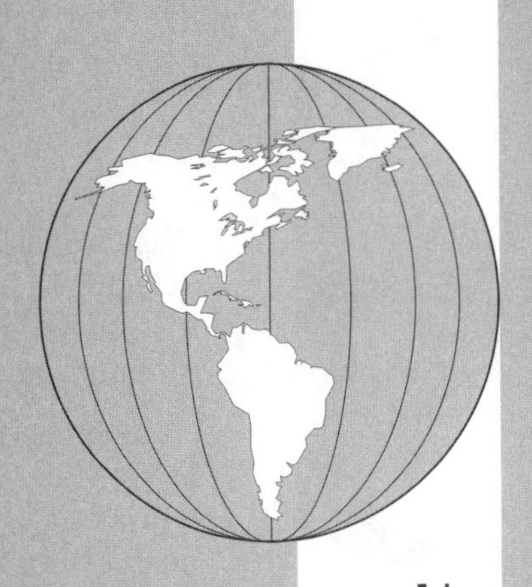

CHAPTER 4

The effective international operation of telecommunications requires some organizations that transcend national boundaries. In this chapter, we examine the needs of international telecommunications that such organizations can solve. We survey some of the important organizations. We conclude by considering how well these organizations cover the needs of the international telecommunications industry and users.

Particular Needs for International Organizations

For companies and individuals to communicate across national borders, they must overcome a host of technical, political, and regulatory barriers. An example that illustrates the type of barriers involved is early use of the telegraph. Before nations reached international agreements regarding the use of Morse code and the technical requirements for interconnecting national telegraph systems, telegrams stopped at national borders. They were then hand carried across the border to a telegrapher who completed the telegram message. To expedite the process, agreements had to be formulated so that a code acceptable and understandable to all could be used.

Standards had to be accepted governing the interconnection of one nation's telegraph system with another's. The International Telegraph Union, a precursor of the International Telecommunications Union (ITU), was one medium through which such agreements and arrangements were made.

Today, telecommunications technologies are much more complex than the early telegraph. International networks carry voice, data, and video. The number and type of telecommunications equipment have proliferated. As fiber-optic technology increases, so does the range of available bandwidths and speeds. If all these networks are interconnected and all these different types of equipment are universally functional, all—or at least most—nations must develop and agree to standards. Standard setting takes place through the workings of international organizations.

Because each nation has its own approach to structuring and regulating telecommunications, the efforts of international bodies often enhance the success of international communications networks. One example is the treatment of private lines. In some nations, private line services aren't even available. In other nations, users aren't permitted to connect private lines to the public switched network (PSN). In others, such interconnection is allowed. A telecommunications manager faces a range of regulatory options and prohibitions. International organizations provide a forum to reach compromises and attain some form of international uniformity.

International organizations, such as the International Satellite Organization (Intelsat), for example, serve another necessary function. For those communications technologies that require a large amount of investment and a coordinated effort, an international organization provides the mechanism for accumulating needed capital and allocating resources in an orderly fashion. The deployment of international satellite telecommunications is an example of such efforts.

Finally, international telecommunications depends on resources that don't belong to any one nation. The spectrum (or range of frequencies available for communications) is an example of such a resource. So too is the geosynchronous orbit (the orbit 22,300 miles above the equator in which satellites move at the same speed with the earth). Through the workings of international organizations, a coherent, accepted approach can be adopted for the allocation of these scarce world resources.

For all these reasons, nations have formed and joined a range of organizations designed to facilitate international telecommunications. The ITU, the European Telecommunications Standards Institute (ETSI), the International Standards Organization (ISO), and Intelsat are some examples of such groups.

Organizations

In this section, we provide background information on the most important organizations in international telecommunications. Additional organizations exist in some regions; many are modelled on the organizations discussed here. Organizations must always change to respond to variations in the environment with which they are concerned and to the mutable demands of the members whom they serve. Therefore, the description and discussion of organizations in the following sections is necessarily dated. We hope you can interpret these changes in the context we provide here.

The International Telecommunications Union

The ITU plays a significant and unique role in international telecommunications. Its parent organization, the International Telegraph Union, was founded in 1865. According to George Codding, Jr. and Anthony Rutkowski in their 1982 study of the history and function of the ITU, the International Telegraph Union was the "first genuine international, intergovernmental organization to see the light of day."[1] James Savage describes the ITU as the "sole global body empowered to supervise and manage the international telecommunications order."[2]

These descriptions of the ITU are by no means overstated. The ITU emerged as the first organization to struggle with the technical issues presented by new communications technologies and the problems inherent in balancing the interests of a divergent international membership. Today, though organizations like Intelsat, the General Agreement on Tariffs and Trade (GATT), the European Community (EC), and the Organisation for Economic Cooperation and Development (OECD) have emerged, the ITU is still the only organization with a mandate for the oversight of global telecommunications.

The 163 members of the ITU are part of a specialized agency of the United Nations. The membership consists of national governments. The purposes of the organization are set out in its basic document, the *International Telecommunications Union Convention*:

> *...to maintain and extend international cooperation between all Members of the Union for the improvement and rational use of telecommunications*

> *of all kinds, as well as to promote and to offer technical assistance to developing countries in the field of telecommunications;...to promote the development of technical facilities and their most efficient operation with a view to improving the efficiency of telecommunications services, increasing their usefulness and making them, so far as possible, generally available to the public...to harmonize the actions of nations in the attainment of those ends.*[3]

In other words, the ITU's overriding purpose is to facilitate international communication.

In pursuit of these goals, the ITU fulfills the following functions:

- Coordinates the allocation of radio frequencies and registers radio frequency assignments in order to prevent interference
- Coordinates the assignment of positions in the geosynchronous orbit
- Recommends international technical standards
- Defines telecommunications services
- Collects and publishes information
- Promotes telecommunications development efforts in developing nations.

An underlying consideration in all ITU activities is, as is stated in the Preamble to the organization's Convention, the full recognition of "the sovereign right of each country to regulate its telecommunications." This consideration is often a complicating factor because the ITU membership represents a full range of regulatory approaches, from almost total competition to total monopoly. Defining services and recommending uniform standards and approaches to the provision of services is no easy task with such a divergent membership. The standards-setting process is further complicated by the swift pace of technological development, which the standards-setting process often can't match.[4]

Because the ITU deals with the allocation of spectrum and geosynchronous "parking spaces," both fixed resources, the organization must balance the needs of developed nations that are ready and anxious to use those resources today with the needs of developing nations that want assurances that resources will be available when they are technically prepared to use them. Much debate arises because of the wide technology gap between the developed and developing nations that compose the ITU. Developing nations push for the ITU to play a larger role in development efforts, a role the developed nations are reluctant to endorse.

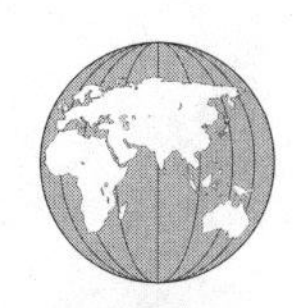

Since its beginning, the ITU has adapted its structure and operations in response to technological changes. The organization has also maintained some balance among the various interests of its diverse membership in order to advance the continued development of international telecommunications.

A Short History of the ITU

The ITU has always been willing to change to meet technical and political developments. During its long history, the organization expanded to encompass emerging technologies, attracted a steadily increasing number of members, and modified its organizational structure to accommodate its increasingly divergent membership.[5]

The International Telegraph Union was formed in Paris in 1865, at a conference of the Austro-German Telegraph Union and the Western European Telegraph Union convened by Napoleon III. This Paris meeting was productive. In addition to drafting an International Telegraph Convention (treaty) and annexed telegraph regulations, those in attendance identified the need for regular meetings to address issues of concern. The next such meeting was held in Vienna in 1868. At the Vienna meeting, a secretariat (administrative body) was formed for the Union. Located in Berne, Switzerland, this secretariat was called the Berne Bureau. In 1906, the name was changed to the International Bureau of the Telegraph Union.

By 1868, the organization had put the necessary elements of operation in place. It had established a pattern of meetings, a constitution, and an administrative arm. The Union held 11 conferences between 1865 and 1932. The primary focus of these conferences tended to be the telegraph. This isn't surprising because the telegraph was the dominant form of land-based communication at the time. The major concern addressed was telegraph rates, rather than technical issues. The emphasis on rates was troublesome for the United States because, while most other telegraph systems were government-owned and operated, the U.S. system was privately owned. While other governments could discuss and commit to rates, the United States could not.

The issue of adding telephone regulation to the Union's Convention was brought up in 1885 but was met with objections from several member nations that didn't want many constraints on this new and growing technology. Some cursory language was added to the regulations; however, it wasn't until 1925 that the telephone was brought under the authority of the Union. In 1923, the French Postal Telephone and Telegraph (PTT) administration formed a preparatory technical committee to deal with international telephone usage in Europe. The group created a secretariat and agreed to annual meetings.

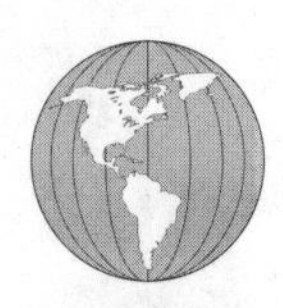

At the Paris Telegraph Conference in 1925, the Union incorporated the telephone committee (the International Telephone Consultative Committee) and also created an analogous committee for telegraphy (the International Telegraph Consultative Committee). Both committees maintained a loose connection with the Union. In 1947 when the ITU became an agency of the United Nations, both committees were more closely aligned with the ITU. Both committees functioned with plenary assemblies (general meetings at which agenda for study were adopted), working groups, and the adoption of nonbinding recommendations.

With the Paris Conference of 1925, the definition of telecommunications began to expand to include both telegraphy and telephony. Radio, or wireless, technology hadn't yet been brought into the organization. Indeed, radio communications developed much differently from telegraphy. Radio communications was privately developed, with its inventor, Guglielmo Marconi, in virtual control of its development and much of its deployment.

Because of interconnection problems caused by Marconi's unwillingness to connect with other brands of equipment, the Germans called for a Berlin Conference in 1903 to deal with these issues. The goal of interconnection among differing systems wasn't achieved until 1912 at a conference in London. A Convention and annexed radio regulations were agreed upon at a 1906 Berlin Conference with 29 nations in attendance. These actions marked some progress toward the establishment of international coordination for radio communications.

The United States, which hadn't yet been greatly active in these efforts, hosted two conferences in Washington during the 1920s. In 1920, the United States pushed for the creation of a Universal Electrical Communications Union, an idea that met with resistance. In 1927, all major nations except the USSR adopted and signed a new expanded radio convention and a set of general and supplemental regulations. An advisory committee (the International Radio Consultative Committee or CCIR) was also established at that time. It was to meet every two years and formulate recommendations regarding radio communications. At this conference, some suggestions were made that the group merge with the Union.

The merger of telegraph, telephone, and radio communications was achieved in 1932 at the Madrid Conference. In Madrid, at a joint meeting of the 4th Radiotelegraph Conference and the 13th International Telegraph Conference, a new organization was formed and named the International Telecommunications Union. Membership in the new ITU was based on adherence to either radio, telegraph, or telephone regulations.

World War II interrupted the work of the ITU. When the war ended, the ITU once again went through a transformation. At an Atlantic City conference in 1947, the ITU became a specialized agency of the United Nations. In its new form, the ITU included a newly created Administrative Council, which would be elected by the membership at its Plenipotentiary Conferences and serve as a governing body between conferences. The Berne Bureau, which had been funded entirely by Switzerland, became a secretariat funded by ITU membership and run by a secretary-general elected at the Plenipotentiary Conferences. With these actions, the organization became more international in scope.

The advisory committees were made permanent organs of the ITU, though they still kept their former operational structure. The CCIT, CCIF, and CCIR all continued to accomplish their work through plenary assemblies and study groups. The directors of each committee were elected by their respective plenary assemblies. At the time, the International Frequency Registration Board (IFRB) was also created. As the Board's major function, it registered frequency assignments and later assumed the task of assigning orbital positions for satellites. The IFRB originally had eleven members. The first members were Argentina, Australia, China, Cuba, Czechoslovakia, France, India, South Africa, the United Kingdom, the United States, and the USSR. This number was reduced to five at a subsequent Plenipotentiary Conference (Montreux, 1965).

The activities in Atlantic City in 1947 created the basic ITU structure and the operating scheme that was to last for over 40 years. The advisory committees on telegraph and telephone were merged into one committee, the CCITT, in 1956.

Steps Toward Creation of the ITU

Development of Telegraphy and Telephony

- **Paris, 1865**: Formation of the International Telegraph Union at a joint meeting of the Austro-German Telegraph Union and the Western European Telegraph Union.

  Drafting of a telegraph convention and annexed telegraph regulations.

- **Vienna, 1868**: Formation of the Berne Bureau.

- **Paris, 1925**: Addition of telephony through the International Telephone Consultative Committee.

  Creation of consultative committee for telegraphy.

Development of Radio Communications

- **Berlin, 1903**: Conference to deal with interconnection problems.
- **Berlin, 1906**: Drafting of radio convention and annexed radio regulations.
- **Washington, 1927**: Expansion of radio convention and creation of consultative committee for radio.

Merger and Continued Development

- **Madrid, 1932**: Merger of telegraphy, telephony, and radio at joint meeting of Radiotelegraph Conference and International Telegraph Conference.

  Formation of new organization named the International Telecommunications Union.

- **Atlantic City, 1947**: Transformation of the ITU into an agency of the United Nations

During those 40 years, the ITU has dealt with issues of increasing importance and complexity. One important issue is proper planning for the use of the radio spectrum. During the first half of the century, spectrum use was allotted on a first-come, first-served basis. Whoever used a frequency registered that frequency in order to avoid interference. Those developing new services needing spectrum and current uses of satellite communications advocate the continuation of this first-come, first-served approach.

Developing nations, however, fearing that the developed nations would monopolize frequencies and noting that frequency and geosynchronous orbital positions are scarce resources, proposed that frequencies and orbital positions be allotted to all countries, even if some can't yet make use of those resources. (For a further discussion of the geosynchronous orbit and why it is so attractive to satellite providers, see Chapter 5, "Technical Issues." Because there are a finite number of spots—called "parking spaces"—in an orbit, these parking spaces are a scarce commodity. The ITU membership debated the issue

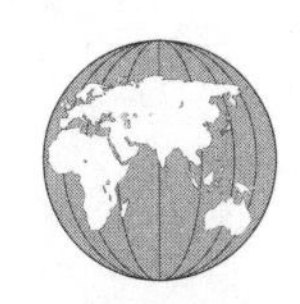

of equitable access to the spectrum and to geosynchronous orbits during several World Administrative Radio Conferences (WARCs). The focus of the WARCs is the allocation of frequency bands to different types of radio communication, including space communication. The development of satellite technology has introduced a whole range of new services requiring spectrum.

WARCs and Their Accomplishments

- **1959 WARC**: Allocated frequencies for space research.

  Suggested a special conference to discuss frequency needs for emerging categories of space communications.

- **1963 EARC (Extraordinary Administrative Radio Conference)**: Allocated frequencies for satellite communications.

  Determined that radio frequencies for satellite services would be allocated on a first-come, first-served basis.

  Formulated a recommendation that the principles of justice and equity be followed in the use of frequency spectrum for space communication.

- **1971 WARC**: Extended the first-come, first-served principle to use of the geosynchronous orbits. However, it also adopted a recommendation that no permanent priority be created for any one nation, nor any obstacle presented for any nation.

  Defined new services, including broadcasting satellite service.

- **1977 WARC**: Determined that the best method for allocating broadcasting satellite service was through an *a priori* planning method, which deferred for the Americas at the request of the United States and later was replaced by a more flexible approach at the 1983 Regional Administrative Radio Conference.

- **1979 WARC**: Debated the merits of *a priori* planning versus the first-come, first serve method.

  Called for the conference to address equitable access to the geosynchronous orbit and frequency bands.

- **1985 and 1988 WARCs** debated the use of the geosynchronous satellite orbit and the planning of the space services utilizing it (Space WARCs).

  ORB-85: Decided on *a priori* planning for new frequencies.

  Coordinated efforts for C-band and Ku-band frequencies in use.

  Continued first-come, first-served procedures for other frequencies.

  ORB-88: Drafted specific plans to implement 1985 recommendations.

- **1992 WARC:** Addressed allocation of frequencies for geosynchronous mobile satellite service and for land mobile telecommunications service.

A source of recurring discussion has been the ITU's role in closing the large gap between the telecommunications infrastructure in the developed world and that found in developing nations. At the 1965 Plenipotentiary Conference in Montreux, Switzerland, a group of developing nations proposed that a special ITU development assistance fund be created, that regional ITU offices be established to help local governments, and that a new ITU committee, similar to the CCITT and CCIR, be created to deal with development issues.[6] None of the three suggestions were adopted in Montreux. However, at the 1973 Malaga-Torremolinos (Spain) Plenipotentiary Conference, a fund was established. Further progress on this issue was made at the Nairobi Plenipotentiary Conference in 1982.

In Nairobi, the ITU approved regional offices to provide assistance at a regional level, decided to fund technical assistance activities from the ITU's regular budget, and created the Maitland Commission to examine the needs of developing countries and suggest ways in which to meet those needs. Also at the Nairobi Plenipotentiary Conference, the ITU adopted a change in procedure designed to broaden developing nations' involvement in ITU activities. The directors of the CCITT and the CCIR would now be elected by the total ITU membership at Plenipotentiary Conferences, instead of by the Plenary Assemblies of the CCITT and the CCIR.

The divergent regulatory schemes of ITU members have complicated the ITU's efforts toward global cooperation in the provision of telecommunications services. These complications were evident in the early days of the ITU,

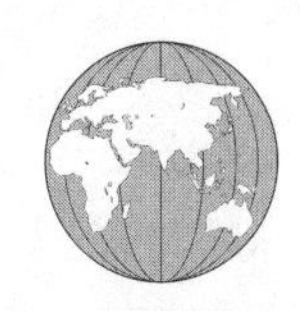

when the United States didn't sign the telegraph or telephone regulations because they were binding and dealt with pricing and accounting methods. In 1973, the ITU changed those regulations to recommendations of the advisory committees, and so the United States could sign the Convention. Complications continue as nations allow varying degrees of competition into their telecommunications services. While some nations (such as the United States, Japan, and the United Kingdom) have encouraged the introduction of competition into virtually all areas of telecommunications, other nations, like France and Germany, have proceeded much more slowly. While the United States, for example, allows the resale of services, other nations, like Singapore, don't. Balancing such divergent approaches is difficult.

In its 1984 set of recommendations, the CCITT suggested that operators of telephone networks could restrict resale and could require owners of private networks to return to the public network as soon as comparable services were available. The recommendations futher supported a ban on the connection of private networks to the public network and continued control by the telecommunications providers over terminal equipment. Though such recommendations were contrary to the developments in Japan, the United Kingdom, and the United States, they agreed with the views of other nations that sought to maintain monopoly control over telecommunications services.

The CCITT formulates recommendations that aren't binding on its members. World Administrative Telephone and Telegraph Conferences (WATTCs) generate and review Telephone and Telegraph Regulations, which are binding. The 1988 Melbourne WATTC addressed the issues of monopoly and competition at a meeting which, for various reasons, is called a pivotal event in the recent evolution of the ITU.[7] At the Melbourne meeting, efforts were made to assert the supremacy of regulation, rather than competition, in the telecommunications market. Some feared that the regulations would extend to private entities using international networks and that enhanced service providers would be required to operate within ITU rules.

A compromise was drafted by then ITU Secretary-General Richard Butler. As a result, private entities using the network wouldn't face compulsory coverage by ITU regulations. Nations gained permission to pursue bilateral innovations in the provision of telecommunications services, and rules weren't written to reinforce monopoly. New obligations were made explicit for operators of the network. Operators were to provide users with access to the network and with their choice of terminal equipment. They were to provide satisfactory quality of service, provide for internetworking, and set prices closer to cost. In this approach to the Telephone and Telegraph Regulations, the ITU recognized and responded to changes in the telecommunications environment.[8]

The Melbourne WATTC answered yet another need to recognize and respond to a changing telecommunications environment. As the pace of technical change increased, the cumbersome standards-setting procedure of the CCITT didn't. The process of plenary assemblies and study groups hasn't kept pace with the development of new services and technologies. The following section describes this process in more detail. In Melbourne, the assembly developed an accelerated procedure to approve new recommendations and revise old ones between plenary assemblies. Instead of waiting for an assembly, the voting countries can use a streamlined postal ballot process. Invoking the streamlined procedure requires a unanimous vote. A 70 percent majority is required to pass a recommendation. This move to accelerate the workings of the CCITT can improve a process that caused many to complain that the organization could delay the introduction of new services or could result in the organization being perceived as irrelevant if standards were adopted by other methods.[9]

Although the Melbourne WATTC made some progress toward streamlining the ITU's procedures, the 1989 Plenipotentiary Conference in Nice, France, appears to have made even greater strides toward making the ITU responsive to current needs. In Nice, ITU members dealt with many issues of continuing concern to the ITU. The Nice Plenipotentiary Conference continued to deal with the needs of developing nations by establishing a Telecommunications Development Bureau and by lowering the monetary contribution required of an ITU member.

The Nice Plenipotentiary Conference also took steps to make the ITU a more efficient body. It called for the establishment of a committee to examine the functions and structure of the ITU and to make recommendations. At this writing, a special Plenipotentiary Conference in Geneva has been scheduled to discuss these recommendations.

Significant ITU Conferences Since 1947

- **Montreux (Switzerland), 1965**: Lesser developed nations suggested the creation of a special development assistance fund.

  A new committee, equal to other consultative committees, formed to address development issues.

  Regional offices were created.

  IFRB was retained and reduced to a five-person board.

- **Malaga-Torremolinos (Spain), 1973**: A development fund was established.

  Telegraph and telephone regulations moved from Convention to CCITT recommendations, allowing the United States to sign the Convention.

- **Nairobi, 1982**: The Maitland Commission was formed to study the needs of developing nations.

  The formation of regional offices was approved.

  CCITT and CCIR directors will be chosen by the entire ITU membership.

- **Melbourne, 1988**: The rights of member nations to pursue competitive and innovative approaches to telecommunications were recognized.

  Accelerated standards-setting procedures were adopted for the CCITT.

- **Nice (France), 1989**: The Telecommunications Development Bureau was created.

  A committee formed to examine structure and functions of the ITU.

- **Geneva, 1992** (Extraordinary Plenipotentiary): Proposals for structural and procedural changes will be reviewed.

Overview of the Current Organization

As currently structured, the ITU is a highly complex organization.[10] It achieves its objectives through conferences, councils, technical bodies, and a general secretariat. The governing document of the ITU is the *International Telecommunications Convention*. This document was first drafted in 1865. In 1989, the Nice Plenipotentiary Conference changed its form. Instead of one document, which was difficult to change, the document now consists of a Constitution and a Convention (or basic regulations). A two-thirds majority vote is required to change the Constitution, but only a simple majority is required to change the Convention. The Constitution can only be changed at plenipotentiary conferences.

The main governing body of the ITU is the Plenipotentiary Conference, which is held every few years and is attended by the full ITU membership. During the Plenipotentiary Conference, the membership:

- Elects the Secretary-General, the Deputy Secretary-General, the five IFRB members, and the directors of the CCITT and CCIR.
- Reviews the Constitution, and if necessary, amends it.
- Determines general policies and priorities of the organization.

Voting occurs on a one-nation, one-vote basis. This voting method leads to bloc voting and gives the developing nations a strong voice in the organization.

Between Plenipotentiary Conferences, the governing body of the ITU is the Administrative Council, the members of which are elected by the Plenipotentiary membership. The Administrative Council implements the decisions of Plenipotentiary Conferences and other conferences (like the WARCs and WATTs) and formulates the agenda for the Plenipotentiary Conferences. The Council decides on matters of technical assistance, budgeting, and relations with the UN.

Originally, the Council consisted of representatives from 18 nations. The Geneva Conference of 1959 raised that number to 25; the Montreux Plenipotentiary Conference of 1965 raised it to 29; the 1976 Malaga-Torremolinos Plenipotentiary Conference raised it to 36; the Nairobi Plenipotentiary Conference in 1982 raised the number to 41; the Nice Plenipotentiary Conference of 1989 lifted the total to 43.

The ITU is responsible for maintaining telephone and telegraph regulations and also radio regulations. These regulations are reviewed and changed at administrative conferences. World Administrative Radio Conferences (WARCs) deal with issues of radio communication; World Administrative Telephone and Telegraphy Conferences (WATTCs) deal with issues of telephony and telegraphy.

Although administrative conferences handle binding regulations, the advisory committees, the CCITT and the CCIR (known together as the CCIs), formulate nonbinding recommendations. The CCIs hold preparatory meetings before WARCs and WATTCs. The CCIs function through plenary assemblies (which establish the areas to be studied) and through study groups and working parties, all of which work to formulate recommended global standards. Because of complaints that this process is too slow to accommodate the rapidly changing telecommunications environment, the Melbourne WATTC created an accelerated procedure for the CCITT. Instead of going through the

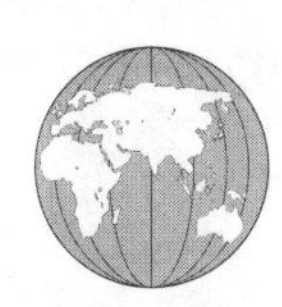

onerous process of establishing interim recommended standards that then must be formally passed at the assemblies, recommendations now can be passed by a 70 percent majority of the voting members using a streamlined postal balloting procedure. All voting nations must agree to use this procedure.

Another organ of the ITU is the International Frequency Registration Board (IFRB). The IFRB, a five-member board, records and publicizes every radio frequency used in the spectrum. The IFRB advises members regarding frequency assignments and other technical matters. The developing nations regard the board as a useful and trusted resource. For this reason, the developing nations fought off efforts to disband the IFRB at the Montreux Conference in 1965, the conference at which the IFRB was decreased from 11 members to 5. The IFRB acts as an arbitrator for nations faced with bilateral or multilateral frequency disputes, and it plays a significant role in preparations for the WARCs.

The ITU's Secretary-General and Deputy Secretary-General oversee the operation of the ITU. The secretariat is responsible for organizing conferences, producing ITU documents and publications, and overseeing the organization's technical cooperation and training programs. Figure 4.1 presents an overview of the ITU organization prior to any actions taken by the Geneva Plenipotentiary Conference.

Before the Atlantic City Plenipotentiary Conference, during which the ITU became a specialized agency of the UN, the official language of the organization was French. At the Atlantic City Plenipotentiary Conference, five official languages were established: French, Chinese, English, Russian, and Spanish. Only French, Spanish, and English were authorized as working languages for purposes of translation. Russian and Chinese would be translated at the expense of the requesting nation. The Nairobi Plenipotentiary Conference added Arabic to the list of official, but not working, languages. The Nice Plenipotentiary Conference placed all six on an equal basis for creating and publishing documents and texts of the Union and for reciprocal interpretation during conferences, plenary assemblies, and meetings of the Union. However, because of the costs involved, full implementation of this measure will take some time. Chinese, Arabic, and Russian, therefore, remain secondary languages.[11]

The membership defrays the expenses of the ITU through a system of unit classes of payments. The total expenses of the ITU are divided by the total number of units subscribed to by the membership, thus arriving at a cost per unit. Members pay according to the unit class they have elected. The Atlantic City Plenipotentiary Conference established eight class units, ranging from a first class unit of 30 to an eighth class unit of 1. In 1973, a new half class unit was created. The Nairobi created fourth unit and eighth unit classes. The Nice Plenipotentiary Conference created a sixteenth unit class for those nations that were determined by the UN or the Administrative Council to be "least developed."[12] Smaller unit classes shift the greater burden of ITU expense to the more developed nations. This reduction also makes ITU membership affordable to the poorest nations.

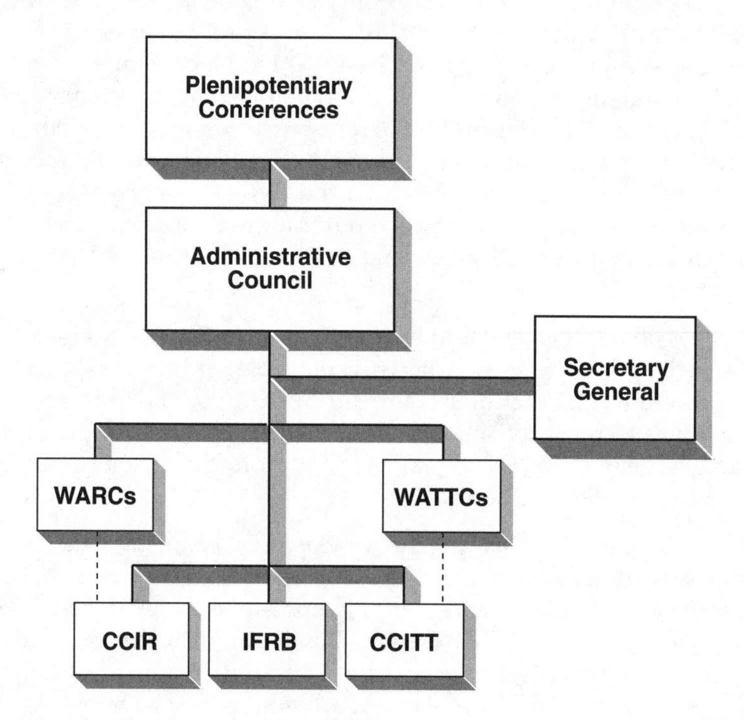

Figure 4.1. *The current structure of the ITU.*

Proposed Changes

At the Nice Plenipotentiary Conference, the ITU membership called for the creation of a committee to study the current ITU structure and to develop recommendations for making the ITU more cost-effective, efficient, and responsive to change. At the end of November 1989, the Administrative Council created a High-Level Committee of 21 countries to conduct a study and generate recommendations. Members of the High-Level Committee included:

- Algeria
- Australia
- Brazil
- Canada
- China
- Colombia
- Denmark
- France
- India
- Indonesia
- Japan
- Mali
- Morocco
- Saudi Arabia
- Senegal
- West Germany (before Unification)
- The United Kingdom
- The United States
- The former USSR
- The former Yugoslavia
- Zimbabwe

The High-Level Committee delivered its report to the Council in May 1991. Its recommendations were, at this writing, scheduled to be addressed during an Extraordinary Plenipotentiary Conference in Geneva in late 1992.

In addition to suggestions regarding the internal management of the ITU, the Committee suggested substantial organizational changes. Its proposals include maintaining the Plenipotentiary Conference as the supreme body of the ITU and scheduling a conference every four years. The Administrative Council would be called the ITU Council and would play a more strategic role for the organization.

The major functions of the ITU would be divided into three sectors: Development, Standards, and Radiocommunications. The work of the CCITT and the standards work now done by the CCIR would be combined. The other workings of the CCIR and the work of the IFRB would be combined under Radiocommunications. The five full-time IFRB positions would be replaced with nine part-time positions. Development work would encompass the current Telecommunications Development Bureau and other development functions.

Each of the three sectors would have a world conference as its supreme body supported by study and working groups. World conferences would be held between Plenipotentiary Conferences. Directors for each sector would head bureaus at ITU headquarters and chair advisory groups formed to aid the sector in developing strategies and approaches. Each sector would have its own budget. The secretary-general would remain the chief officer of the ITU, with a key role in planning, managing, and coordinating ITU activities. The Secretary-General would create a new unit for strategic policy and planning and would form a business advisory forum in order to obtain input from business leaders.[13]

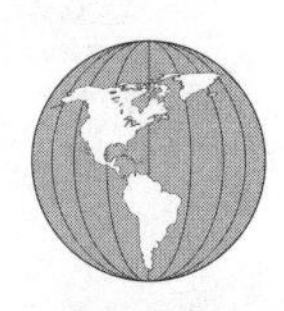

Proposed Changes to the ITU

- Plenipotentiary Conferences will remain the supreme body of the ITU, and regular conferences will be held every four years.
- The Administrative Council will be renamed the ITU Council.
- The Secretary-General will create a new strategic policy and planning unit and form business advisory forum.
- ITU functions will be organized into three sectors:

  Development—to encompass all development activities of the ITU and the Telecommunications Development Bureau

  Standards—to include CCITT activities and standards-setting activities of the current CCIR

  Radiocommunications—to include nonstandards-setting activities of the CCIR and activities of current IFRB
- Each sector will include:

  World conferences

  Study and work groups

  Directors and bureaus at ITU headquarters

  Advisory groups

One recommendation of the High-Level Committee was established before further plenipotentiary action. A World Telecommunication Advisory Council (WTAC) was formed in mid-1992 to give the ITU "strategic advice from the public and private sectors on the telecommunications environment and how, in the light of its dynamic nature, the Union's principal activities could be carried out more effectively."[14]

The 18-person WTAC selected Karlheinz Kaske, Chairman and CEO of Siemens AG of Germany, as its first chair. The other 17 members are senior officials of major international telecommunication companies, including Motorola, AT&T International, NEC, Bell Canada, and representatives from Kuwait, Nigeria, India, and other nations.

Initial Members of the WTAC

Leaders of Private Companies

Karlheinz Kaske; Chairman and CEO, Siemens AG (Chairman of WTAC)

George M. C. Fisher; Chairman, Motorola, Inc.

Randall T. Tobias; Chairman, AT&T International

Haruo Yamaguchi; Chairman, NTT

Tadahiro Sekimoto; President, NEC Corporation

Jean C. Monty; Chairman, Bell Canada

Pierre Suard; Chairman, Alcatel N.V.

Bjorn Svedberg; Chairman, LM Ericcson

A.C.R. Gil; Chairman, SID Telecom of Brazil

G.G. Kudriavtzev; Chairman, Intertelecom of Russia

Representatives of nations

Australia

Costa Rica

Cote d'Ivoire

India

Kuwait

Nigeria

People's Republic of China

United Republic of Tanzania

The High-Level Committee attempts to clarify the ITU's role and standardize its operations through regularly scheduled meetings. The committee also serves to focus the ITU's efforts on strategic and planning efforts and regularly include user and business input as part of the organization's workings.

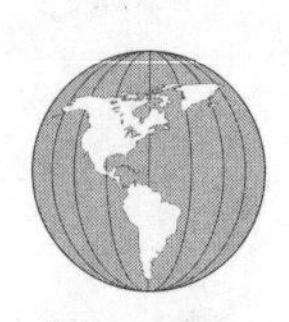

Telecommunications Standards Development in the CCITT

Officially, the CCITT and CCIR produce *Reports* and make *Recommendations* to their members. The recommendations of the CCITT and CCIR are the standards produced by each of these organizations. The production of recommendations is the most visible and impactful of the CCIs' activities. Standards development in the CCIs has undergone significant reform since the Melbourne Plenipotentiary Conference in 1988. Further changes can be expected in the future as the CCIs become more responsive to the needs of its members and the industry in general.

Historically, the CCIs defined the work of its *Study Groups* (SGs) by broadly worded *Study Questions* that defined the scope of the TCs authority and activities.[15] These questions were addressed during the course of a four-year *Study Period.* The CCITT and CCIR voted on the recommendations and reports generated by the technical committees, respectively, at their plenary assemblies, which occurred at the end of each four-year Study Period. Today, as then, Recommendations are adopted only by unanimous vote.

SGs have their own plenary meetings and establish working parties, working groups, and rapporteurs to carry out their work. *Rapporteurs* are individuals appointed to carry out specific activities, such as drafting documents or conducting liaison activities. They are assisted by the CCITT secretariat staff located at the ITU headquarters in Geneva.

The recommendations and reports are developed through the consideration and discussion of contributions. A *contribution* is a technical document proposing a specific solution to a specific problem. Prior to the technical debate, the Technical Committee must agree on the *Terms of Reference* for the standard being developed. The Terms of Reference outline the technical approaches that the committee may take. These are obviously crucial to the outcome of the committee process, and may, in themselves, generate considerable debate.

This four-year cycle of study periods and plenaries was generally satisfactory in the CCITT's early years, but as the rate of technological change increased, it became burdensome. In addition, the unanimity requirement and the dominance of the Administrations in the CCITT's decision-making process (instead of the manufacturers who were doing much of the technical work in standardization) were increasingly inappropriate.

At the 1984 CCITT Plenary Assembly in Malaga-Torremolinos, the CCITT established a committee, called Special S, to review the CCITT's organization

and procedures. The results of this committee called for only minor changes. As a result, additional calls for changes to the CCITT's procedures have come from within the industry. By the time of the next Plenary Assembly in Melbourne in 1988, the United States had established its Committee T1 to handle telecommunications standards development in the United States, Japan had formed the Telecommunications Technology Council (TTC), and the European Telecommunications Standards Institute had been founded. Collectively, these regional bodies were perceived as threatening the CCITT's role in international standardization.[16] At the 1988 Melbourne Plenary, the traditional four-year cycle for the approval of standards was replaced with an approvals procedure that allowed recommendations to be approved as they were ready. Under these new procedures, a recommendation needed unanimous approval in the SG, but only a 70 percent approval from the CCITT membership. Members were required to respond to these ballots within 60 days. The constructive spirit that prevailed at the Melbourne Plenary and enabled these changes to take place is sometimes referred to as "the spirit of Melbourne."

In the wake of these radical changes that revitalized the CCITT as an international standards-making body, practical agreements had to be established between the CCITT, ETSI, Committee T1, and the TTC. These were established at the Fredericksburg (Virginia, United States) Conference in 1990. As a result of the Fredericksburg Conference, the procedures for coordinating among the three bodies direct that:

- The TTC, Committee T1, and ETSI would meet periodically in relatively informal meetings and compare their plans for standardization
- They would compare and identify the common elements of their plans
- The directors of the CCITT and CCIR would use the common elements to establish their work priorities[17]

The CCIs differ in other important aspects from other standards setting organizations, such as ISO and the IEEE, because the telephone administrations of the member countries play an important role. Their role has significance in two respects.

First, because the CCITT is a treaty organization and the participating telephone administrations are the legal entities governing telephony and telegraphy, the reports and recommendations of the CCITT often have the force of law. This means that equipment not conforming to the relevant CCITT recommendation may not be attached to the telephone network in those countries. Therefore, the standards developed by a voluntary consensus standards committee become *de jure* standards.[18] This gives the CCITT considerable

power as a standards-setting organization and the national administrations a very important role in the CCITT.

This role of the national administrations is the second way in which the CCITT differs from the others. In most other standards organizations, users are underrepresented. In the CCITT, the administrations (effectively user organizations) help define which standards efforts will be pursued within the CCITT, and which ones will not. In addition, the representatives of administrations often take an active role in the development of the technical standards, even though they have no product development capability. This is troublesome for manufacturers, because they have little power to articulate their agenda and vision for the future. As we argued elsewhere, the manufacturers are almost always in a better position to make the necessary cost versus functionality tradeoffs because they ultimately have to produce products. While the user's opinions are useful in determining the future agenda, a bias as strong as that found in the CCITT is generally not desirable.

The CCITT has a somewhat unique membership structure consisting of four membership classes. Administration membership (open to any of the ITU's 168 members) and Recognized Private Operating Agencies (RPOAs) (open to basic telecommunications service providers)[19] have full voting rights at the plenary meetings. Scientific and Industrial Organizations (SIOs) such as IBM have a different type of membership. These organizations influence the standards process heavily, although they do not have a vote at the plenary meetings. Each SIO must be approved by their national government (such as U.S. Department of State for a U.S. firm). There are currently 159 SIO members of the CCITT. Finally, other standards organizations, such as ISO, may attend meetings to ensure coordination among different standards setting organizations. In 1988, 39 organizations fell into this membership category.

Standards setting in the CCITT will continue to change in the future. The current Director, Dr. Theo Irmer, has set the following objectives for the CCITT:[20]

- Rapid standards production
- Flexible, uncomplicated procedures
- Market orientation
- Coordination with other regional organizations
- A stronger role for nonadministrations, such as some form of power sharing between administrations and nonadministrations

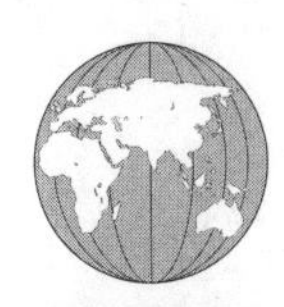

He proposed an 18-month production cycle of standards, a reorganization of study groups, decentralization of work and parallel production efforts with the ISO, and the International Electrotechical Committee (IEC) as mechanisms for achieving these objectives. In addition, he proposed the consolidation of the standards related to the CCITT and the CCIR for the purpose of improving efficiency. If these changes are implemented, they will radically change the CCITT's structure and behavior. Based on pressures from regional bodies and ISO/IEC, the CCITT may have no choice but to implement changes such as these.

Issues and Problems

Whatever final structure emerges from the Extraordinary Plenipotentiary Convention in Geneva, the ITU will continue to face the challenge of balancing the often divergent interests of its membership. The organization must strike a balance between the development needs of some members and the willingness of other members to pay for those needs. In crafting its regulations and formulating recommended standards, the ITU continues to experience friction between its members who want to encourage competition and those who favor regulation and monopoly. In its treatment of the spectrum and orbital assignments, issues of equitable access continue to arise.

As technological change continues at an increasing pace, the ITU must respond quickly in its standards-setting process and continue to work with other entities—GATT, ETSI, OSI, and others—to facilitate the continued development of international telecommunications.

International Standards Organization

The ISO was founded in 1947 as a voluntary, nontreaty organization. The purpose of ISO is to achieve worldwide agreement on international standards. The membership currently includes about 90 countries. Since its inception, the ISO has prepared and approved many thousands of standards, covering topics from screw threads to data processing.

The ISO is divided into 160 technical committees, each containing several subcommittees. Each Technical Committee (TC) is responsible for developing standards in a relatively broad area. TC97, for example, is charged with "Standardization, including terminology and definitions, in the area of

information systems, including computers and office equipment." This general area is subdivided and assigned to subcommittees (SCs). These subcommittees have general responsibility in a narrow area. JTC1/SC6, for example, is concerned with telecommunications and information exchange between systems.[21]

The ISO comprises the representative standards organizations of any country wishing to participate. The ISO certifies these bodies to carry out standards activities. Each of these *member bodies* is entitled to participate and exercise voting rights on any Technical Committee of the ISO. Some developing nations don't have standards organizations; they are considered *correspondent members*. What follows are some countries and their respective member organizations:

Country	*Member Committee*
Canada	The Canadian Standards Association (CSA)
France	Association Française de Normalisation (AFNOR)
Germany	Deutsches Institute für Normung (DIN)
Japan	Japanese Industrial Standards Committee (JISC)
The United Kingdom	British Standards Institute (BSI)
The United States	The American National Standards Institute (ANSI)

The structure of the ISO differs from that of ITU organizations. The ISO holds plenary meetings at closer intervals, which the member bodies determine. Unanimity isn't required to carry a motion, as is necessary in the ITU, although it is normally the objective of the committee chairmen to achieve unanimity where possible. ISO publishes *International Standards* as opposed to the *Recommendations* and *Reports* of the ITU.[22]

The process of adopting a standard in ISO has several well-defined phases. A *project formal description* (FD) is developed that defines the scope of the project. The appropriate Subcommittee of the Technical Committee charged with such projects develops this FD. The standard first exists as a *Working Draft* (WD), then as a *Draft Proposal* (DP). As further technical details are worked out and agreed upon, this becomes a *Draft International Standard* (DIS). Finally, it is adopted as an *International Standard* (IS).

Unlike the CCITT, ISO is not a treaty organization, so a governmental agency does not need to coordinate and represent its activities. In some countries,

like the United States, involvement in ISO activities can occur without government involvement. The *General Assembly* (GA) meets every three years and establishes policy in the ISO. The Council organizes and supervises these policies; its chairperson is also the president of the GA, which also comprises the GA Vice President, the treasurer, and 18 additional officials elected from the member bodies. The Council meets annually and determines the technical structure of ISO, accepting approved standards for publication and appointing members of the Technical Board and Executive Board. The Executive Board assists the Council on administrative and organizational matters, whereas the Technical Board advises the Council on matters related to the standards and establishment of committees in the absence of the Council. The Council contains six Standing Committees (on Planning, Certification, Consumer Policy, Development, Information, and Reference Materials) that focus on the impacts of standards, rather than the standards themselves. The Council is supported by a Central Secretariat, which is located in Geneva. The Technical Committees (TCs) that perform the actual standards developments are supported by secretariats from the member bodies.

The European Community

As this is written, the single internal market of the European Community (EC) is scheduled to come into existence at midnight on December 31, 1992. The EC intends to create an internal market devoid of barriers that block the free movement of people, goods, and money across the national borders of the 12 EC member states. A major tool in moving toward an internal market is telecommunications. In pursuit of the internal market, EC policymakers have worked toward the development of Europe-wide telecommunication networks and services.

When envisioning a barrier-free community, EC policymakers regard telecommunications as a vehicle for unity and an important prerequisite for economic growth. To facilitate the development of an EC telecommunications infrastructure, policymakers focused on initiatives to build infrastructure and liberalize the telecommunication environments in all the EC nations. Their efforts have implications for the development of telecommunications in all parts of the globe.

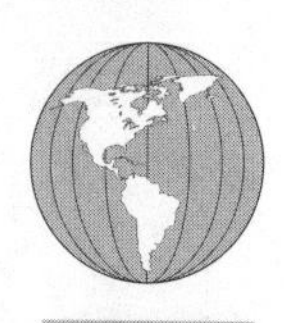

The History of the European Community

The introduction of a single internal European market in 1993 marks one milestone in a continuing series of unification efforts that focus on matters of trade, but often go beyond trade to matters of defense and diplomacy. The internal market will initially encompass the 12 members of the European Economic Market:

Belgium	Italy
Denmark	Luxembourg
France	the Netherlands
Germany	Portugal
Greece	Spain
Ireland	the United Kingdom

Conditions indicate that other nations are waiting to join the EC. This continues a unification process that began in the chaotic conditions following World War II.

Faced with the devastation following World War II, the nations of Europe realized that a "peaceful process of integration on a voluntary basis was the only way to full recovery and to provide Europeans with a better future."[23] Their response formed a variety of organizations. The Council of Europe, founded in 1949, focused on social issues and human rights. The Western European Union—consisting of Belgium, France, Italy, Luxembourg, the Netherlands, the United Kingdom, and pre-Unification West Germany—dealt with issues of defense. Euratom was formed to address atomic energy concerns.

Several organizations were created to handle trade and economic matters. The Economic Commission for Europe was formed to facilitate relief after World War II, to work toward the removal of East-West trade barriers, and to help foster European cooperation in research. The Organization for European Economic Cooperation was formed in 1948 to channel aid from the United States to European nations in need of assistance. The initial membership of this organization included 16 Western European nations, with the United States and Canada functioning as associate members. That organization later changed its focus to facilitate cooperation on common economic and development issues and changed its name to the Organisation for Economic Cooperation and Development, or OECD. Belgium, France, Germany, Italy,

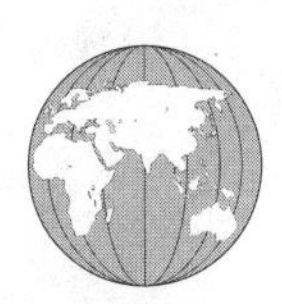

Luxembourg, and the Netherlands formed the European Coal and Steel Community in 1951 to create a tariff-free market for coal, coke, iron ore, scrap iron, pig iron, and steel among the six member nations (the Six).

Against the backdrop of these various efforts at unity, the European Economic Community (EEC) was formed with the signing of the Treaty of Rome in 1957. The Treaty of Rome provided for the removal of trade barriers for oil, natural gas, and electricity among the Six and added to the progress of the Benelux nations (Belgium, the Netherlands, and Luxembourg) in creating their own customers' union in 1948.[24] In 1965, a treaty was signed merging the EEC, the European Coal and Steel Community, and Euratom. The merger went into effect July 1967. The foundation was now in place for a broader attempt at European union.

After changing its position on membership, Great Britain finally joined the EC in 1973, along with Denmark and Ireland. Having expanded from six to nine members, the EC in the 1970s also sought to move toward what has been called macroeconomic integration or the movement toward monetary union and eventual political union.[25] The increase in oil prices and recessionary conditions during the 1970s hampered EC efforts toward integration. During this period, nations placed a greater emphasis on national interests and the creation of protectionist measures, rather than movement toward a unified market.

By the early 1980s, the perception grew that the creation of barriers and the increased costs caused by duplication of efforts across national boundaries was causing the EC nations to fall behind the United States and Japan in economic growth and market performance.[26] This perception resulted in the drafting of a White Paper outlining a plan for a single Europe. Lord Cockfield, the White Paper's drafter, notes that, during the mid-1980s, talk abounded concerning the of need to "relaunch" the European Community and to do so through an internal market.[27]

The White Paper proposed that a single market be created, free from physical, technical, and fiscal barriers.[28] The member states adopted Lord Cockfield's White Paper in a meeting of the European Council in December 1985 and drafted the Single European Act (SEA) to accomplish the goal of an internal market by the end of 1992. The SEA was ratified by the 12 EC members during 1986 and 1987 and finally came into effect July 1, 1987.

The SEA set the stage for the creation of a single European market among the EC members on midnight December 31, 1992. The creation of that market involves the discussion and acceptance by the member states of almost 300 measures designed to create a market of:

- No customs duties
- Free movement of people, goods, and capital
- Some harmonization of tax rates
- An EC-wide single market for telecommunications

Progress toward the goal of a single market in 1993 was evident, with the acceptance of over three-fourths of the necessary measures by early 1992.[29]

Conditions indicate that the single internal market might extend beyond the 12 EC members. Several members of the European Free Trade Association (EFTA), which opted for a looser free trade area rather than a customs union in the 1960s, are now considering the possibility of joining the European Community in the 1990s. After the turn of the century, several Eastern Europe nations also are likely candidates for membership.[30]

The tendency toward greater unity recently extended beyond the internal market envisioned by the SEA. During a December 1991 summit in Maastricht, the Netherlands, EC leaders drafted a treaty envisioning European union through the adoption of a single currency by 1999, the encouragement of common EC stands on foreign policy and defense issues, and the adoption of common standards for working conditions. The Danish defeat of a referendum to adopt the treaty and the French passage of a referendum on Maastricht by a close margin place the future of the union in question. However, regardless of the Maastricht Treaty's outcome, the single internal market envisioned by the White Paper and outlined in the SEA is, at this writing, expected to emerge on January 1, 1993.

The Organization of the European Community

The major bodies of the EC include the Commission, the Council of Ministers, the European Parliament, the Court of Justice, and the Economic and Social Committee. Each of these bodies plays a specific role in the governance of the EC.[31] The activities of all these groups are based on the provisions of the Treaty of Rome and the SEA.

The decision-making body of the EC is the Council of Ministers, which usually meets in Brussels. Those on the Council represent their own countries; this isn't the case in the other units of the EC, in which supranational

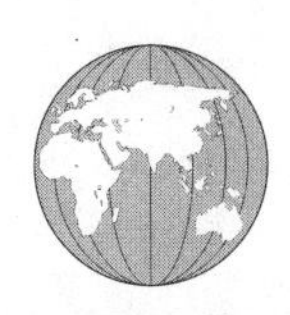

interests are represented. The presidency of the Council rotates among the member nations every six months. The Council has its own secretariat to handle administrative matters. Voting strength on the Council is in proportion to the size of its member nations: France, Germany, Italy, and the United Kingdom have ten votes each; Spain has eight votes; Belgium, Greece, the Netherlands, and Portugal each have five; Denmark and Ireland have three each; and Luxembourg has two. The total of all possible votes is 76. A qualified majority is equal to 54 votes.

From 1965 until the SEA came into being in 1987, the Council followed a process of requiring a unanimous vote on significant issues. The unanimity requirement was decided upon during a meeting of the Council held in Luxembourg following France's refusal to move toward majority voting; the resulting decision to seek unanimity in voting was therefore called the Luxembourg Compromise. Qualified majority voting rather than unanimity for most issues was introduced with the SEA.

The Commission of the EC has several functions, including:

- Proposing new policies
- Exercising executive powers when implementing policies of EC treaties
- Administering EC funds and Euratom research programs
- Serving as guardian of the treaties
- Representing the EC in international organizations

Commissioners are appointed for four-year terms; the President of the Commission is appointed for one two-year term, which is usually renewed for an additional term. The Commission also has six vice-presidents.

The Commission is organized into directorates-general, which are usually identified by number and assigned specific areas of concern. The directorates-general are headed by directors general; heads of divisions serve beneath the directors general. The Commission has a fairly large staff (around 14,000 in 1989), and nine working languages.

Since 1979, the European Parliament (EP) has been selected by direct elections. Membership in the Parliament is proportional by size of member state. France, Germany, Italy, and the United Kingdom each have 81 members of Parliament; Spain has 60; the Netherlands has 25; Belgium, Greece, and Portugal each have 24; Denmark has 16; Ireland has 15; and Luxembourg has 6. The various national political parties of the member states are represented in the Parliament. The SEA increased the EP's powers in 1987. The EP participates

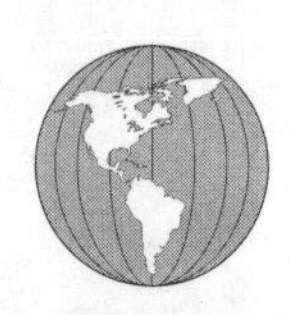

in a cooperative procedure in which it gives opinions on Commission proposals and Commission proposals adopted by the Council. If the EP rejects a Council proposal, the Council must be unanimous in overriding the EP's rejection.

The Court of Justice, which sits in Luxembourg, consists of 13 judges, with each appointed for a six-year term and possible reappointment. The members of the Court, like the Commissioners and the EP members, are expected to represent EC interests and not the interests of their respective nations. The Court ensures that EC laws are followed.

The Economic and Social Committee (ESC) of the EC serves an advisory function. The ESC comprises 189 representatives of interest groups in the member states. Those interest groups encompass labor, employers, and other groups. The ESC is consulted on issues involving social policy, training, agriculture, and other matters and has the right to initiate consideration of issues of importance by the bodies of the EC.

The Treaty of Rome provides several instruments that the Council and the Commission can use in implementing policies. These legal instruments include:

- Application of the Treaty of Rome rules. The Commission can ask for enforcement of these rules in the Court of Justice.
- EC directives, which are binding on member states regarding the goal of the directive. The Council issues directives in response to proposals from the Commission. The Commission itself can issue directives in those cases where member states grant exclusive or special rights to specific organizations.
- EEC regulations, which the Council usually issues and are binding in all member states.
- EEC decisions, which are individual measures directed toward a specific nation or person. The Council usually issues decisions.
- EEC recommendations, which are not legally binding and are issued either by the Council or by the Commission.[32]

All these legal instruments have been applied in matters involving telecommunications in the EC. Indeed, by the end of 1991, EC efforts to promote an advanced telecommunications infrastructure had involved nine directives, one regulation, five decisions, three recommendations, and seven Council resolutions.[33]

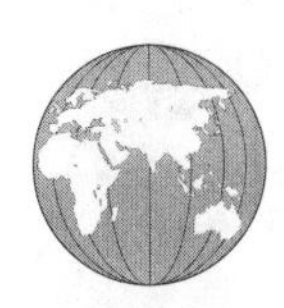

Telecommunications Initiatives of the European Community

Telecommunications wasn't a concern of pre-EC organizations, nor of the EC in its early years. In the early 1950s, efforts by the Council of Europe to form a European postal and telegraph union were rejected. Discussions of forming a postal and telegraph union along the lines of the European Coal and Steel Community bore no results. Efforts in the EC also failed to see tangible results. The PTTs of EC member states met in 1964 and again in 1977, but addressed only postal matters. Telecommunications at that time was regarded as a "public infrastructure service," and as such, wasn't perceived as having "any direct relevance for trade or even industrial policy."[34]

Efforts by the Commission and the EP to move toward a unified telecommunication policy in the EC didn't result in Council activities during the 1970s and 80s. Not until the mid-1980s, when concern regarding AT&T's potential activities after divestiture in the United States, fear of dominance by IBM, and a growing understanding of the importance of telecommunications as an economic force moved the Council, were some actions taken.[35]

In December 1984, the Council approved an action program to address five areas:

- Formulation of a coordinated plan for network and telecommunications services development in the EC that would include common infrastructure projects like the introduction of ISDN, the development of cellular communications, the development of broadband networks, and the development of community-wide videocommunication services.
- Creation of a community-wide market for telecommunications equipment.
- The launching of a program to develop the technologies needed to establish broadband technologies (the Research and Development for Advanced Communications in Europe or RACE program).
- Establishment of assistance for lesser developed regions of the EC—special telecommunications action for regional development (STARS) program.
- The coordination of negotiating positions in international organizations dealing with telecommunications issues.[36]

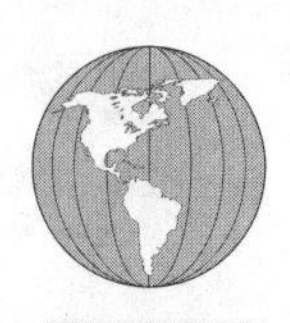

The SEA recognized the importance of telecommunications to the economic growth of the EC and to the process of unification when telecommunications was identified as an important component in the single market initiative.

The importance of telecommunications to EC efforts was explicitly expressed in the Green Paper, the Commission's program for telecommunications:

> *A technically advanced, Europe-wide and low-cost telecommunications network will provide an essential infrastructure for improving the competitiveness of the European economy, achieving the internal market and strengthening Community cohesion—which constitute priority Community goals reaffirmed in the Single European Act. Telecommunications [has] a great influence not only on services in general, such as financial services, transport and tourism, but also on trade in goods and on European industrial cooperation.*[37]

The Green Paper, which was proposed by the Commission in 1987 and carried the basic objectives that were approved by the Council in 1988, presented a series of proposals designed to create an open and robust telecommunications environment. The Green Paper's proposals included:

- Continuation of PTT monopolies only in the provision of network infrastructures and basic services.
- Unrestricted provision of competitive services, including value-added services, in member states and between member states for own use, shared use, or provision to third parties.
- Creation of EC-wide interoperability through strict adherence to standards for the network infrastructure.
- A clear definition of requirements for interconnection and access to the network (open network provisioning, or ONP).
- An unrestricted market for terminal equipment.
- Separation of regulatory and operational activities.
- No cross-subsidization of services or manufacturing activities by telecommunications administrations.
- Creation of the European Telecommunications Standards Institute (ETSI).[38]

The Commission is active in making the Green Paper proposals a reality although the Treaty of Rome doesn't explicitly give the Commission oversight in the area of telecommunications. The Commission bases its activities in this area on various articles of the EEC treaty, including Article 235, which allows

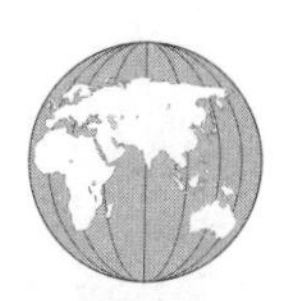

the Commission to address policy fields not provided for by the treaty if actions in such policy fields are necessary for the realization of the common market.[39] The Commission also places reliance on Articles 85, 86, and 90 of the Treaty of Rome, which deal with issues of competitiveness, the abuse of dominant positions, and the activities of organizations granted special rights by member states.

The EC has made progress on several issues raised by the Green Paper. In the area of terminal equipment, a 1988 directive abolished monopoly rights. A 1990 directive dealt with mutual recognition of type approval across the Community. *Type approval* is the process in which a governmental entity within a nation certifies that a piece of equipment may be connected to the public network in that country. The 1988 directive was challenged in the Court of Justice by Belgium, France, Germany, Greece, and Italy. However, the Court found that the Commission had acted properly under Article 90 of the Treaty of Rome.

The EC moved in the area of services as well, adopting in 1990 a directive calling for competition in all value-added services and progressive competition in data communication services. Simple resale of nonvoice services is allowed as of January 1, 1993, with extensions granted until 1996 to countries with underdeveloped data communication markets. Belgium, France, Italy, and Spain have argued that the Commission exceeded its authority in liberalizing the services area as well. In the area of ONP, the EC issued a directive addressing the basic principles of providing access and interconnection to the network through harmonized standards, usage conditions, and technical interfaces. Subsequent implementing directives regarding network security, tariffs, interoperability, and other issues will be required. In late 1990, the EC also issued a paper on satellite communication, calling for liberation of the earth segment, unrestricted access to space segment capacity, commercial freedom for space segment providers, and harmonization measures to facilitate Europe-wide services.[40]

EC efforts continue to introduce greater degrees of competition into telecommunications in the EC. The 1990 directives on service competition and ONP required that the Commission review the status of liberalization and the need for further action in 1992. As part of that review, the Commission adopted an advisory paper that provided four options for the future:

- Freezing the liberalization process as it is (maintaining the status quo).
- Introducing extensive regulation of tariffs and investments at the Community level to remove bottlenecks and the surcharge on intra-Community tariffs.

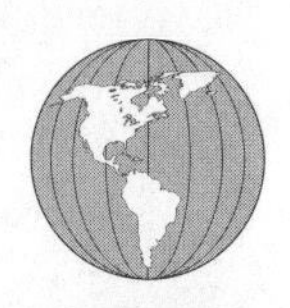

- Liberalizing all voice telephony for international and national calls.
- Taking the intermediate step of opening voice telephony between member states to competition.[41]

Of the four options presented in the advisory paper, which was drafted by Sir Leon Brittan (EC Commission Vice President in charge of competition) and Filipo Maria Pandolfi (Telecommunications Commissioner), the fourth is most likely to be considered, especially since intra-Community calls are higher than national long distance rates.

Recent discussions regarding the possible reorganization of responsibilities dealing with telecommunications issues suggest that the EC will continue to focus on telecommunications matters. The possibility exists that Directorate General XII (research) will merge with Directorate General XIII (information technology and telecommunications). Instead of two, three directorates now involve telecommunications issues, including a new and larger directorate for telecommunications policy and postal services. The management of EC research programs in the telecommunications sector is split between two directorates, one dealing with advanced telematics and the other dealing with mobile, broadband, and security issues.[42] Whatever final organizational structure emerges, the Commission will continue to address the proposals outlined in the Green Paper.

The Significance of the European Community to International Telecommunications

At a very basic level, the EC is significant to international telecommunications because of its size. With 320 million inhabitants, the EC member states represent a potential market of consumers more plentiful than the United States with its 250 million. Beyond pure population considerations, the EC represents a large market for telecommunications. According to some studies, while the United States comprises 35 percent of the world's telecommunications market and Japan accounts for 11 percent, the EC nations as a whole represent over 20 percent of the global market.[43] As the second largest telecommunications market, the EC has influence on the global marketplace.

The EC's recognition of the importance of telecommunications as a vehicle for economic growth and unity can't help but have an effect on other nations

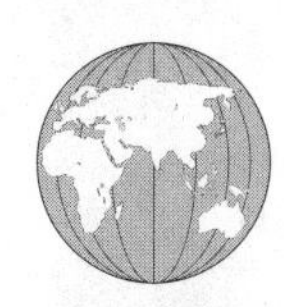

and their view of telecommunications as a means of economic development. It is surely no accident that all the recently liberated Eastern European nations view telecommunications infrastructure development as a necessary step toward growth. It is also no accident that these nations approach telecommunications infrastructure with an eye toward competition and privatization. While the United States and Japan provide examples of the benefits of both competition and privatization, the example presented by the EC is much closer to home and perhaps more compelling. This influence is strengthened by the EC's commitment to lending technical assistance to its eastern neighbors. According to Rudolf Meijer, a principal administrator of one of the Commission's directorates general, the EC budgeted $1.25 billion for technical assistance to eastern nations during 1992.[44]

The EC's Green Paper, with its neatly presented proposals, can serve as a blueprint, not just for EC nations, but for EC neighboring states as well. The EC's emphasis on liberalization, the avoidance of cross-subsidies, ONP, and the separation of operation from regulation provide yet another strong voice for change from the old ways of providing telecommunications services. The provisions of the Green Paper and the emphasis on telecommunications as a vehicle for economic progress lend credence to the perception that telecommunications is no longer a publicly provided utility, but an integral element of trade and economic growth.

The Conference of European Posts and Telecommunications Administrations

The CEPT was founded in 1959 for the purpose of coordinating the activities of European Postal and Telecommunications Administrations. Its membership is limited to European countries and now consists of member states of the EC and the European Free Trade Association, along with Turkey and the former Yugoslavia—26 states in all. Membership might also be available to the Eastern European states. Formally, CEPT regulations are established in the founding Act of the Conference. CEPT is independent of other pan-European organizations, both politically and economically.

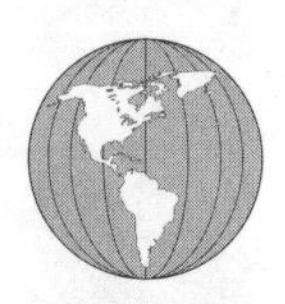

CEPT was founded to support:

- The development of services of common interest
- The exchange of information
- The study of questions concerning the organization, technical aspects, and functioning of services
- Simplification and improvement of postal and telecommunications services
- Examination of proposals presented at conferences

CEPT performs its work through Plenary Assemblies and Committees. The Plenary Assembly meets every two years and is the supreme authority of CEPT and deals with all organizational issues. At the Plenary, members nominate one of their peers to be the Managing Administration, the obligations of which encompass presidential and secretarial duties, supervision of ongoing work between Plenaries, and the organization of the next Plenary. A permanent staff, the Liaison Office, is located in Berne, Switzerland, to support the Managing Administration.

CEPT has two Commissions, one to deal with postal matters and the other to deal with telecommunications matters. Each of these commissions studies service provision, tariffs, and technical issues. The commissions can appoint committees to study specific topics, such as satellite and transatlantic communications. CEPT ultimately had a very strong influence in the CCITT standards-setting process, because CEPT positions are normally supported by all the European administrations.

Historically, CEPT meetings have been closed to anyone but members. The rationale was that privacy made it possible to reach compromises and take politically unpopular positions without paying the price. If they wanted, members could coordinate their positions with the telecommunications industry in their own countries; however, most countries did so only with selected suppliers in closed meetings rather than with the industry as a whole in open meetings. This limited the kinds of input they received and the decisions they made. Clearly, a strong incentive existed to make decisions that perpetuated the monopoly positions of member administrations, regardless of user input.[45]

CEPT and the EC formally recognized each other in 1975.[46] The recognition came by way of an EC request for the harmonization of telecommunications standards among countries. CEPT created the Coordination Committee on Harmonization (CCH) to support this effort. Because much of CEPT's work

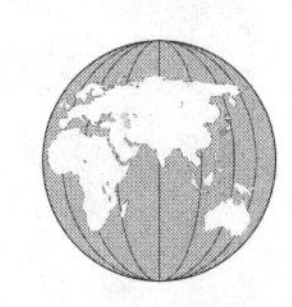

wasn't performed openly, equipment manufacturers were distressed at their inability to participate. Furthermore, member administrations weren't bound to adopt CEPT recommendations, which led to the limited success of CEPT efforts at harmonization.

As the EC became more powerful, the influence of its actions became greater, so that a framework for cooperation between the EC and CEPT was necessary. This need was met by the Copenhagen Memorandum of Understanding (MoU) in 1985.[47] The MoU was partly the result of several member states investigating the redefinition of telecommunications service provision and of complaints between CEPT members.[48] These problems led to the creation of the Technical Recommendations Application Committee (TRAC). With the MoU, CEPT recommendations became binding on members. The procedure for developing these European standards, called NET (Normes Europeennes des Telecommunications), was quite involved. As a result, harmonized regional standards were developed too slowly for the needs of the EC. Consequently, the EC included in its Green Paper the separate ETSI that could develop standards more quickly and in a more businesslike manner.

CEPT continues to exist after ETSI, but its global influence is diminished because it no longer has control over the standards development process. In addition, the EC's directorate for competition policy (DG IV) found that CEPT's tariff plans were a cartel action under Article 85 of the Treaty of Rome.[49] This caused CEPT to reorganize and further change its focus. For instance, it created a Commercial Action Committee to deal with the concerns of users. Many CEPT members formed the European Telecommunications Network Operators group (ETNO) as an additional industry organization to represent their interests outside CEPT.

The European Telecommunications Standards Institute

The ETSI was established in 1988 as a direct result of the EC's Green Paper to assume control over the regional telecommunications standards development process. In contrast to CEPT, ETSI is open to anyone who has an interest in developing European standards. The ETSI has five membership classes:

- Administrations, administrative bodies, and national standards organizations
- Public network operators
- Manufacturers
- Users
- Private service providers, research bodies, consultants, and others

The EC and the EFTA also hold memberships in ETSI. In committees, voting is a weighted national scheme, where larger countries have more voting power than small countries. This is unique in standards development, where votes are based either on unanimity or majority rules.

ETSI has three major organizational elements: the General Assembly, the Technical Assembly, and a Secretariat with a Director. Power is distributed among these three components. The ETSI General Assembly is responsible for administration only. It doesn't make technical decisions on the content of standards. Rather, the General Assembly is responsible for organizational decision-making, such as accepting members, budgeting, and procedures. The General Assembly elects the Director. The technical work of ETSI takes place in the Technical Assembly. The General Assembly creates and manages the Technical Committees, which in turn manage the work of Project Teams.[50] Project Teams are unique in that they consist of individuals who are dedicated full time to a standards effort and who are paid from the ETSI Annual Work Program Budget.[51] In addition, the Technical Assembly supervises the work of the ISDN Standards Management (ISM) Committee, the Strategic Review Committee (SRC), and the Intellectual Property Rights Committee (IPRC).

1990 ETSI Technical Committees

NA	Network Aspects
BT	Business Telecommunications
SPS	Signalling, Protocols, and Switching
TM	Transmission and Multiplexing
TE	Terminal Equipment

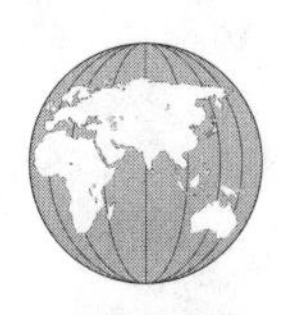

EE	Equipment Engineering
RES	Radio Equipment and Systems
GSM	Special Mobile Group
PS	Paging Systems
SES	Satellite Earth Station
ATM	Advanced Testing Methods
HF	Human Factors

1990 Selected ETSI Project Teams

PT5	Group 4 Class 1 Facsimile at 64kbps over ISDN
PT10	Digital European Cordless Telephone (DECT)
PT14	Intelligent card and card related aspects of a terminal for telecommunications use
PT2V	ISDN Administrative Coordination Team
PT6V	Telecommunications Facilities for People with Special Needs

ETSI maintains close ties with other regional and international standards-setting organizations. Largely because of the emergence of ETSI, then Secretary-General of the ITU, Richard Butler, became concerned that other standards organizations would bypass the CCITT and CCIR in the development of global standards. These standards organizations include the ETSI in Europe, Committee T1 in the United States (with the secretariat that is the Exchange Carriers Standards Association or ECSA), and the Telecommunications Technology Council (TTC) of Japan. This concern stimulated the 1990 Fredericksburg Conference that defined the relative roles of the regional standards bodies and the CCITT/CCIR.

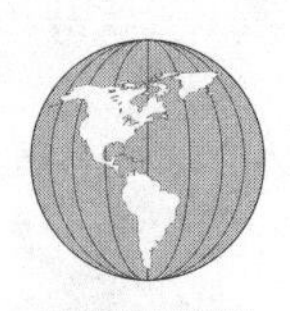

The International Telecommunications Users Group

The International Telecommunications Users Group (INTUG) is an association of telecommunications users dedicated to promoting their interests at the international level.[52] At a 1974 meeting in Brussels, INTUG was formed to ensure that the concerns of telecommunications users were represented at the international level, including global and regional bodies. To that end, INTUG became a member of the CCITT in 1979. INTUG's formal objectives are to address user concerns with the following major issues:

- **Monopoly Authority and the Rights of Users**—INTUG believes that regulations and telecommunications service provision should be user oriented, prices should be fair, and high levels of network performance should be maintained. In particular, INTUG believes that competition is beneficial and should be promoted wherever possible.
- **Free Access to Telecommunication Networks**—INTUG promotes open access to all national and international network services with the intent of eliminating restrictions wherever possible.
- **Freedom in User Choice of Equipment and Services**—INTUG supports competition and choice wherever possible so that users have the largest range of alternatives possible.
- **Constructive Cooperation Between Public Authorities and Users**—INTUG supports open dialogue between the user community, regulatory bodies, and network service suppliers.

A full-time staff and an executive director support INTUG's activities. INTUG is governed by a council consisting of up to three delegates from each full member. The Council approves policy positions and elects new council members. A subset of the Council is the Executive Committee, which consists of the Chairman of the Council, the Executive Director, the Treasurer, the Vice Chairman for Administration, the Vice Chairman for Regulatory Affairs, the Vice Chairman for International Relations and Development, and the Vice Chairman for Strategic Planning. The Vice Chairmen also serve as liaisons for regional user groups in Europe, the Americas, and the Pacific Basin. Currently, INTUG has members in Australia, Austria, Belgium, Canada,

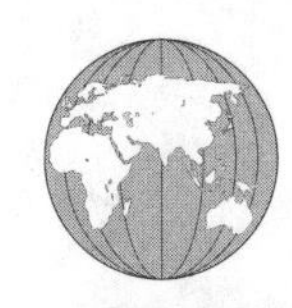

Denmark, Finland, France, Hong Kong, Japan, the Netherlands, New Zealand, Norway, South Africa, the United Kingdom, and the United States.

INTUG has three classes of membership:

- **Full Members** who are national and international user organizations that have objectives consistent with INTUG's
- **Associate Members** who are significant users of international telecommunications facilities
- **Individuals** who are interested in telecommunications developments such as academics and consultants

Policy positions in INTUG are developed at regular meetings, which are held at various locations throughout the world. At these meetings, national delegates report on the status of relevant issues to update their members. In addition, working groups, established to deal with specific issues, make progress reports. Finally, working groups receive new projects, or new working groups are formed, as current developments dictate.

The *INTUG News*, which is published quarterly, frequently publicizes these reports from national representatives and working groups. INTUG's policy positions are then articulated and represented in international organizations such as the CCITT, the OECD, and the EEC.

Intelsat

Satellite systems have played a major role in international telecommunications since the mid-1960s. After people recognized that three satellites, placed in geosynchronous orbit 120 degrees apart, could provide communications coverage to the whole world,[53] communications linkage among all nations became a viable undertaking. Two questions had to be addressed before international satellite communications became a reality:

> How can such a costly enterprise be financed?
> Who should control this significant international project?

The answer to both questions was Intelsat, an international satellite cooperative which, at least in its early years, was heavily dominated by the United States. This organization continues to feel strong U.S. influence.

Intelsat is an international consortium of 122 nations that owns and operates 18 satellites, with which it provides a range of communications services to its

membership and to some 60 nonmember nations, territories, and protectorates. The history and organization of Intelsat presents an interesting study because it reflects the early and continuing influence the United States has exercised in the international telecommunications arena. This influence is based not only on technological development but also on the disproportionately large share of international traffic generated by the United States.

Intelsat's history is also instructive because it reflects the impact of new technologies and emerging competitive forces on the old-line monopoly approach to the delivery of telecommunications services. As happens in other areas of telecommunications, new players, like PanAmSat, enter the satellite communications market while users ask for new services and more flexible access to the communications resource. The former monopoly providers respond by moving to de-average rates, lower prices, and liberalize their service offerings. Intelsat is no exception.

A Short History

Intelsat was created as the result of a U.S. initiative. After much discussion on the part of the U.S. government about whether space communications should be a public or private undertaking,[54] the U.S. Congress passed the Communication Satellite Act of 1962. The Communication Satellite Act stated that U.S. policy was

> *. . . to establish, in conjunction and in cooperation with other countries, as expeditiously as practicable a commercial communications satellite system, as part of an improved global communications network, which will be responsive to public needs and national objectives, which will serve the communication needs of the Unites States and other countries, and which will contribute to world peace and understanding.*[55]

The United States, then, set up a policy to create a satellite system that would be operated on a commercial basis and established, not by one nation, but by a group of nations. In addition to providing communications, the system would also fulfill the foreign policy objective of contributing to world harmony.

In pursuit of these goals, Congress created in the Communication Satellite Act a "private corporation, subject to appropriate governmental regulation,"[56] called the Communication Satellite Corporation, or Comsat. Comsat is actually a quasi-private entity. It was originally half-owned by carriers like AT&T and ITT and by stockholders. Its board of directors consisted of three individuals selected by the carriers, three selected by the stockholders, and three appointed by the government. AT&T and ITT eventually divested themselves

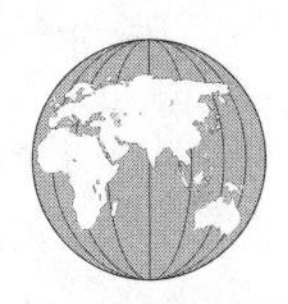

of Comsat ownership, and of their right to select three of its directors, now selected by stockbrokers. The government still appoints three directors.[57]

Comsat's mission was to work with other countries in setting up a satellite system. Negotiations with the newly formed CEPT and with other nations resulted in the creation of a consortium, which was the precursor to today's Intelsat. Comsat was chosen to be the manager of the consortium, which was based on two separate agreements, one signed by the various member governments and the other by the telecommunications administrations of the member nations. These arrangements represented a five-year interim agreement which, in addition to naming Comsat as consortium manager, created an Interim Communication Satellite Committee (ICSC) to function as the decision-making body. Signatories with membership shares of 1.5 percent or more were represented on the ICSC; voting was by majority for most issues. However, substantive matters required a majority representing more than 12.5 percent of the largest share percentage. A majority was impossible to attain without U.S. support because the U.S. share at that time was over 50 percent.

On August 20, 1964, 11 nations signed the Intelsat interim agreements: Australia, Canada, Denmark, France, Italy, Japan, the Netherlands, Spain, the United Kingdom, the United States, and the Vatican City.[58] Neither developing nations nor the USSR were involved in these early developments.

Discomfort with U.S. dominance of the interim arrangements led to an ICSC draft report late in 1968 that recommended Comsat be replaced with a more independent and broadly representative entity. The draft report also suggested a change in the voting formula that would reduce U.S. influence. A final draft report was approved in 1971. The result was a multilateral Intelsat agreement signed in August 1971 and activated in February 1973. That agreement established the current Intelsat structure.

The Intelsat agreement also included a provision that paved the way for the introduction of competition in the international satellite arena. At the insistence of the European nations, Article XIV(D) was approved. This article allowed the creation of separate satellite systems, provided those systems posed no economic harm to Intelsat. Article XIV(D) is the subject of much debate and discussion, and, as we'll discuss later, might no longer be a useful or necessary component of the Intelsat agreement.

Structure And Organization

Membership in Intelsat is open to any nation that is a member of the ITU. Participation by the membership takes place at two levels. The governments

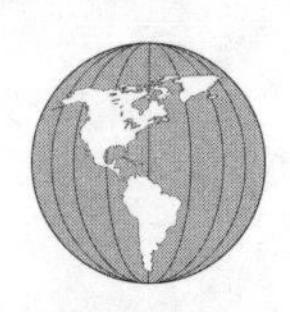

of Intelsat member nations are parties to the multilateral agreement that establishes the organization and delineates its governing principles. The member nations' telecommunications administrations, or other designated representatives, are signatories to an operating agreement that sets forth technical, operational, and financial guidelines. This unique two-part structure reflects the need to address both the political issues involved in dealing with such a large scale international undertaking and the technical matters that must be resolved.

Intelsat is a cooperative; its owners bear the costs of the undertaking and share in the resulting revenues. The percentage of costs borne and revenues realized by each member is determined by that nation's investment share. Each member's investment share in Intelsat is based on its percentage of usage of the system. That share is determined annually on March 1 and is based on each member nation's portion of the preceding six months' total satellite use.

Table 4.1. Intelsat ownership shares, in descending order.

Country	Signatory	Investment Share
United States	Comsat	21.864515
United Kingdom	British Telecom	12.057422
Japan	KDD, Inc.	4.500624
France	France Telecom	4.240515
Germany	Federal Ministry for Post and Telecommunication	4.197863
Australia	OTC Limited	2.787103
Italy	Società Telespazio	2.475500
Spain	Telefónica de España, S.A.	2.442714
Canada	Teleglobe Canada, Inc.	2.253752
Singapore	Telecommunication Authority of Singapore	2.007955
India	Videsh Sanchar Nigam Ltd.	1.877491
Saudi Arabia	Govt. of Saudi Arabia	1.870299
Brazil	Empresa Brasileira de Telecomunicaçoes S.A.	1.799084

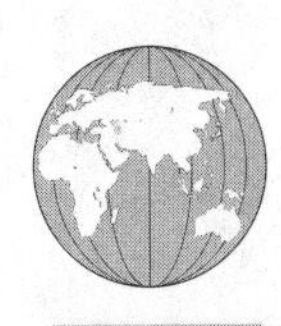

Country	Signatory	Investment Share
People's Republic of China	Ministry of Posts and Telecommunications	1.590957
Republic of Korea	Korea Telecom	1.421507
Argentina	Comisión Nacional de Telecomunicaciones	1.313707
Colombia	Empresa Nacional de Telecomunicaciones	1.282447
Netherlands	PTT Nederland NV	1.219379
South Africa	Telkom SA Ltd.	1.144476
United Arab Emirates	Ministry of Communications	1.130136
Switzerland	Direction Générale de l'Enterprise des Postes, Téléphones et Télégraphes	1.091446
Thailand	Government of Thailand	.999491
Kuwait	Ministry of Communications	.950249
Venezuela	Venezuelan Telephone Co.	.938148
Nigeria	Nigerian External Telecommunications, Ltd.	.894590
Peru	Empresa Nacional de Telecommunicaciones del Peru	.812826
Belgium	Régie des Télégraphes et des Téléphones	.777604
Iran, Islamic Republic of	Telecomm Company of Iran	.773224
Mexico	Govt. of Mexico	.745446
Malaysia	Syarikat Telekom Malaysia Berhad	.729971

continues

Table 4.1. continued

Country	Signatory	Investment Share
Pakistan	Government of the Islamic Republic of Pakistan	.726073
Norway	Norwegian Telecom	.699644
Israel	Israel Telecomm Corp. "BEZEQ"	.665429
New Zealand	Telecom Corp of New Zealand	.661044
Chile	Empresa Nacional de Telecomunicaciones S.A.	.654975
Egypt	Government of the Arab Republic of Egypt	.649338
Sweden	Swedish Telecom	.643207
Philippines	PHILCOMSAT	.640919
Portugal	Companhia Portuguesa Radio Marconi	.619760
Indonesia	PT INDOSAT	.575707
Denmark	Tele Danmark A/S	.549720
Greece	Hellenic Telecom Org. (OTE)	.495059
Austria	Government of Austria	.489891
Turkey	Government of Turkey	.399796
Ecuador	Instituto Ecuatoriano de Telecommunicaciones (IETEL)	.393215
Jamaica	Jamaica International Telecommunications (JANITEL)	.379821
Cameroon	Société des Télécommunications Internationales du Cameroun	.345282
Kenya	Kenya Posts and Telecom. Corp.	.328774
Algeria	Govt. of the Democratic and Popular Republic of Algeria	.313398
Qatar	Qatar Public Telecom. Corp.	.295682

Country	Signatory	Investment Share
Dominican Republic	Compania Dominicana de Teléfonos, C. por A.	.289647
Jordan	Government of the Hashemite Kingdom of Jordan	.279862
Yugoslavia	Community of the Yugoslav Posts, Telegraphs, and Telephones	.277053
Iraq	Government of the Republic of Iraq	.246610
Bangladesh	Telegraph and Telephone Board of Bangladesh	.242658
Cyprus	Cyprus Telecom. Authority	.234556
Oman	Sultanate of Oman	.208125
Cote d'Ivoire	Government of the Republic of Cote d'Ivoire	.200060
Ireland	Irish Telecommunications Board	.199898
Mauritius	Overseas Telecom. Services Co.	.190338
Finland	General Directorate of Posts and Telecom. of Finland	.182260
Morocco	Office National des Postes et Télécommunications	.179084
Iceland	Government of Iceland	.171861
Yemen	Govt. of the Republic of Yemen	.161255
Paraguay	Administración Nacional de Telecomunicaciones	.161000
Bolivia	Empresa Nacional de Telecomunicaciones (ENTEL)	.158412
Zaire	Office National des Postes et Télécommunications du Zaire	.153676

continues

Table 4.1. continued

Country	Signatory	Investment Share
Togo	Office des Postes et Télécommunications du Togo	.149696
Libya	Government of the Great Socialist People's Libyan Arab Jamahiriya	.148688
Haiti	Télécommunications d'Haiti S.A.	.141279
Ethiopia	Telecommunications Service, Provisional Military Govt. of Socialist Ethiopia	.129015
Angola	Empresa Pública de Telecomunicações (EPTEL)	.127391
Sudan	Govt. of the Democratic Republic of the Sudan	.119662
Zambia	Govt. of the Republic of Zambia	.118595
Malawi	Dept. of Posts and Telecom.	.112693
Senegal	Govt. of the Republic of Senegal	.109776
Mali	Télécommunications Internationales du Mali	.103231
Bahamas	Bahamas Telecom. Corp. (BATELCO)	.100456
Nepal	Nepal Telecom. Corp.	.090267
Ghana	Ministry of Transport and Communications	.089954
Syria	Government of the Syrian Arab Republic	.086840
Burkina Faso	Office des Postes et Télécom. du Burkina Faso	.068010
Tanzania	Posts and Telecom. Corp.	.064473

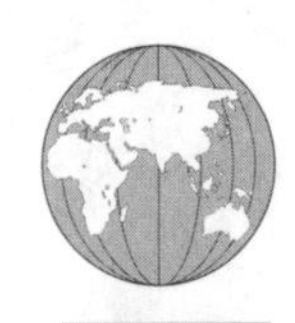

Country	Signatory	Investment Share
Uganda	Ministry of Power, Post, and Telecommunications	.064038
Mozambique	Empresa Nacional de Telecomunicações	.062527
Guinea	Secrétariat d'Etat aux Postes et Télécommunications	.059877
Afghanistan	Ministry of Communications	.050000
Azerbaijan	Aserbaidschan PhoneSat Communication-System Ges.m.b.H.	.050000
Barbados	Barbados External Telecom. Ltd.	.050000
Benin	Office des Postes et Télécom. de la République Populaire du Benin	.050000
Cape Verde	C.T.T.-Empresa Publica dos Correios e Telecomunicações	.050000
Central African Republic	Government of the Central African Republic	.050000
Chad	Société des Télécommunications Internationales du Tchad	.050000
Congo	Government of the People's Republic of the Congo	.050000
Costa Rica	Instituto Costarricense de Electricidad	.050000
El Salvador	Administración Nacional de Telecomunicaciones (ANTEL)	.050000
Fiji	Fiji International Telecomm. (FINTEL)	.050000
Gabon	Société des Télécommunications Internationales Gabonaises	.050000

continues

Table 4.1. continued

Country	Signatory	Investment Share
Guatemala	Empresa Guatemalteca de Telecomunicaciones	.050000
Honduras	Empresa Hondureña de Telecomunicaciones	.050000
Lebanon	Government of Lebanon	.050000
Liechtenstein	Government of the Principality of Liechtenstein	.050000
Luxembourg	Government of Luxembourg	.050000
Madagascar	Société des Télécommunications Internationales de la République Malgache	.050000
Mauritania	Government of the Islamic Republic of Mauritania	.050000
Monaco	Government of the Principality of Monaco	.050000
Nicaragua	Instituto Nicaraguense de Telecomunicaciones y Correos	.050000
Niger	Government of the Republic of Niger	.050000
Panama	Instituto Nacional de Telecomunicaciones (INTEL)	.050000
Papua New Guinea	Post and Telecom. Corp	.050000
Romania	Ministry of Posts and Telecom.	.050000
Russian Federation	Ministry of Posts and Telecom.	.050000
Rwanda	Ministère des Transports et des Communications de la République Rwandaise	.050000
Somalia	Ministry of Posts and Telecom.	.050000

Country	Signatory	Investment Share
Sri Lanka	Government of Sri Lanka	.050000
Swaziland	Posts and Telecom. Corp.	.050000
Trinidad and Tobago	Trinidad and Tobago External Telecommunications Company	.050000
Tunisia	Adm. for Post, Telegraph, and Telephone of Tunisia	.050000
Uruguay	Administración Nacional de Telecomunicaciones	.050000
Vatican City State	Government of the Vatican City State	.050000
Viet Nam	Direction Générale des Postes et Télécommunications de la République Socialiste du Viet Nam	.050000
Zimbabwe	Government of Zimbabwe	.050000

Source: Intelsat Annual Report, 1991-92.

Investment shares range from 22 percent, held by the United States, to the smallest allowable share, .05 percent. In 1991, almost 47 percent of the investment shares were held by five countries: France, Germany, Japan, the United Kingdom, and the United States.[59] A large number of Intelsat members are developing nations (as much as 80 percent, according to Heather Hudson).[60] However, despite the large proportion of membership from the developing world, their influence is mitigated by the manner in which the organization's Board of Governors is selected. Representation on that most powerful unit of the organization is based on ownership shares.

The organization has a four-tier structure. Governments, or representatives of governments, that are parties to the Intelsat Agreement meet every two years in the Assembly of Parties. At these meetings, matters of policy are discussed and the general objectives of Intelsat.

Each year, telecommunications administrations that are signatories to the Operating Agreement or their representatives convene a Meeting of Signatories. The signatories consider financial, technical, and operational matters.

The Board of Governors holds the real power in the organization. Meeting four times a year, the governors are representatives of those member nations that hold a minimum investment share, as determined by the Board of Signatories. In 1989 that minimum share was 1.539350.[61] Two or more members with shares less than that minimum amount may pool their shares to reach the minimum and be represented by one governor. Also, one governor may represent five or more members in an ITU region, regardless of their investment share. There can be only one such governor per ITU region, and only five total governors selected in this manner. In March 1989, the Board comprised 29 governors, representing 102 of the 117 member nations; in 1992, the Board comprised 28 governors representing 103 member nations.

The administrative work of the organization is handled by the Executive Organ. Originally, a Director-General, who reported to the Board, and three Deputy Director-Generals headed the Executive Organ. In a recent effort to "reflect a new, more commercial orientation" for Intelsat, the Board approved a management restructuring in which the Deputy Director assumed the additional title of Chief Executive Officer, and the Deputy Directors were replaced by four Vice Presidents. In addition to an executive vice president for operations and services and a vice president and chief financial officer, now vice presidents head engineering, research, information, and administration.[62] In early 1992, the Executive Organ, now referred to as Intelsat Management, had a staff of 778. Intelsat headquarters are in Washington, D.C. According to the Headquarters Agreement between the United States and Intelsat, the organization and its employers are accorded the rights and privileges granted a foreign embassy.[63]

The Structure of Intelsat

Participation at Two Levels

- **Intelsat Agreement**—Established organization; delineated governing principles; signed by governments.
- **Operating Agreement**—Defines technical, operational, and financial guidelines signed by telecommunication administrations.

Four Tier Structure

- **Assembly of Parties**—Meeting of Parties to the Intelsat Agreement; meetings every two years; consideration of policy matters.

- **Meeting of Signatories**—Meeting of Signatories to the Operating Agreement; annual meetings; consideration of technical, financial, and operational matters.
- **Board of Governors**—Representation based on ownership shares; meetings four times per year; chairman and vice chairman; four advisory committees oversee work of the organization.
- **Executive Organ (Intelsat Management)**—Performs administrative functions; Director General and CEO; four vice presidents:

  Executive VP for Operations and Services

  VP and Chief Financial Officer

  VP for Engineering and Research

  VP for Information and Administration

Services Offered

Intelsat provides service to its 122 members and to almost 60 other countries, territories, and dependencies. Services on Intelsat are acquired, in most nations, by going through the telecommunications administration. In the United States, those entities ordering Intelsat services, including carriers like AT&T, order through Comsat, the U.S. signatory to the Intelsat agreement. The British have allowed Cable & Wireless to bypass the UK signatory and order directly from Intelsat, but such arrangements aren't the norm. In most instances, carriers, or private entities, if allowed in a particular nation, must go through the PTT. The German Ministry of Posts and Telecommunications in May 1992 added yet another wrinkle to possible arrangements by allowing satellite communication licensees to acquire satellite capacity from any authorized signatory, not just through the German signatory.[64]

The largest percentage of Intelsat business remains international voice telephony, with over 65 percent of the organization's revenues coming from that source in 1988.[65] That percentage, however, dropped to 43 percent in 1991.[66] The organization also offers digital services, including Intelsat Business Service (IBS), an integrated digital service designed to handle voice, data, telex, fax, and videoconferencing. Intelnet is Intelsat's digital data service designed for use with very small aperture terminals (VSATs). Intelnet is used for financial and news distribution services and for interactive transactions such

as banking. Intelsat is also heavily used for television, both on a full-time and an occasional basis. Intelsat is also used for cable restoration service, or back-up service in case of cable outage. During 1991, Intelsat was used to restore four major fiber-optic cables, including TAT-8 (Atlantic), TCS-1 (Caribbean), and HAW4/TPC3 and NPC (Pacific).

In addition to international services, Intelsat is used for domestic and regional communications by nations who either don't have domestic satellite systems or have communications needs beyond the capability of their domestic system. In South America, five countries are using Intelsat domestic leases and transponders for unrestricted use (TUU) service for a regional Andean satellite system. In Africa, the Regional African Satellite Project (RASCOM) will enable its 51 members to meet regional needs by merging their domestic Intelsat leases. In Europe, Germany is using Intelsat domestic and regional services to link east and west communications.

Many nations that use domestic services include developing countries. Intelsat provides other services for remote areas and low traffic routes as well as developing nations. Project Access provides free use of space segment capacity for education, health, and other social services, along with free use of capacity for special global television events that are humanitarian in nature. A videoconference for World AIDS Day was the first such global event.

Since the launch of its first satellite, the Early Bird, in 1965, Intelsat's capacity to provide services has grown immensely. The Early Bird carried 150 channels. In 1988, Intelsat's 13 satellites could accommodate almost 120,000 channels. Those 13 satellites included the fifth generation of Intelsat satellites. The sixth series of satellites is now operational, and Intelsat is considering bids for its seventh generation.

Table 4.2. Intelsat spacecraft, present and future.

Series	Year of First Launch	Lifetime (Years)	Lifetime Capacity
INTELSAT V	1980	7	12,000 Circuits, and 2 TV
INTELSAT V-A	1985	7	15,000 Circuits, and 2 TV
INTELSAT VI	1989	13	24,000 Circuits, and 3 TV
INTELSAT K	1992	10	Up to 32 high-quality TV channels

Series	Year of First Launch	Lifetime (Years)	Lifetime Capacity
INTELSAT VIII	1993	10–15	18,000 Circuits, and 3 TV
INTELSAT VIII-A	1995	10–15	22,500 Circuits, and 3 TV

Source: Intelsat Annual Report, 1991-92

The Impact Of Competition

Like most other communications monopolies, Intelsat faces challenges from potential competitors. Some of those challenges are based on technological advances, notably transoceanic fiber-optic cable providers; others are based on the erosion of former regulatory or policy approaches.

The seeds for erosion of the single satellite system policy were planted in the document that established Intelsat itself. Article XIV(D) of the Intelsat Agreement states:

> *To the extent that any Party or Signatory or person within the jurisdiction of a Party intends individually or jointly to establish, acquire or utilize space segment facilities separate from the INTELSAT space segment facilities to meet its international public telecommunications services requirements, such Party or Signatory, prior to the establishment, acquisition or utilization of such facilities shall furnish all relevant information to and shall consult with the Assembly of Parties, through the Board of Governors, to ensure technical compatibility of such facilities and their operation with the use of the radio frequency spectrum and orbital space by the existing or planned INTELSAT space segment and* to avoid significant economic harm to the global system of INTELSAT. [emphasis added][67]

Early applications of Article XIV(D) created very little controversy and appeared to offer no danger to the maintenance of a one-system policy. Until the early 1980s, the Intelsat Board of Governors accommodated requests for five satellite systems to operate outside Intelsat. However, these tended to be regional systems that had little to do with the huge U.S. market. As Aronson and Cowhey have noted:

> *Until 1983 Intelsat strategy was to accommodate regional satellite systems that did not impinge on the U.S. market. Everything else was*

secondary. Regulating the U.S. market set the terms for competing around the world. If Intelsat could limit entry to the United States, its future was reasonably secure.[68]

Inevitably, given the huge U.S. market, potential separate system providers emerged. In 1983, a U.S. firm named Orion filed an application with the FCC for permission to establish a separate transatlantic satellite system. The Orion application, and four more applications that followed, resulted in a protracted controversy in the U.S. government, with the procompetitive forces (notably the Department of Commerce) urging the benefits of competition and the more hesitant forces (the Department of State) worrying about the foreign policy implications of such a move.

In late 1984, President Ronald Reagan notified the Secretaries of State and Commerce that "separate international communications satellite systems are required in the national interest," that the United States would fulfill its obligations under the Intelsat agreement, and that the Secretaries were directed "jointly" to inform the FCC of "criteria necessary to ensure the United States meets its international obligations and to further its telecommunications and foreign policy interest." [69]

A letter from the Secretaries to the chairman of the FCC resulted from this presidential memorandum. In their letter, the Secretaries directed the FCC to enforce two conditions on any separate system:

(1) each system is to be restricted to providing services through the sale or long-term lease of transponders or space segment capacity for communications not interconnected with public-switched message networks (except for emergency restoration service; and,

(2) one or more foreign authorities are to authorize use of each system and enter into consultation procedures with the United States Party under Article XIV(d) of the INTELSAT Agreement to ensure technical compatibility and to avoid significant economic harm.[70]

So, the U.S. one-system policy gave way to a more open environment, but with stipulations. The systems couldn't connect to the public network, and system providers had to receive permission to operate from another authority, in addition to the FCC. The FCC included these stipulations in its Separate Systems decision in 1985 and granted five companies authority to operate, including Orion and PanAmSat. PanAmSat has been the most vigorous proponent of a more liberal interpretation of Article XIV(D) and even of rescinding it entirely. Recent developments, as we discuss shortly, suggest that this will soon be the case.

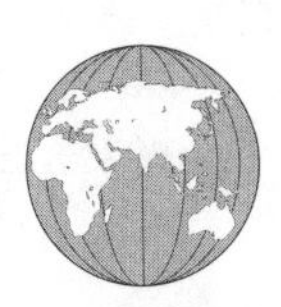

In late 1990, Intelsat agreed to some limitations on the consultation procedures called for under Article XIV(D), notably the economic harm provision. Under this provision, operators of separate systems had to show that their existence would pose no economic harm to Intelsat's operation. The limitations agreed to exempted providers of separate systems from offering less than 30 36-MHz equivalent transponders for international traffic not connected to the public switched network (in other words, private lines) from the economic harm provision. At the same time, Intelsat also agreed to exempt separate systems that offered less than 100 64-kbps equivalent circuits for traffic that did interconnect with the public network.[71]

In the summer of 1992, the Intelsat Board accepted a working study group recommendation to further liberalize interpretation of the economic harm provision. The recommendation was presented to Intelsat's Assembly of Parties in Sydney, Australia, November 2-6, 1992, and was approved. This recommendation eliminated the requirement that potential economic harm be considered for all separate satellites providing private line services, regardless of transponder equivalence, even if the lines were connected to the public switched network. Furthermore, the recommendation called for raising the upper limit on the exemption from 100 64-kbps per separate system to 1,250 circuits per satellite. The Assembly adopted a four-to-six year transition period, with higher thresholds being set in 1994 and again in 1996, and a total phaseout of harm tests by the late 1990s.[72]

Complete elimination of Article XIV(D) provisions will clear the way for competition in all international services available through satellite communications. If the competing satellites are allowed unlimited connection to the public network, Intelsat will lose its current unique position. Indeed, PanAmSat claims that the unique position is already gone. In a policy White Paper, PanAmSat points to recent changes in Intelsat's Executive Organ, Intelsat's decision to seek debt financing, and the privatization of many of Intelsat's owners as proof that the organization is no longer a nonprofit consortium. Rather, according to PanAmSat, the organization is a "private enterprise" and as such should be stripped of its treaty.[73]

The situation has changed a good deal since the United States spearheaded the creation of one satellite system for all international traffic. With U.S. concurrence, actions are underway to replace protected monopoly with competitive entry. The international situation is reminiscent of what is happening in many nations, notably the United States. It is difficult not to compare PanAmSat's arguments and actions with MCI's incursions into the U.S. long distance market. PanAmSat will persevere for many of the same reasons MCI was successful. Intelsat's pricing policies, the growing amount of potentially

lucrative traffic, and increasing user demands for more flexible services are pushing toward the opening of the satellite communications market.

Transoceanic Cable Systems

Intelsat faces more than competition from alternative satellite systems. It also faces competition from an alternative technology, transoceanic fiber-optic cables. Submarine cables linking continents predated satellite technology by a century. The first telegraph line between Canada and the United Kingdom was laid in 1858. The first telephone cable, TAT-1, was built in 1956 and had a 36-circuit capacity. The Early Bird satellite, with its total capacity of 480 telephone channels, was larger than the largest submarine cable of that time, which could only carry 256 channels.[74]

Satellites offered greater capacity than conventional submarine cables. Satellites also offered greater coverage of territory. Submarine cables didn't extend to all continents, and messages still had to be deployed inland by some means. Satellites offered coverage to any location with an earth station. Satellites still offer the benefits of potentially universal coverage. However, because of fiber-optic technology and the growth of traffic on certain routes, the economic picture has changed somewhat.

Fiber-optic cables have significant capacity. TAT-8, the first fiber-optic submarine cable, had a maximum capacity of 37,800 voice circuits.[75] Fiber's large capacity, when coupled with the significant rate of growth on high traffic routes, makes the laying of fiber-optic submarine cables an economically attractive undertaking. Figure 4.2 illustrates the deployment of submarine cables.

Fearing that submarine cables might pose a threat to the continued viability of Intelsat, the FCC followed a policy of *balanced loading*, in which carriers balanced traffic between satellite and cable facilities. Since the mid-1980s, the FCC has moved away from a balanced loading policy, not requiring any balanced loading from non-AT&T carriers and liberalizing these requirements for AT&T.[76] The relationship between submarine cables, deployed on high traffic routes, and Intelsat, the provider of service to all, is reminiscent of the bypass or "cream skimming" arguments heard in regard to terrestrial traffic. The same forces are at play in the international arena as well.

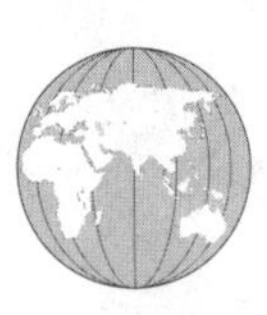

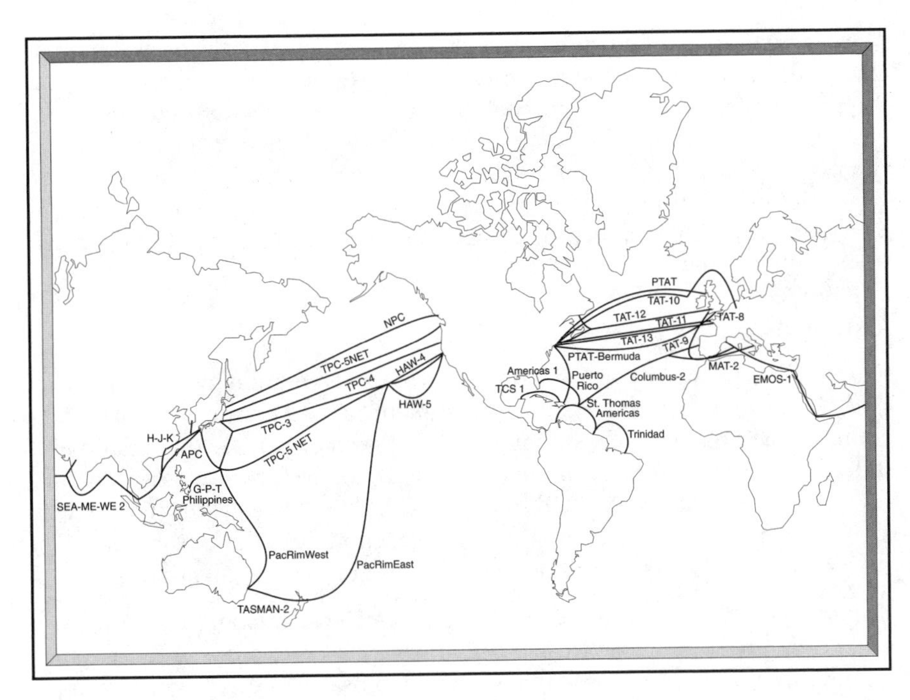

Figure 4.2. Global digital fiber optics network.

Pricing and Bypass

Intelsat's pricing strategies have been similar to those of other monopoly service providers with a major goal of deploying service as widely as possible. As a consortium, Intelsat provides service to all its members and other entities. To make service affordable to all members, Intelsat has used an average pricing strategy. Regardless of the traffic density of a route, Intelsat users pay the same per annum lease fee for all routes. As a result, high traffic routes subsidize low density routes. The high traffic routes are, in effect, overpriced, providing an opportunity for competitors to enter the market and offer service at lower prices but still at a high profit.

Intelsat has made efforts to combat the resulting bypass of its facilities. Its 1988 annual report pointed out that, by the beginning of 1989, the organization had lowered rates 14 times and that the price of a voice channel had fallen from $32,000 per annum in 1965 to $1,000 in 1988. New specialized offerings like IBS and bulk capacity services are, in effect, a means of deaveraging services and offering users more flexibility.

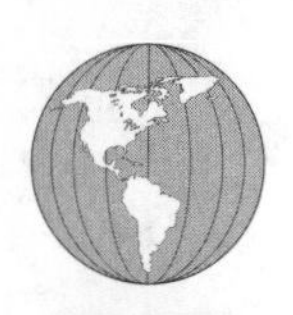

The organization still has what amounts to a universal service role. Submarine cables connect developed nations, not underdeveloped continents. The organization's ownership approach also complicates its ability to plan. Because ownership is based on usage, not forecasts of usage, low-volume users can make inflated forecasts of future use with impunity. The result could be overbuilding, resulting in higher rates. Intelsat isn't overseen by any regulatory body; restraints on its pricing structure come from competitive forces.

Comsat

Examining what has happened to Comsat since its creation in 1962 is instructive. Unlike other nations, where government-owned entities provided telecommunications, the United States had a system of private ownership and didn't have a telecommunications administration to function as a signatory to Intelsat. Comsat was created in part to fulfill that role. Comsat is the U.S. signatory to Intelsat. The company takes direction from the U.S. government in determining its position on issues that arise in the organization.

Comsat was originally crafted as a carriers' carrier. For example, a company seeking international service would approach a carrier like AT&T; AT&T would then purchase access to Intelsat from Comsat. The FCC regards Comsat as a dominant carrier and regulates its activities. In a series of actions in the mid-1980s, the FCC determined that it wasn't beneficial to allow carriers to bypass Comsat and gain direct access to Intelsat. However, the FCC determined that customers could bypass carriers and order directly from Comsat. The FCC further determined that Comsat could offer end-to-end international communications service in competition with AT&T and other carriers, but had to do so through a separate subsidiary. PanAmSat recently requested that the FCC strengthen those subsidiary requirements, citing Comsat's peculiar situation of being both a signatory to the Intelsat agreement and a potential competitor to the new separate satellite systems.

Comsat stands in a odd position. As a signatory to Intelsat, it has a voice in pricing and deployment decisions that affect the competitive position of Intelsat with separate satellite systems like PanAmSat. As the provider of end-to-end communications service, Comsat is itself a competitor to these separate systems. Additionally, Comsat is both the only means by which U.S. carriers obtain access to Intelsat and a competitor of those same carriers.

Comsat's difficulties might not be unique. As other nations privatize their telecommunications administrations and allow competitive carriers to enter the market, they, too, create situations in which entities both sell service to

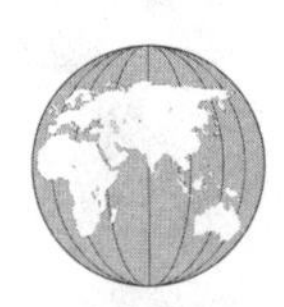

and compete with their own rivals. As more players enter the telecommunications market, the old structures might not be able to accommodate the new relationships and the new rules of the game.

Inmarsat

The maritime counterpart to Intelsat is Inmarsat. Established in 1979, and operational in early 1982, Inmarsat provides satellite communications for ships and offshore enterprises. Like Intelsat, Inmarsat is a cooperative organization, financed by its signatories. The organization was created from a series of conferences held by the United Nation's International Maritime Organization.

Inmarsat, which consists of 44 member nations, comprises an Assembly, a Council, and a Directorate. The Assembly of signatories to the agreement meets every two years to address policy issues. The Council, which is the most powerful organ in Inmarsat, consists of representatives from the 18 signatories with the largest investment shares and four representatives from the other member nations, who are elected by the Assembly based on geographical representation. The organization's Directorate, headquartered in London, carries out its Administrative duties. The Directorate is headed by a Director General, who is appointed by the Council. Members and nonmembers purchase service from Inmarsat. Services include telex, voice, fax, and data circuits up to the T1 level. Inmarsat leases space on other satellite systems, including Intelsat.

Domestic and Regional Satellites

Intelsat and Inmarsat might be the only truly international satellite consortia; however, a large number of domestic and regional satellites provide a range of broadcast and point-to-point services. For example, the Palapa system in Southeast Asia provides services to Indonesia (its country of origin), Malaysia, Singapore, Thailand, and the Philippines. Arabsat, which is a consortium of Arab League members, provides services to Arab nations.

The Soviet and Eastern European counterpart to Intelsat is Intersputnik, which had 14 member nations and operated in 17 nations.[77] With the breakup of the Soviet Union, Intersputnik is under the control of the Commonwealth of Independent States. A growing number of circuits on Intersputnik are being made available for communication with the United States and the rest of the

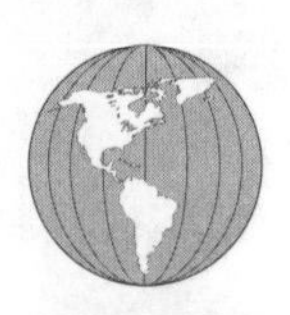

West. In Western Europe, Eutelsat provides a range of services to its 26-nation membership.

These satellite organizations, and others like them, reflect both the benefits of cooperative efforts in launching and maintaining expensive satellite systems and the widely recognized advantages of satellite technology in deploying a range of services.

Issues and Conclusions

An analysis of Intelsat's history and changing position uncovers increasingly familiar trends in the international telecommunications market. The early dominance of the United States and the lessening of that dominance; the tension between new services and technologies requested by users in developed countries and the need to provide services to underdeveloped nations; the impact of new technologies on old service arrangements; the replacement of protected monopolies with competitive entry; and questions about the continued viability of old institutions in their current form are all secondary themes. The impact of separate satellite systems on international communications and on organizations like Intelsat and Comsat will be interesting developments during the 1990s.

Critique/Analysis of Organizations

As is the case with all aspects of telecommunications, the forces of competition and technological change affect international organizations. As a result, their continued effectiveness depends on their ability to adapt to an increasingly complex and rapidly moving environment.

The proliferation of transoceanic cables and the development of private satellite systems question the continuing role of Intelsat, for example. Technological advances such as fiber optics have changed the economic dynamics of transoceanic cables, making them a viable alternative to satellites. Procompetitive policies have boosted new players onto the stage, such as PanAmSat, who are anxious to provide service. Intelsat's position as the dominant provider of international connectivity might change.

As technologies develop at a faster rate, the standards-setting processes set up by international organizations have a difficult time keeping pace. Equipment manufacturers haven't waited for standards, as is evident by the creation of de facto standards. The creation of such bodies as the ETSI points to the continuing regional pressures that affect international telecommunications. As telecommunications becomes a greater economic force, the stakes grow. The pressures of regionalism increase as nations, and groups of nations, seek to maximize their economic position in such areas as equipment manufacture. A stronger role in the standards-setting process is key to success in equipment manufacturing and service deployment.

In many ways, the needs of large, multinational users point out the shortcomings of international organizations. Large users who function in a variety of nations seek to establish multipoint networks. They want to find universal services provided under similar conditions across borders. International organizations are still groups of individual countries, and these countries follow a pattern of formulating bilateral telecommunications agreements. The international settlement arrangements discussed in Chapter 3, "Economic Issues," illustrate this type of agreement. International organizations can't persuade their individual country members to adopt universally accepted practices and procedures. The CCITT recommendations regarding the resale of private line services are an example of this continuing pattern. The CCITT recommendations allow a great deal of latitude on the part of national administrations regarding the existence and the conditions surrounding resale.

Endnotes

1. George A. Codding, Jr. and Anthony M. Rutkowski, *The International Telecommunication Union in a Changing World* (Dedham, Mass.: Artech House, 1982), p. 3. This work provides an excellent, in-depth look at the history of the ITU, as well as an analysis of the ITU's organization and functioning in the early 1980s.

2. James G. Savage, *The Politics of International Telecommunications Regulation* (Boulder, Colo.: Westview Press, 1989), p. 10.

3. International Telecommunications Union, International Telecommunications Convention, as quoted in "Trade in Telecommunications Services, Notes by the Secretariat," *GATT Secretariat*, MTN.GNS/W/52 (May 19, 1989).

4. International Telecommunications Union.

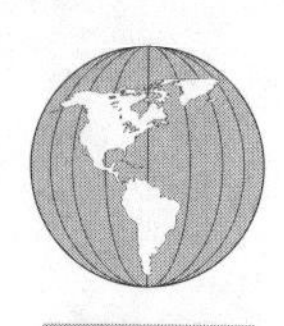

5. I am indebted to George Codding's works for much of the following history of the ITU.

6. George A. Codding, Jr., "The Nice Plenipotentiary Conference," *Telecommunications Policy*, vol. 14, no. 2 (April 1990), p. 140.

7. Peter Cowhey and Jonathan D. Aronson, "The ITU in Transition." *Telecommunications Policy*, vol. 15, no. 4 (August 1991), p. 305.

8. Cowhey, p. 306.

9. Stanley M. Besen and Joseph Farrell, "The Role of the ITU in Standardization: Pre-eminence, Impotence or Rubber Stamp?" *Telecommunications Policy*, vol. 15, no. 4 (August 1991), p. 311.

10. The complexity of the ITU structure makes it difficult to understand at first glance. James Savage, in *The Politics of International Telecommunications Regulation*, uses an analogy to a private corporation to try to explain the interworkings of the various ITU sections, but he admits that such an analogy is "imperfect" (p. 15ff.).

11. Codding, *Telecommunications Policy*, pp. 147-148.

12. Codding, *Telecommunications Policy*, p. 148.

13. "Tomorrow's ITU: The Challenges of Change," *Telecommunications Policy* (August 1991), pp. 269-270.

14. Quoted in "World Telecom Advisory Council Established to Assist ITU Efforts," *Telecommunications Reports International*, vol. 3, no. 11 (May 29, 1992), p. 7.

15. As of May 1992, there were 15 Study Groups.

16. Stephen Temple, *A Revolution in European Telecommunications Standards Making* (ETSI, 1991), p. 68.

17. Temple, p. 71.

18. In fact, in several countries manufacturers sell not to the end-users, but to the administrations, who then lease the equipment to the end-users. The administrations therefore have the power of a monopoly in their countries.

19. As of May 1992, there were 65 RPOAs. These consist of companies such as AT&T, MCI, France Telecom, British Telecom, and so on.

20. From Dr. Irmer's remarks at the 1992 Annual Conference and Exposition of the International Communications Association, Atlanta, Georgia.

21. Joint Technical Committee 1 (JTC1) is a committee jointly formed by ISO and the International Electrotechnical Commission (IEC) for the standardization of information technology.

22. In practical terms, there is no difference between a CCITT Recommendation and an ISO International Standard. CCITT Recommendations are functionally equivalent to standards. CCITT Reports, on the other hand, aren't standards, nor do they attempt to be. They are the result of an investigation undertaken by an approved committee to acquire facts and other information about an issue of interest to the CCITT.

23. Jeffrey Harrop, *The Political Economy of Integration in the European Community* (Aldershot, U.K.: Edward Elgar Publishing, 1989), p. 7. Harrop provides an excellent review of the precursors to the EC and the process leading up to the creation of the EC.

24. Harrop, p. 12.

25. Harrop, p. 13-14.

26. *1992-Planning for the Information Technology Industries*, researched and compiled by Eurofi plc., IBI International Business Intelligence Series (London: Butterworth, 1989), pp. 4-5.

27. Interview with Lord Cockfield in "The Chief Architect of 1992, Lord Cockfield," *Europe* (January/February 1992), p. 13.

28. *Information Technology Industries*, p. 5.

29. Bruce Barnard, "The Countdown Continues," *Europe* (January/February 1992), 17-18.

30. "On the Way to the Forum," *The Economist* (July 11, 1992), pp. 14, 17-18.

31. See Harrop, pp. 23-40, for a more detailed overview of the EC structure.

32. Herbert Ungerer and Nicholas Costello, *Telecommunications in Europe: Free Choice for the User in Europe's 1992 Market; the Challenge for the European Community* (Luxembourg: EC, 1988), pp. 133-134.

33. *Telecommunications Reports International* (September 6, 1991), p. 9.

34. Volker Schneider and Raymund Werle, "International Regime or Corporate Actor? The European Community in Telecommunications Policy," *The Political Economy of Communications: International and European Dimensions*, ed. by Kenneth Dyson and Peter Humphreys (London: Routledge, 1990), p. 88.

35. Schneider and Werle, pp. 88-93.

36. *Information Technology Industries*, pp. 45-46.

37. Quoted by Ungerer and Costello, p. 23.

38. See *Towards a Dynamic European Economy — Green Paper on the Development of the Common Market for Telecommunications Services and Equipment*, COM(87) 290 (June 30, 1987).

39. See Schneider and Werle, pp. 96-97.

40. Herbert Ungerer, "Liberalization of European Telecommunications," *Transnational Data and Communications Report* (January/February 1991), pp. 17-21.

41. *Telecommunications Reports International* (October 30, 1992), pp. 1-3.

42. *Telecommunications Reports International* (November 13, 1992), pp. 13-14.

43. Cited by Ungerer and Costello, p. 32.

44. *Eastern European and Former Soviet Telecom Report*, vol. 3, no. 7 (July 1, 1992), p. 7.

45. Eli Noam, *Telecommunications in Europe* (New York: Oxford University Press, 1992), p. 300.

46. Gerd Wallenstein, *Setting Global Telecommunications Standards: The Stakes, The Players, and the Process* (Norwood Mass.: Artech House, 1989), p. 203.

47. Stephen Temple, *European Telecommunications Standards Institute: A Revolution in European Telecommunications Standards Making*, ETSI, 1991.

48. Noam, pp. 305-308.

49. Noam, p. 301.

50. *1990 ETSI Annual Report*, p. 6.

51. Temple, *European Telecommunications Standards*, p. 28.

52. You can obtain more information on INTUG by contacting the Executive Director at this address: G.G. McKendrick; Executive Director; International Telecommunications User Group; 31 Westminster Palace Gardens; Artillery Row; London SW1P 1RR; England; +44-71-799-2446 (telephone); +44-71-799-2445 (fax).

53. Arthur C. Clarke, the British physicist, made this discovery and explained it in, "Extra-Terrestrial Relays: Can Rocket Stations Give World-Wide Radio Coverage?" *Wireless World* (October 1945), pp. 305-308.

54. See Heather Hudson, *Communication Satellites, Their Development and Impact* (New York, The Free Press, 1990), pp. 1-38, for an overview of these discussions.

55. Communications Satellite Act of 1962, Public Law 87-624, 76 Stat. 419, August 31, 1962.

56. 76 Stat. 419.

57. Hudson, p. 26.

58. See *Intelsat Annual Report for 1988-89* (Washington D.C., 1989), p. 45.

59. See *Intelsat Annual Report for 1991-92* (Washington D.C., 1992), pp. 32-36.

60. See Heather Hudson, *Communication Satellites: Their Development and Impact*, p. 148.

61. According to the *Intelsat Annual Report 1988-89* (Washington, D.C., 1989), p. 49.

62. *Telecommunications Reports* (March 9, 1992), page 32.

63. International Telecommunications Satellite Organization Headquarters Agreement, 28 UST, 2249-2257, November 22, 24, 1976.

64. *Telecommunications Reports International* (May 15, 1992), pp. 1-2.

65. All specific numbers cited are from the *Intelsat Annual Report 1988-89*, (Washington, D.C., 1989). Also, the description of services are based on information provided in that report.

66. *Intelsat Annual Report for 1991-92*, page 26.

67. Agreement Relating to the International Telecommunications Satellite Organization, 23 UST 3854, August 20, 1971.

68. Jonathan David Aronson and Peter F. Cowhey, *When Countries Talk: International Trade in Telecommunications Services* (Cambridge, Mass.: Ballinger Publication, 1988), p. 121.

69. Presidential Determination No. 85-2 (November 28, 1964).

70. Letter of Secretaries of State and Commerce, Attachment B to *Notice of Inquiry and Proposed Rulemaking*, CC Docket No. 84-1299, 100 FCC 2d.

71. *Telecommunications Reports International* (November 23, 1990), pp. 12-13.

72. *Telecommunications Reports* (November 13, 1992), p. 14.

73. *Telecommunications Reports International* (May 29, 1992), pp. 16-17.

74. Joseph Pelton, *Global Communication Satellite Policy: Intelsat, Politics, and Functionalism* (Mt. Airy, Md.: Lomond Books, 1974), p. 46, cited by Hudson, p. 29.

75. Leland Johnson, "International Telecommunications Regulation," *New Directions in Telecommunications Policy*, ed. Paula R. Newberg (Durham, N.C.: Duke University Press, 1989), p. 117.

76. According to Leland Johnson, AT&T has agreed to allocate 34 percent of its growth to satellite systems until 1995, p. 110.

77. Hudson, p. 131.

Technical Issues

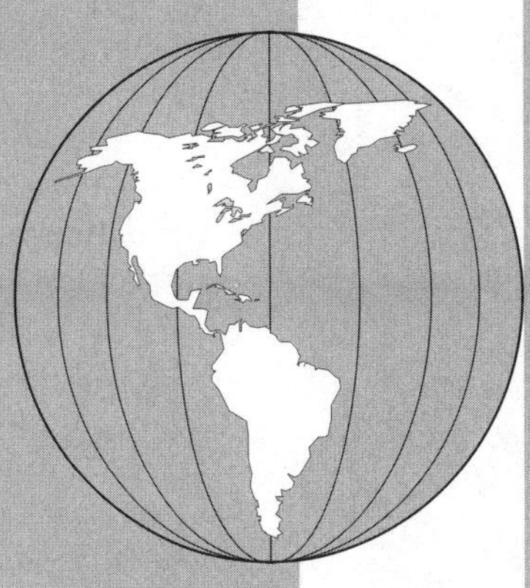

CHAPTER 5

In this chapter, we introduce the key technologies that affect international telecommunications. We intend this chapter only as an introduction because a thorough treatment of the technologies would require many volumes. We refer the interested reader to more detailed sources for further information.[1] This chapter is organized first according to the major types of communications systems, voice and data. Video transmission is emerging as an important applications technology and is briefly discussed as well. We then arrange the subject matter to discuss the major classes of functionality required in a telecommunications system: user systems, switching systems, and transmission systems. Obviously, these areas interact: changes in user systems can drive changes in transmission systems, for example. The primary advantage of organizing the chapter in this way is that the unique characteristics of voice and data communications systems are preserved. Within each category, we will follow the historical development of each kind of system.

Voice Communications

Though the telegraph, which is essentially a form of data communications, was the first application of telecommunications, voice communication over the telephone was the first that was easily used by the general public. Telegraph operation requires a mastery of Morse code, which is sufficiently formidable to be daunting to most users.[2] Therefore, select individuals mastered Morse code and became the key operators who performed the telecommunications

function for the end-user. For a telegraph system to operate efficiently, key operators had to be busy much of the time. This meant that, from time to time, they had a backlog of messages to transmit. Similarly, most end-users did not have the message traffic to justify a private key operator, so users had to share the key operator in a central location. If users wanted to exchange information simultaneously, as in a conversation, both parties needed to be present at the operator's station and tolerate significant delays (on the order of minutes), between a message and its response. Thus, simultaneous, real time communication between two parties was not feasible. The telegraph actually serves as a device to transmit messages that are roughly equivalent to letters, but at a much higher speed than is practical with the postal system.

The invention of the telephone changed all that. With the telephone, two parties could hold a conversation in real time over a great distance. Users didn't need to master a special code or develop the dexterity to operate a telegraph key. They could operate many aspects of the telephone system themselves, eliminating the intermediary (the key operator). The nature of the telephone channel eliminated all significant delays between a message and the response to that message.

In the early days of the telephone, users needed a separate telephone instrument for each party with whom they communicated. The wire that was installed served only those two instruments. The cost of this arrangement made it impractical for networks of any appreciable size. With the introduction of the switchboard operator, a single line and telephone instrument could connect to and be used for communication with any other user. The essence of this approach is facilities sharing: in addition to simplifying the user's interface and making that interface universal, it made the provision of telephone service more economical. This economy occurs because many users could employ the facilities, such as transmission wires and operators, in recognition of the fact that users typically are not on the system all the time. When users' demands exceed a component's capacity, blockage occurs: a user's call attempt is not successful. Determining the appropriate capacity is a difficult task.

With these innovations, the telephone came to dominate the telegraph as a means of communication, especially among wealthier citizens. As the network grew in size (and as technology matured), it became increasingly affordable, to the point where most developed countries can now provide economical telephone service to virtually all their households.

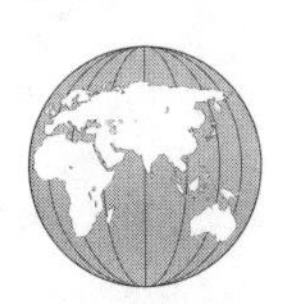

The Structure of Voice Systems

Technological advances replaced many manually operated telephone exchanges with automatic ones. These traditional operators have been replaced by switches that might be electromechanical or electronic, with operators relegated to customer assistance and directory assistance roles. Figure 5.1 illustrates the components of a voice network and how they interact with each other.

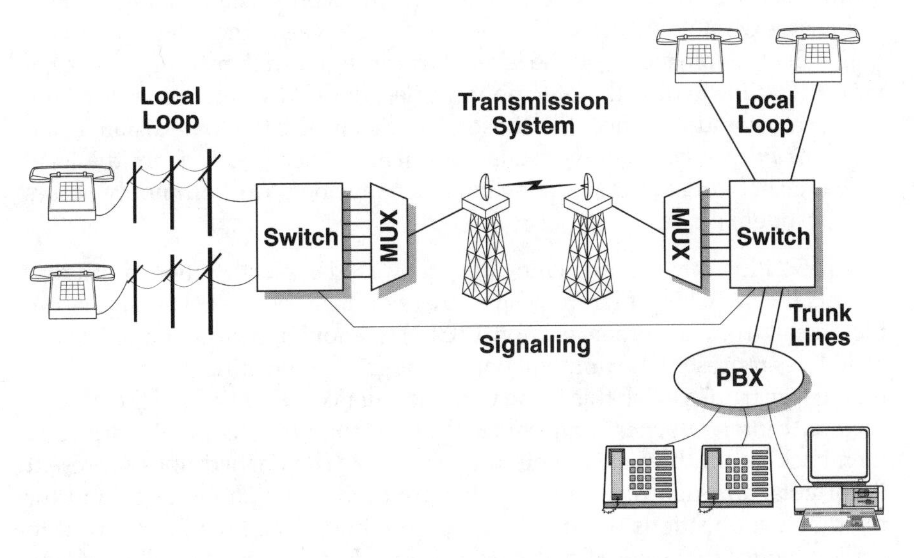

Figure 5.1. Elements of the voice communications network.

Figure 5.1 represents the structure of a typical voice communications network. This approach may vary in many ways, but all must have the same functionality. The user interacts with the network by way of the telephone instrument. The telephone instrument is normally connected to a switch, in a *central office* (CO), with a pair of wires called the *local loop*. COs serve several thousand to 100,000 users. Because this clearly doesn't represent the entire universe of users, other COs must also exist. These COs could be interconnected on a point-to-point basis, but just as this was infeasible for the individual telephone user, it is not feasible for COs; interconnecting so many COs in a country would not be cost-effective.

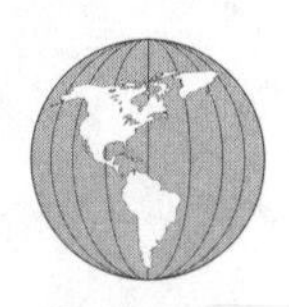

Instead, these COs are connected to a tandem office, or toll center, using *trunk lines*. The *tandem office* acts as a central office for central offices: it switches traffic between COs. In most countries, interconnecting tandem offices with other tandem offices is necessary. You can easily see that this is done efficiently with yet another higher level of switching. This must continue until interconnecting all switches with each other is efficient.

In the United States, the typical model for the switching network has five levels of switching. At the highest level, the so-called Class 1 offices, all switches are interconnected. Lower level switches are interconnected with higher level switches, except when direct interconnection between other switches is economically feasible based on the level of traffic between them. For instance, it may be feasible to directly interconnect adjacent COs as well as connect them to the same tandem office. For those countries where the long distance network, or *interexchange* network, is administered separately from the local exchange network, access into the interexchange network commonly occurs at the tandem offices.

Figure 5.2 illustrates the switching hierarchy used in North America and represents the switching hierarchies in other countries as well. This figure depicts a call from customer location 1 (CL1) to another customer at CL2. The solid lines represent the nominal path through the switching hierarchy: if no high usage trunks exist, this is the path the call takes. It travels to the nearest regional center, which is connected to the corresponding regional center, and then back down through the hierarchy to CL2. The dashed lines represent "shortcuts" through the hierarchy that are used when enough demand for the direct connections occurs. These are the high usage trunks referred to earlier. Notice that several paths are possible. In this case, the call travels to the tandem office and is connected directly to the CO serving CL2. This provides for a much shorter path through the network, reducing the call setup time and the cost of providing the telephone service to those customers. In fact, in the United States, most major interexchange carriers have changed the way they select a route for calls through the network so the calls use the lowest possible level of the hierarchy.

How are voice services provided? What are the user devices? What are the elements of switching and transmission technologies in use today and in the future? What implications do these have for international telecommunications? These are some of the questions we address in subsequent sections.

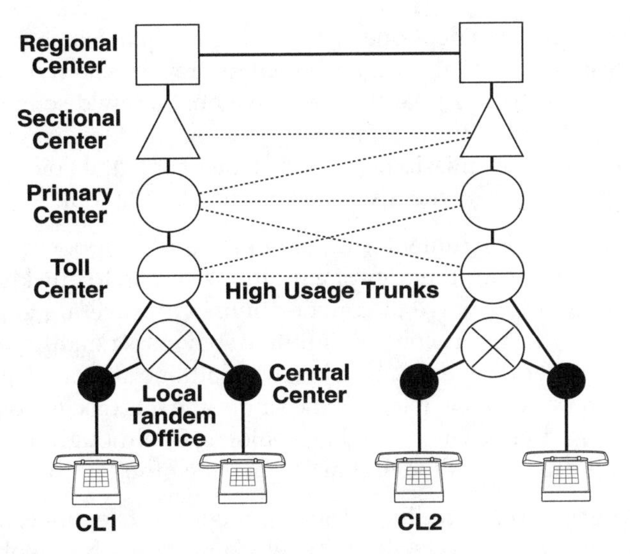

Figure 5.2. North American switching hierarchy.

User Devices

The common end-user device for voice communications is a telephone. The telephone set has undergone several evolutionary cycles. The telephone works by converting sound waves on the speaker into electrical signals that are transmitted on a wire. The speaker in the handset converts the electrical signal into sound waves that the listener can hear. In addition to these functions, the telephone instrument must perform several signaling functions.

By far the most common type of telephone is an analog, single line telephone instrument. A telephone instrument is *analog* if the voice is sent by a continuously varying electrical signal. The amplitude of the signal is a function of the volume of the user's voice, and the frequency of the signal is a function of the pitch of the user's voice. A loud, booming voice (basso) has a high amplitude dominated by low frequencies; whereas a quiet, high voice (alto) has a low amplitude dominated by high frequencies. The telephone channel normally limits the range of frequencies it carries so the network can be used as efficiently as possible without significantly deteriorating the quality of the transmitted speech.

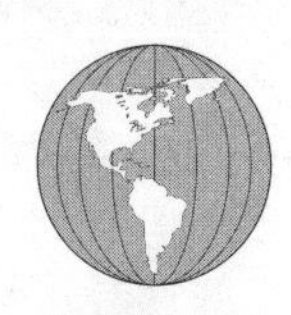

The alternative type of telephone instrument is a digital telephone, where the user's voice is converted to a digital signal for transmission. A digital signal consists of an electrical signal that can have only a few discrete levels. The user's voice, which is fundamentally an analog signal, is converted into a digital format for transmission. At the distant station, the voice is converted back into analog form so the distant user can understand it.

Business telephone instruments may simultaneously support multiple lines: the possibility exists for the user of a multiline station to quickly switch between lines so as to engage in conversations with several people at once. Fundamentally, the technology of multiline stations is not different from the single line sets, which we describe below. Multiline stations are simply expanded and repackaged versions of the single line sets discussed previously. This applies to digital sets as well as analog sets, although the differences between single and multiline digital sets may be more profound.

When placing a call on a traditional analog single line telephone, the user first lifts the handset from the cradle of the telephone base (the telephone industry jargon refers to this as "going off hook"). The instrument signals the switching equipment in the CO that the user wishes to place a call. The switch then provides the user with a dial tone to indicate it is ready to receive input. The user dials the number of the call, and the telephone instrument signals that number to the telephone switch. When the switch receives the call, it exchanges signals with all the other switches involved in the call, and the call is established. The distant user is alerted by a ring, which is placed on the local loop by the distant user's CO. When the user who receives the call picks up the telephone handset, the telephone instrument signals the switch that someone is receiving the call, and the conversation begins. When the conversation ends, both users place their handsets on the cradle (the users "go on hook"), which causes the telephone instruments to signal to the switch that the conversation has ended. The switch then releases the trunks used for the call, so other users can use them. Figure 5.3 illustrates the flowchart of a telephone call at a high level. This flowchart can be specified at a greater level of detail.

A telephone instrument requires electricity to operate. Early telephone instruments contained batteries or a small, hand-operated generator. Maintenance problems caused by the many batteries motivated the telephone companies to power the telephone instruments directly from the COs. In manual exchanges, users must first establish a connection with the telephone operator. In the early days, a caller accomplished this by producing a ringing voltage with a hand-generated crank. The caller spoke the destination telephone number to the operator, who registered the source and destination of the call, manually connected the two lines, and manually rang the destination

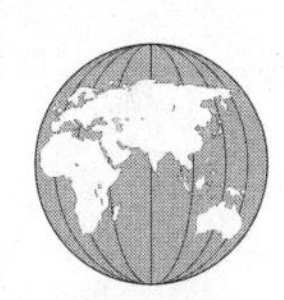

telephone. If more offices were involved, the operator had to contact the next office directly to determine the status of the destination. If the destination was available, both operators connected the call over an agreed-upon trunk. The operator had to monitor the status of the connection to determine when the callers terminated the call. The relative inefficiency of this method motivated the development of automatic exchanges.

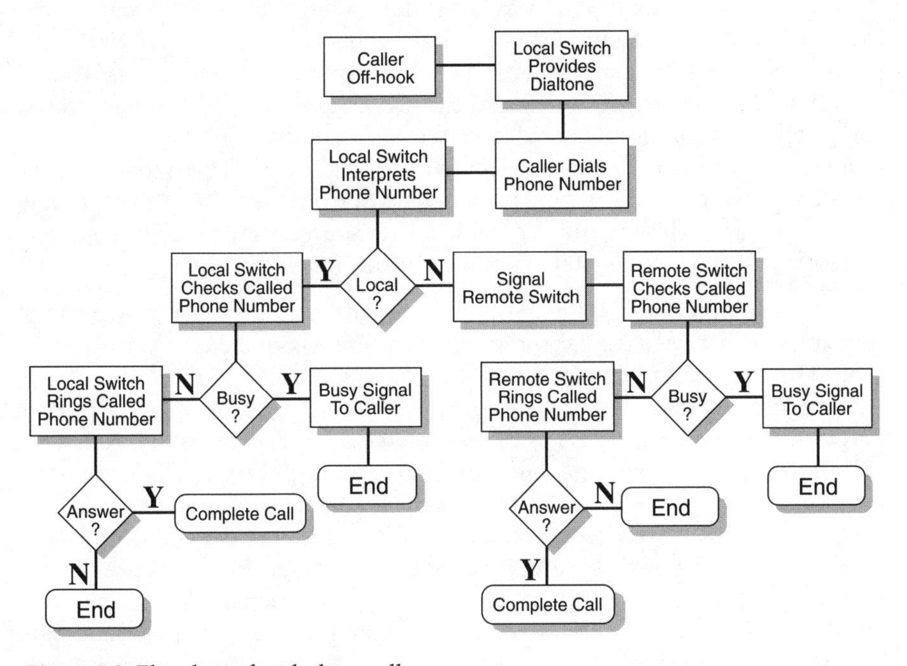

Figure 5.3. *Flowchart of a telephone call.*

In automatic exchanges, signaling between COs may be done by controlling the electrical current that powers the telephone instrument. When the handset is resting on the cradle (the telephone is *on-hook*), no electrical current flows over the wire. When the user picks up the handset (the user's telephone goes *off-hook*), current begins to flow, signaling to the switch that the user wishes to place a call. The caller dials the destination number by pulsing the electrical current on and off (in a rotary dial telephone) or by generating tones in the telephone instrument (in a tone dial telephone). In each case, the switch in the CO registers the selected number and proceeds to establish the connection.

Digital telephones require a digital local loop to work. The digital local loop technology is specified in a set of standards, referred to collectively as the *Integrated Services Digital Network* (ISDN). Digital loops have been relatively slow to emerge, because of the large and costly conversion problem faced by the telephone companies, as well as the relative absence of useful applications for these loops. Some countries are more strongly committed to converting their infrastructure to digital local loops than others; many European countries are investing heavily in this technology, as is Japan. With digital telephones over a digital loop, signaling happens quite differently. In this case, both the telephone instruments and the switches in the CO contain microcomputers that communicate with each other over the digital loop to set up a call, transmit dialed digits, indicate a ring, and so on. These instruments are often powered locally by the user instead of from the CO (the way analog instruments are). This requires a reliable power source and a reliable telephone network to have a reliable telecommunications network.

Most local loops throughout the world remain analog, which means data transmission between digital computers requires a conversion device to succeed. This conversion device is a modulator-demodulator, or *modem*. These devices transmit data signals from computer to computer, over the telephone network. Modems support a larger variety of data transmission rates. As a rule of thumb, the higher the data transmission rate supported by a modem, the more costly the modem is. Modems can be purchased to operate over dial-up connections or over leased analog circuits. These modems use different technologies and generally cannot be used interchangeably.[3] Because the capacity of analog local loops is limited, the speed at which modems can transmit data is also limited. Without coding and data compression, modems can transmit at a theoretical maximum of about 26,000 kbps. With data compression, these speeds can be boosted to about 60,000 kbps. As the quality of the telephone line deteriorates, so does the speed at which the modem can transmit. In many developing countries, the maximum practical modem speed is 2,400 bps.

One advantage of a digital loop and digital telephones is that modems are unnecessary. One might still require a connection device, but that device is much simpler in nature than a modem and therefore costs much less. The possibility exists that one could simultaneously transmit voice and data over the same digital local loop, a feat not possible on a wide scale with an analog loop.[4] We discuss the transformation of the network to digital beyond the local loop in more detail in subsequent sections.

Data Communications

Data communications differs in some fundamental ways from voice communications. On the surface, they differ because the native form of voice is analog, and the native form of data is digital. If this were the only significant difference, we'd have no need for different kinds of networks. More important differences between voice and data communications stem from the nature of traffic between voice and data and the requirements placed on the communications channel by each application in terms of delay and errors.

At the conversation level, two people on a telephone call occupy approximately 40 percent of the capacity of a circuit. This is because, in general, only one person talks at a time, and pauses occur when nobody is talking. Voice conversations are generally symmetric (the channel is used equally in both directions, on average) and take an average of two minutes to complete. Voice conversations are generally tolerant of noise on the communications channel. Users normally have enough experience with their native language that it's possible for them to distinguish words even on noisy lines. On the other hand, users generally don't tolerate delays very well. This disrupts the flow of conversation and can result in the breakdown of the communications channel. (For this reason, CCITT Rec. E171 recommends the use of at most one satellite line in a connection. Satellite systems introduce significant delay.)

Data communication has quite different characteristics. In most cases, data can tolerate reasonable delays, but not errors. In interactive, terminal-oriented communications, the traffic is characterized by asymmetric data flows, where data flow in one direction is heavier than in the other. This traffic often occurs in bursts: data flows for short periods, followed by long periods of no activity. Finally, sessions may last a long time compared to voice conversations. In data transfer applications, a substantial amount of data must often be transferred at one time. This normally results in a heavily asymmetric communication that lasts as long as it takes to transfer the data. During this time, the channel is heavily occupied. Communication with characteristics such as these suggests the type of channel suitable for data differs significantly from the type of channel appropriate for voice. Table 5.1 summarizes the different characteristics of voice and data traffic.

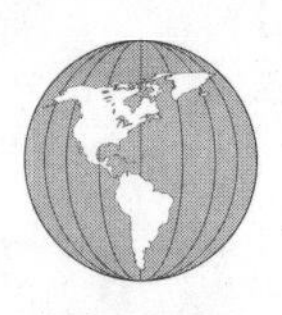

Table 5.1. Voice and data traffic characteristics.

	Voice	Data
Noise Tolerance	High	Low
Delay Tolerance	Low	High
Symmetry	High	Low
Burstiness	Low	High

Historically, data communication over long distances occurred over leased telephone lines using modems developed for that kind of line. Because this reserved a telephone channel 24 hours a day and 365 days a year, this was economical only for high-volume communications. Not all applications required this level of data transfer; additionally, some applications required location flexibility. To satisfy these users, modems were developed to work over dial-up telephone lines. As we discussed previously, data communications sessions generally take up much more time than voice communications; if the session is interactive, the line is idle much of the time. Data communications managers soon determined that this was not cost-effective and sought alternatives.

Packet switching was developed in response to this need. In contrast to circuit switching, where a circuit of dedicated bandwidth is established for exclusive use by a customer, packet switching packages the customer's message into packets, and transmits these packets over a shared communications line: many customers share the same bandwidth, even though this is transparent to them. This method was economical because no single customer needed the entire channel at all times. In fact, packet switching could dynamically allocate the bandwidth among the customers; customers who needed more bandwidth because they had more data to transmit simply transmitted more packets. Packet switching quickly became the preferred approach for data communications because it not only made more efficient use of a communications channel, but it also provided some value added features. These features include error checking and correction, databases, and so on. In fact, packet-switched services are often referred to as *value added network* (VAN) services.

Figure 5.4 illustrates a simple packet-switched network. This network consists of three packet switches, four user terminals (A through D), and three host computers (1 through 3). The packet switches are interconnected with a single bidirectional line as shown. We can assume Terminal A has a session

with Host 1, Terminal B with Host 2, Terminals C and D with Host 3; in addition, Hosts 1, 2, and 3 exchange information with each other. Because the packet switches are connected with a single line, all the devices must share that line with each other. The packet switch selects the output line and manages the sharing process. Information from the terminals and hosts is broken into packets, each of which is transmitted independently. In Figure 5.4, each of the packets is labelled with its source device; in reality, both source and destination device addresses aren't necessary. To share the line, the packet switch transmits packets sequentially, interleaving packets from various devices with each other. If a device must send multiple packets, they may be interleaved with other packets from other devices that also must use that line.

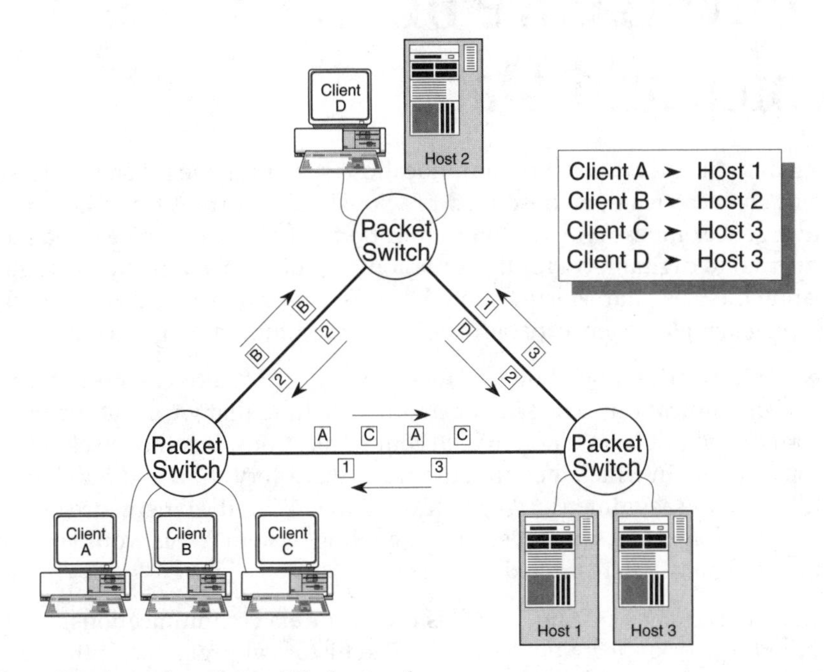

Figure 5.4. A simple packet-switched network.

Practical networks are substantially more complex than the simple network illustrated here. In addition, a device called a *packet assembler and disassembler* may be required to construct packets if the user devices (the terminals and hosts in Figure 5.4) don't do this automatically. In fact, one of the services that many commercial packet-switched networks provide is packet assembly and disassembly so that users don't need to do this explicitly before using the network.

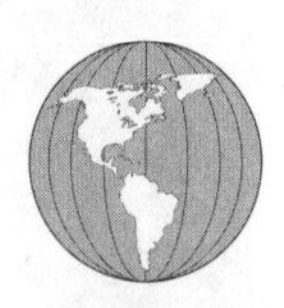

Devices that use a data communications network can vary quite a bit. Historically, they consisted primarily of terminals and host computers. In addition to terminals and host computers, customers today use devices such as minicomputers, personal computers, point of sale terminals, and so on. The range of applicable devices has indeed become quite large. Packet switches themselves are specialized computers. They have been optimized to process input and output quickly and efficiently. Their architecture has often been modified to make this possible.

Convergence of Voice and Data

Even though voice and data communications systems have substantially different characteristics, as described previously, the current trend is toward convergence of these types of communication. When analog voice signals are converted to digital form for transmission, you don't need much to imagine treating them as data. The question then arises how one would go about developing a single system capable of efficiently accommodating voice and data.

We previously described how circuit switching is not suited to most types of data communications. The kind of circuit switching that we are accustomed to thinking about is unsuited to data communications. If someone could develop a system in which the call setup times were very short and the channel could be made asymmetric, it could be suitable. With the emergence of *common channel signaling systems*, such as Signaling System 7, this would be possible, particularly in restricted geographic areas.

Similarly, packet switching, which is used for data communications, can be adapted to work effectively for voice. The historical problems with packet switching in relation to voice communications are the delays for each packet and the delay variations between packets. With the rise of high-speed networks and fast nodal processors for packets, an emerging technology referred to as *asynchronous transfer mode* (ATM) can handle both voice and data traffic on a single network. To date, only experimental implementations of this technology exist over a wide area, and many of these focus on data communications, not voice communications. Hopefully, this technology will deliver Broadband ISDN (B-ISDN) service when it becomes economically and technically feasible.

Transmission Systems

Transmission systems are the telecommunications system elements that send the user's signal over a distance to other telecommunications facilities. Transmission systems can carry voice, data, or both. Traditionally, transmission systems carried information exclusively in analog form. In countries with a well-developed telecommunications infrastructure, use of analog transmission equipment has almost disappeared. Digital transmission equipment has several advantages. It can easily integrate voice and data; the equipment can be implemented cheaply with large-scale digital integrated circuits; maintenance is easier; and the transmission system can be engineered to an arbitrarily low noise level. These advantages have proven persuasive, even if using analog transmission technology is often more efficient from a capacity utilization perspective.

A large telephone system requires large capacity transmission systems to handle the large number of calls. Systems that can carry many channels simultaneously normally provide high capacity, a capability that results in an analog system with a large bandwidth or a digital system with a high bit rate. To be utilized efficiently, these large capacities must be organized. The method for organizing transmission capacity is called *multiplexing*. A device called a *multiplexer* (or MUX) systematically places individual voice channels into the larger bandwidth of the high capacity transmission system and removes them from the high capacity channel at the distant end.

Media

The media that carries communications signals falls into two distinct categories, cable systems and wireless systems. As the basic distinction between the two, cable systems require a continuous right-of-way between end-users. Wireless systems do not need the right-of-way, but they may require an adequate transmission path in terms of clearance between sender and receiver. Historically, cable systems preceded wireless systems in widespread practice.

Cable

In the earliest systems, a single wire was used, with a path through the earth as the return channel. While this was economical, it limited the rate at which information could be transmitted down the wire, or conversely, the distance that information could travel between relays, or repeaters. The remnants of these systems are still found along old railroad tracks; in some countries, these *open wires* are still in use. By building a complete circuit—by stringing two wires for every circuit instead of one—engineers found that the capability of these systems improved dramatically. By twisting these pairs together, the capabilities of wire pairs improved. Modern telephone loops consist largely of this *twisted pair* cabling, although different loop carrier systems exist to improve the capacity of the loop plant even more.

Coaxial Cables

As the demand for telephone services grew, so did the number of cables needed to satisfy this demand. It soon became necessary to consider transmission systems (which we discuss later) and higher capacity cables. At first, electronics augmented the twisted wire pairs—twisted together to improve their transmission characteristics—to improve the carrying capacity of the wires. The demand for telephone services soon outstripped the capability of twisted pair transmission systems to carry the needed traffic. Engineering measurements showed that another type of cable, *coaxial cable*, provided higher capacity than twisted pair cables. Coaxial cable consists of a wire (the conductor) surrounded by a flexible metallic cylinder, called the shield. Figure 5.5 illustrates the major cable types. Many analog transmission systems using frequency division multiplexing (FDM) soon used coaxial cable. About this time, microwave radio systems began to emerge, so research on cable systems slowed somewhat.

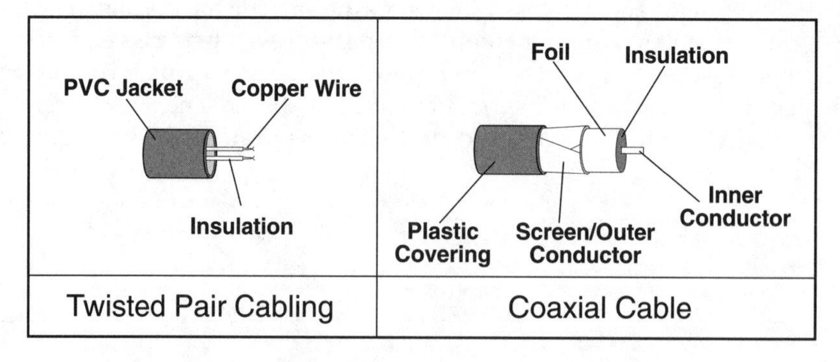

Figure 5.5. Types of cabling systems.

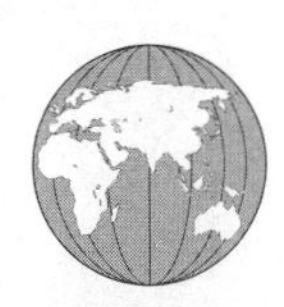

After coaxial systems, which are still in use today, research and development on cable systems turned to millimeter waveguides and optical fibers. *Millimeter waveguides* consist of precisely machined pipes that guide radio waves at low losses. Systems based on this technology never emerged as significant, largely due to the success of optical fibers.

Optical Fibers

Optical fibers consist of thin glass strands that carry pulses of light rather than electrical signals. As a result, they are immune to the electromagnetic interference that plagues most electrical systems. Additionally, the fibers are small and lightweight with a very high capacity for carrying information. The full capacity of the fibers has not been exploited, although researchers are developing systems that make more efficient use of the capacity. The limiting factor in fiber-optic capacity utilization is that the electronic systems at the ends and in the relays must be placed along the fiber at intervals to boost the signal because they cannot operate at the speeds that the fiber can carry. Fiber-optic transmission systems have dominated the industry in the last decade in regard to new installations.

Wireless Systems

The characteristics that differentiate wireless systems from cable systems have caused the former to fall in and out of favor over the last 50 years. Most importantly concerning these characteristics, a physical connection between stations is unnecessary. This fact enables entities to employ wireless systems more effectively than cable systems in some terrains. Wireless systems also allow some applications, notably mobile communications, that would be virtually impossible with cable systems.

Early wireless systems were based on simple transmitters that sent telegraph signals over the Atlantic Ocean. After World War II, much of the technology developed for radar was transferred to the private sector, and microwave communications systems emerged. *Microwave systems* use signals at very high frequencies (above 1 GHz or 1 billion cycles per second) as carriers for information. These signals demand a line of sight between transmitter and receiver, requiring the careful engineering of transmission paths. This system also limits the distance between stations due to the curvature of the earth and other natural and man-made obstacles. When engineering a transmission system, only the acquisition of land and facilities for these stations is necessary; no permissions, beyond frequency use permissions, are required for the intervening acreage, provided a building or some other structure is not in the way.

In remote, mountainous, urban, and otherwise hostile environments, this is a substantial advantage over cable systems.

Wireless systems, however, have signals that are easy to intercept, unlike cable-based systems. While technologies such as data encryption solve this problem to a great extent, this disadvantage remains a vulnerability. Some forms of wireless communication, such as the cordless telephone, may not enjoy the protections found in the wiretap laws of some countries (such as the United States).

Digital technology provides some technological solutions to this problem. To properly receive a digital signal, the receiver must understand the format of the signal being transmitted and know when to expect the signal and at what bit rate. Digital wireless systems can efficiently and effectively obscure the synchronization signals that must be transmitted between the transmitter and the legitimate receiver. Thus, meaningfully interpreting the bits being sent becomes difficult for an intruding receiver. Through encryption, the bits themselves may also be encoded, resulting in a more secure transmission channel.

Microwave Systems

As the technology matured, microwave radio systems began to operate at ever higher frequencies; they have been engineered as analog and digital transmission systems. Microwave systems have recently fallen out of favor as a long-haul transmission technology in many countries because of the emergence of high-capacity, cost-effective, fiber-optic transmission media. Microwave systems remain an important technology for short distance networks, where repeaters are unnecessary, such as linking facilities across a major highway, river, or other obstacle where a dedicated path is difficult or costly to acquire.

Satellite Systems

One important special case of a microwave transmission system is the satellite communications system. Satellite systems use the same range of frequencies that land-based (terrestrial) microwave systems use, but their repeater is located 22,500 miles above the earth, usually in a synchronous, equatorial orbit. Satellites in this orbit retain a constant relative position over the earth, so that, for example, a satellite over Nairobi, Kenya, in geosynchronous orbit is always over Nairobi. Therefore, building inexpensive earth stations is possible; the antennae need not be expensive and failure prone tracking systems. Many current earth stations allow readjustment of the antenna to capture the signal from a different satellite. Once adjusted, the antenna is fixed in its position and need not be moved.

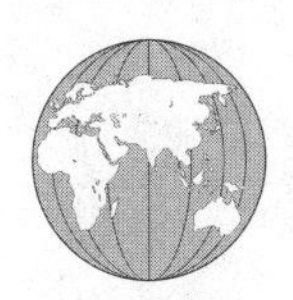

Satellite systems have particular advantages for very long distance communications systems and for geographically dispersed point-to-multipoint systems. A satellite system consists of two earth stations and a transponder on a communications satellite in orbit above the earth. The source earth station transmits signals to the satellite transponder, which receives the signal and retransmits it to all stations within its "footprint," including the receiving earth station. Figure 5.6 illustrates this concept. As the figure shows, the satellite can broadcast the identical message to many users over a wide geographic area. For this reason, satellites are often used for the distribution of television programming.

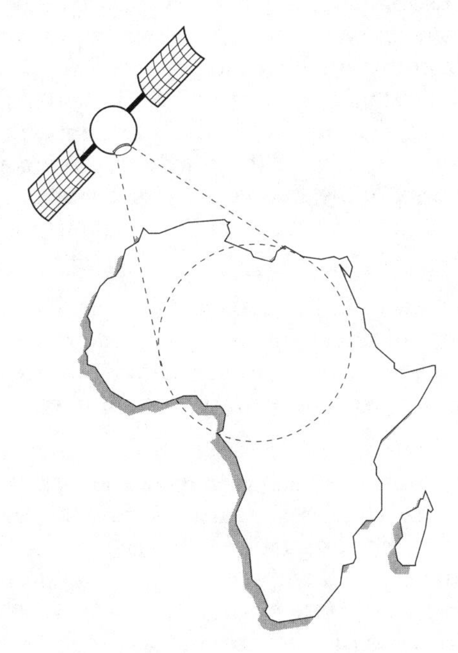

Figure 5.6. *A satellite footprint over Africa.*

As satellite technology improves, so do the types of systems that use satellite communications. Using *very small aperture terminals* (VSAT) is common today for most forms of satellite communications. VSAT systems have the advantage of small earth station antennae that are relatively inexpensive and easy to manage. The small antenna size has been achieved as a result of improvements in earth station electronics and in satellite transceivers with higher transmission power.

Mobile Applications

Mobile applications are the ultimate for wireless transmission systems. Mobile systems have several forms that are relatively familiar. In the simplest form of a mobile system, each user possesses a mobile radio capable of operating on a few channels. These radios incorporate no signaling or contention resolution; the functions are left to the user and are handled by essentially social arrangements, such as, "Meet me on Frequency (or Channel) A at 10:15 a.m." If someone else happens to be using the frequency at that time, the users wait until the channel is free.

The widespread use of automobiles as an important form of transportation motivated the desire for mobile telephone use. In principle, each automobile has a telephone instrument with an attached radio system. The radio system must perform all the signaling functions in addition to the traditional radio functions of transmission, modulation, and reception. The earliest mobile systems had a single transmitter (base station) serving an entire metropolitan area. The capacity of these servers was soon exhausted, motivating the emergence of an alternate form of mobile telephone use.

This alternate form relies on several base stations scattered throughout a metropolitan area. Each station transmits at a much lower power than was used in the single transmitter approach discussed previously. A system of this architecture can satisfy many more users because the frequencies in a service area can be reused some distance away without interference.

This form came to be called *cellular systems* because the metropolitan area was divided into numerous cells, each with its own base station. Figure 5.7 illustrates how a cellular system is structured. As you can see, a single switching office interconnects each base station in the cell sites, enabling users to pass from one cell to another cell without losing the connection. This is also the point at which the cellular system is interconnected with the public switched telephone network. When a subscriber passes across a cell boundary, the call is "handed off" to the next cell. This causes a momentary outage imperceptible to voice users, but that can disrupt mobile data communications over a cellular system.

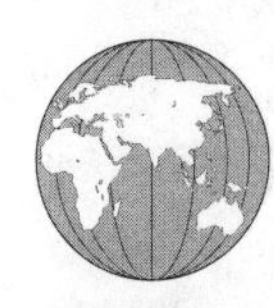

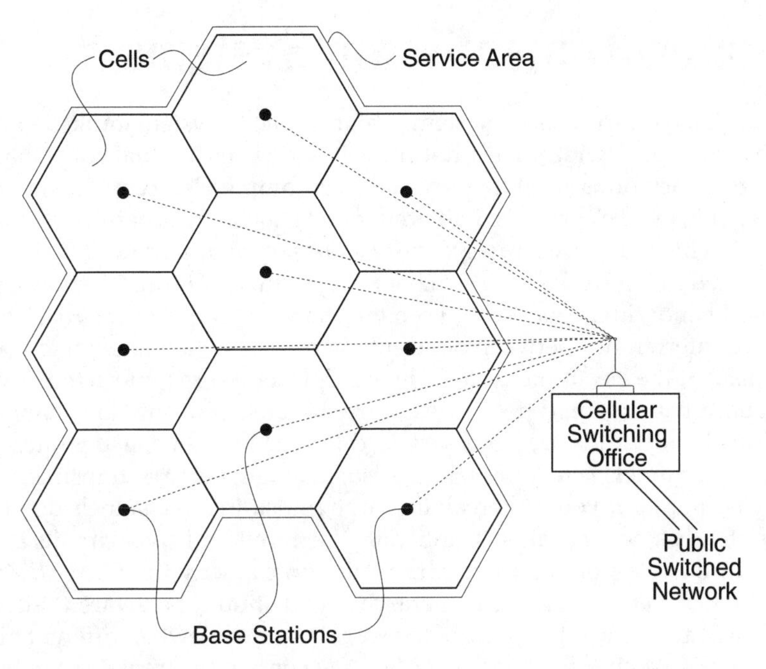

Figure 5.7. A cellular telephone system.

Recent trends have further stimulated demand for wireless communications, such as mobile telephones and wireless *local area networks* (LANs) for data communications. Current industry plans call for the introduction of *personal communications systems* (PCSs), wherein designers take the cellular concept to the limit in the sense that they design the cells to be very small and their transmitters to operate at very low power. This enables designers to build telephones that fit in shirt pockets. These phones would be impossible without continuing advances in wireless technology. Wireless LANs are also emerging as an important technology. Here, computers and terminals do not have a cable-based network connection. Instead, they have a small radio transceiver that uses well-defined protocols to communicate with others. Each room may have a central transceiver connected to a wire-based local network.

Each of the transmission systems can support analog or digital signals. While the trend points toward digital communications systems, understanding the distinction between these kinds of communications systems is important.

Analog Transmission Systems

For an analog transmission system, a high-capacity system implies a system with a large bandwidth. The multiplexer "stacks" individual voice channels on top of each other until the bandwidth is completely occupied: the multiplexer shifts voice channels onto predefined subchannels of the larger bandwidth.[5] This is referred to as *frequency division multiplexing* (FDM) and is illustrated in Figure 5.8 for a six-channel system. As Figure 5.8 shows, some unused bandwith is placed between the channels, called *guard bands*, which reduce interference between adjacent channels (a phenomenon known as *crosstalk*). If the electronic systems in multiplexers were perfect, these wasteful guard bands would not be necessary. Because transmission channels of various bandwidths are necessary in different locations, using only high-capacity channels is inappropriate. Also, managing a large transmission facility is simpler if you can break the large transmission channels down into consistently smaller channels, until only the individual voice circuits are left. We call this way of organizing transmission capacity the *FDM Hierarchy*. CCITT has standardized this hierarchy; you should be aware that North America and some other parts of the world use a somewhat different hierarchy. *CCITT Recommendations G.211 to G.544* contain the transmission hierarchies for various transmission systems.

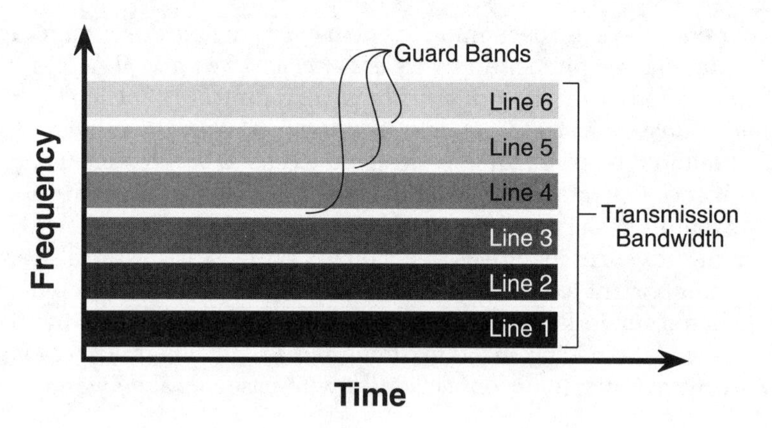

Figure 5.8. *Frequency division multiplexing.*

The first multiplexing level is referred to as a *group*. In a group, 12 voice channels are bundled together into a transmission channel that requires 48 kHz

of bandwidth.[6] Ten groups constitute a *supergroup*, which carries a total of 120 channels and occupies 240 kHz of bandwidth. If a higher capacity is required, the standard specifies that five supergroups be combined into a *mastergroup*, containing 600 channels and occupying 1.232 MHz of bandwidth. The final collection of circuits is referred to as a *supermastergroup*. Actual transmission systems may accommodate multiple supermastergroups, but the supermastergroup is the highest level of organization; each supermastergroup consists of three mastergroups, or 1,800 voice circuits, and occupies 3.872 MHz of bandwidth.

Carrier systems are specific transmission systems engineered for information transmission. Specific carrier systems are designed for specific types of cable and use a particular multiplex hierarchy. Most carrier systems are designed for use by network providers. Three criteria specify carrier systems: the cable type over which they operate, the amount of bandwidth they use (in MHz), and the number of channels they use. Typically, a carrier system with a higher bandwidth over a particular type of cable cannot go as far between repeaters as one with lower bandwidth over the same cable type. *Repeaters* are electronic systems that boost the power level of a signal. They must be placed at regular intervals in a transmission system to meet the transmission quality objectives of the system.

Some CCITT Recommendations for Analog Carrier Systems

G.31x	Carrier Systems for Open Wire (x=1-4)
G.32x	Carrier Systems for Loaded and Unloaded Symmetric Cable Pairs (x=2-7)
G.33x	Carrier Systems for 2.6/9.5mm Coaxial Cable Pairs
G.332	12 MHz System (45 Supergroups)
G.333	60 MHz System (180 Supergroups)
G.334	18 MHz System (60 Supergroups)
G.34x	Carrier Systems for 1.2/4.4mm Coaxial Cable Pairs
G.341	1.3 MHz System (5 Supergroups)
G.343	4 MHz System (15 Supergroups)
G.344	6 MHz System (20 Supergroups)
G.345	12 MHz System (45 Supergroups)
G.346	18 MHz System (60 Supergroups)

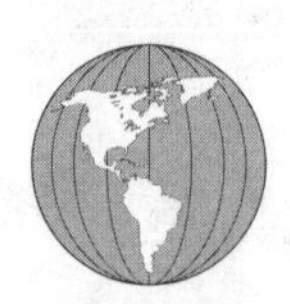

Digital

When the T1 carrier system was developed in the early 1960s, an alternative method of organizing a high-capacity channel became necessary for digital channels. Instead of "stacking" channels in frequency (as is done in FDM), the entire channel is given to one circuit for a short interval on a rotating basis. This is referred to as *time division multiplexing* (TDM). In other words, the circuits are organized sequentially instead of simultaneously. If the sequences are short enough, they appear to make simultaneous use of the bandwidth. So, if six channels share a larger bandwidth, each channel uses one-sixth of the bandwidth on average. Figure 5.9 illustrates this concept. The channel is organized into a frame that repeats regularly, and each circuit occupies a slot in the frame. A framing signal defines the start of frames. The demultiplexer can clearly detect this signal. Using digital systems also eliminates the need for the guard bands that FDM requires.

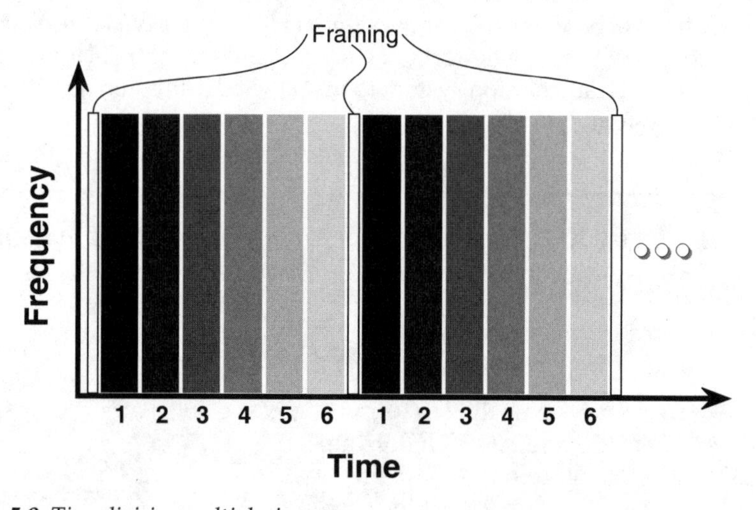

Figure 5.9. *Time division multiplexing.*

A digital transmission system requires a digital signal. For data communications, this is no problem. For voice communications, however, a conversion is necessary because voice is typically an analog signal. The digital representation of a voice signal is called *pulse code modulation* (PCM). To convert a voice signal to a digital format, the voice signal is first sampled at regular intervals. These samples are then converted to digital format for transmission across

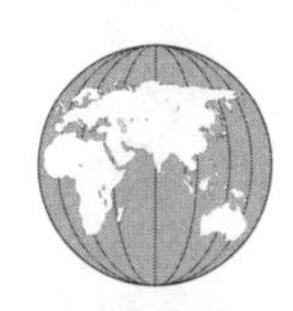

the network. At the receiver, the digitized samples are converted back into analog format. After the reconverted analog samples are filtered, they are transmitted to the user. If the sampling rate is high enough, the users perceive no difference in the quality of the signal. For voice telephone channels, a bit rate of 64 kbps is needed for PCM. To further improve the quality of this conversion from analog to digital and back, the quieter portions of the signal are amplified, and the louder ones are not. The receiver reverses the process exactly, so the listener hears a signal as it should sound. This is a process called *companding* (a combination of the words compression and expanding).

The CCITT standardized a form of companding called *A-Law companding*, in reference to the way companding is performed. Unfortunately, North American countries had begun using a different form of companding, referred to as *µ-Law companding*, prior to the establishment of the A-Law standard. Without *transcoder* devices (which convert A-Law PCM to µ-law PCM), a perceptible loss of fidelity occurs if an A-Law encoded signal is decoded by a µ-Law device and vice versa. For users of dial lines, this conversion is provided as part of the connection service. Users of digital leased lines for voice must be sensitive to the need for this conversion.

Careful study of PCM systems revealed significant efficiencies when encoding the difference between successive samples instead of the samples themselves. This has led to the development of *adaptive differential PCM*, or ADPCM. ADPCM requires only 32 kbps for a voice signal, which allows for a doubling of capacity on existing circuits because the rate is half a standard digital voice channel. ADPCM works very well for voice, but not so well for modem signals because it was optimized for voice. Some users report problems with high-speed dial-up modems over ADPCM channels, although slower modems seem to exhibit few difficulties.

As with FDM, different capacities suit different needs. Having a basic level of aggregation and higher levels of aggregation is necessary. In TDM, the CCITT has standardized the hierarchy. Once again, the North American multiplexing hierarchy differs from the international standard. In North America, the basic circuit is a DS0 channel, which has a capacity of 64 kbps[7] and can carry a single uncompressed voice circuit. In North America, the next level in the multiplexing hierarchy is the DS1 rate (also called T1, after the carrier system that implemented this rate). A T1 can carry 24 DS0 channels and has a carrying capacity of 1.536 Mbps.

In the CCITT hierarchy, the next level is E1, which consists of 30 DS0 channels with a rate of 2.048 Mbps. This rate is useful for many corporate communications networks because it represents a communication capacity and a

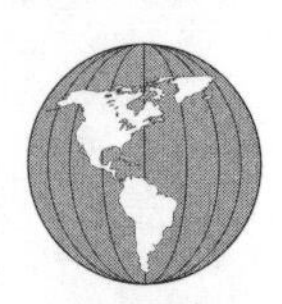

scale economically usable in their networks. Telephone companies find this rate less useful, though, because this channel is fairly small. In the North American hierarchy, the next commonly used channel is a DS3 channel, which has a data rate of approximately 45 Mbps and can support 672 DS0 channels. The telephone companies may use a higher speed, 274 Mbps, which supports 4,032 DS0 channels. In the CCITT hierarchy, the next practical level is approximately 34 Mbps, supporting 480 DS0 channels. Finally the highest level in the CCITT hierarchy supports 1,920 DS0 channels at a rate of approximately 139 Mbps. *CCITT Recommendations G.700 to G.772* define the CCITT digital transmission systems.

North American Digital Hierarchy

DS0: 64 kbps = 1 Voice Channel
DS1: 1.544 Mbps = 24 DS0
DS3: 44.736 Mbps = 672 DS0
DS4: 274.176 Mbps = 4,032 DS0

CCITT Digital Hierarchy

E0: 64 kbps = 1 Voice Channel
E1: 2.048 Mbps = 30 E0
E2: 8.448 Mbps = 120 E0
E3: 34.368 Mbps = 480 E0
E4: 139.264 Mbps = 1,920 E0
E5: 565.148 Mbps = 7,680 E0

This difference in digital hierarchies translates into a difference in transmission systems. Users of single (or DS0) lines or dial-up connections are unlikely to notice a difference between these systems. However, users of higher speed services, such as DS1 (T1) service or E1 service, must be sensitive to these differences when crossing national boundaries where different transmission systems are prevalent. In the United States, you can purchase E1 service for international circuits to facilitate connection, although the reverse is not always true: you may not be able to purchase DS1 circuits in foreign countries to connect to the United States and other North American hierarchy users.

Standards committees have developed a new approach to digital transmission that is being introduced into the telephone company plant. This technique, called the *synchronous optical network* (SONET), makes the management of individual circuits in high-capacity channels much easier and cost-effective. It also allows telephone systems to receive standard interconnection at very high speeds, and therefore, at very high capacities. As its name implies, SONET is intended exclusively for optical transmission systems. In SONET, the lowest rate is STM-1, equivalent to a DS-3 Channel in the North American Digital Hierarchy (although it could carry as many as 810 DS0 channels). Higher levels in the hierarchy are constructed from multiples of these STM-1 channels or from other STM-n channels. SONET represents the future approach in countries conforming to the North American hierarchy (and the CCITT hierarchy), so no further development is anticipated in the traditional digital hierarchies.

SONET Digital Hierarchy

STM-1: 51.85 Mbps
STM-3: 155.52 Mbps
STM-9: 466.56 Mbps
STM-12: 622.08 Mbps
STM-18: 933.12 Mbps
STM-24: 1.244 Gbps
STM-36: 1.866 Gbps
STM-48: 2.488 Gbps

Switching Systems

In previous sections, we discussed the relationships among different kinds of switching devices. At this point, we'll briefly review the kinds of technologies used to perform the switching function. Historically, a switching office comprised many rows of operators, who sat at switchboards connecting calls using cords and plugs. This was called *circuit switching* because it involved the establishment and management of circuits to carry voice communications. As the size of the network grew, this technology became infeasible and introduced an element into the network that retarded competitiveness.

Almon Strowger, developer of the first automatic switch, was originally a mortician by trade. In the town where he plied his trade, his competitor's wife was the local switchboard operator. When a request for a mortician came to the telephone operator, Strowger's funeral home was consistently the second choice. This placed him at such a disadvantage that he developed a method for automatically switching calls, without the intervention of an operator. This invention became so important to the telephone industry, that Strowger formed the Automatic Electric Corporation, which was later purchased by General Telephone and Electric (GTE). Switches based on Strowger's design, patented in 1891, are still in operation in many parts of the world.

These switches were conceptually based on the same idea as a telephone switchboard. The lines and trunks were separated from each other spatially, and the objective called for the interconnection of these physically distinct entities. Several technologies have implemented this design, from Strowger's step-by-step switch through the *crossbar* technology that succeeded it, to the modern implementations using tiny relays or even semiconductor switches. This switching technology is called *space division switching* and is generally well-suited for both analog and digital signals, although some exceptions exist. Figure 5.10 illustrates a space division switch as a matrix of crosspoints that are spatially separated from each other. In this figure, the points where lines cross are crosspoints; telephone A is connected to telephone 3 and telephone C is connected to telephone 2 because the crosspoints are closed. Clearly, telephone A can connect to any telephone 1 through 4. The addition of some paths to the right of the figure allows connections among telephones A through D (a connection between stations B and D is illustrated); similarly, the two additional paths at the top of the figure allow connections among telephones 1 through 4 (although none are used in Figure 5.10).

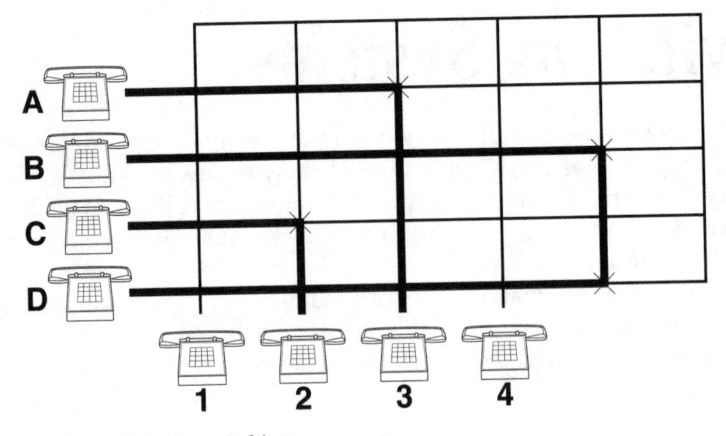

Figure 5.10. *Space division switching.*

In addition to the switching technology, discussed previously, consider how these switches are controlled. In the earliest switches, the human operator provided the control. As the technology became more automatic, the control mechanisms were reconsidered. Control is necessary so that the switch can respond appropriately to user signals. (See Figure 5.3 for an illustration of a user's interaction with a switch.) At first, the control was *hard wired* as part of the switching machine. Any changes to the control mechanism, for whatever reason, required a change in the physical system of the switch. As customers demanded more and varied features and the switches became more complex, an alternate approach to controlling a switch became necessary.

The ideal situation called for a control mechanism that could be changed without requiring a change in hardware. Fortunately, digital computer technology began to emerge as a viable technology at about this time. It provided an excellent method of controlling a switching system. This technology had the advantage of offering an alterable control system without any hardware changes. Switches that use computer technology to control the switches are referred to as *stored program control* switches. Thus, one can have an analog switching system controlled by a digital computer; indeed, many examples of switches like this exist today.

As digital transmission proliferated because of its inherent advantages over analog transmission, designers considered how to build a switch differently so it could take advantage of this transmission technology. As the economic motivation, they could eliminate costly analog-to-digital conversion devices (called *channel banks*) if the digital signal could be switched directly, making the provision of telephone service more cost-effective.

Because digital transmission facilities use TDM to organize their bandwidth, a switching function could be performed by rearranging information among the time slots. Figure 5.11 illustrates two multiplexers and a device called a *time slot interchanger* (TSI) to illustrate that this performs a switching function. As the figure shows, the input lines to the multiplexer on the left (A, B, C, D) end up at different output lines at the multiplexer on the right (1, 2, 3, 4). The symbols in the parentheses represent the information transmitted by each channel. The frame shows the information packaged into a frame for transmission. The arrow indicates the direction of transmission. As the result of the operation, the data is rearranged, as shown by the data associated with output channels. Consequently, a switching function is performed on the information. For example, the information from A (represented by &) travels to 4. This is functionally equivalent to closing the switch between A and 4 in Figure 5.10.

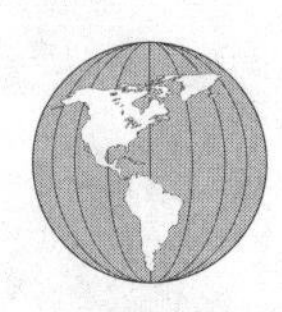

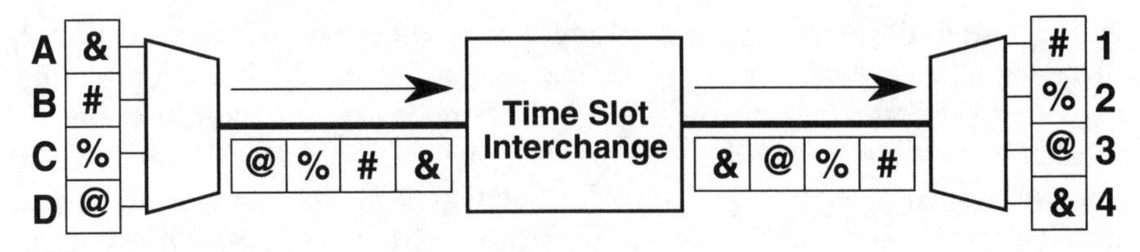

Figure 5.11. Time division switching.

Time slot interchange works by storing all the information of each time slot in memory and then reading the memory in a different order than it was stored. The speed of the memory used by the TSI limits the number of channels that a TSI can support. The faster the memory, the more channels the TSI can support. In practice, a single TSI cannot support more than a few thousand lines. This is far too small for most applications, so switch designers have found it necessary to develop architectures with space stages as well as time stages. In doing so, they have developed switches that can switch digital signals effectively. Switches of this type always use stored program control; all modern switches sold by major vendors are of this architecture.

Signaling

Because these switches are all digital, a new, more efficient form of signaling between switches is desirable. Older space division switches used the actual trunks that the voice channel would use to set up calls. The interoffice signaling was *in-band* and used a modified form of the pulsing technique discussed earlier to pass called number information from switch to switch. Figure 5.12 illustrates the relationship between the signaling channel and a message channel for a simple switching network. As the networks grew in size, the time delays of this arrangement, as well as the reduced trunk utilization efficiency,[8] began to be intolerable. In addition, telephone fraud was much easier with this scheme. Together, these factors resulted in significant revenue losses for telephone companies.

A scheme called *common channel signaling* (CCS) was developed to solve this problem. CCS employs a separate digital signaling channel between switches. All signaling information is transmitted over this single, common channel. This possibility occurs only when all switches used stored program control because the controllers were computers. CCS then amounts to computer-to-computer communications. The network is utilized more efficiently because

no trunks are committed to a call until the path between stations is available, and the destination station is available. Because the signaling rates can be much faster than is possible with in-band signaling, calls can be set up faster, and networks can become larger without efficiency losses.

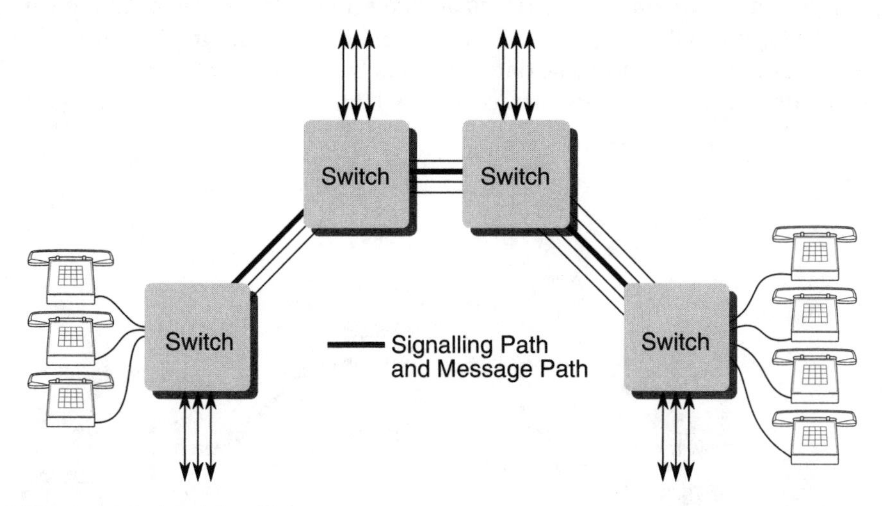

Figure 5.12. *In-band signaling.*

Signaling System Number 7 (SS7) is a common channel signaling method that the CCITT first approved in 1980. In common channel signaling, the call information is sent between switches over a channel dedicated only to signaling: all signals have the same channel in common with each other. The Bell System introduced the *common channel interoffice signaling* (CCIS) system in the 1970s, which was a variant of CCITT Signaling System Number 6. In the meantime, the CCITT began development on SS7.

SS7 defines the protocols and system architecture of a packet network that overlays the circuit-switched voice (or data) network of a common carrier. The nodes of this packet network are called *signal transfer points* (STPs). Several telephone network switches in a region are connected to an STP to pass signaling information. STPs are interconnected with digital circuits. A switch may be connected to multiple STPs for redundancy. When a call is placed on a switch, the SS7 network carries the signaling information from the originating switch to the destination switch by way of the STPs. Unlike the in-band signaling approaches, you can upgrade the capacity of an SS7 network by adding STPs, increasing the data rate of the channels that interconnect STPs or increasing the speed of the STPs themselves.

In addition to providing standard signaling services, SS7 gives users access to enhanced network services. To offer such services, a telephone network must install a *service switching point* (SSP) in the telephone switching offices and service control points (SCPs). The SSPs provide network users with access to network databases, such as an 800 number database, a calling card database, and databases that might define other features in the future. The SCP regulates access to these databases. An SCP enables a third party to ensure only authorized users can access the data.

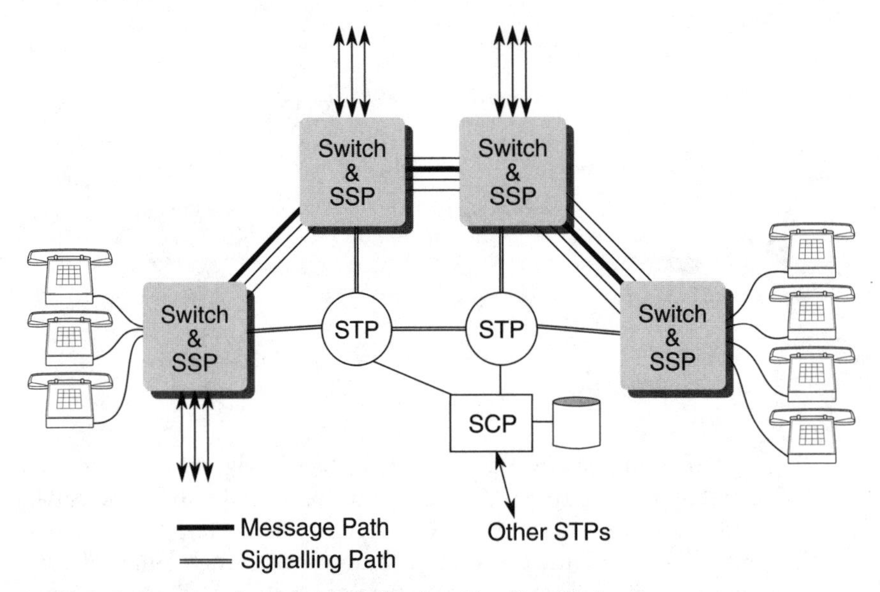

Figure 5.13. *Common channel signaling using Signaling System 7.*

Operating Considerations for International Networks

The operation of international telecommunications networks requires some special considerations and international cooperation. First, a user's telephone number must be globally unique so that a call can be placed easily to any number worldwide. The question is, how can you accomplish this without infringing on the national operating prerogatives of countries? Secondly, given

that globally unique numbers can be assigned, how are calls connected? We will address these issues in the subsequent sections.

International Numbering

The problem with international numbering is assigning globally unique telephone numbers. Countries have historically treated telephone numbering differently and would prefer to continue doing so. To facilitate uniqueness, the CCITT has standardized the international numbering plan in Recommendations E.160 to E.167.

The CCITT specifies that an international number can consist of the following components:

Prefix—This allows the selection of different number formats, transit networks, or services. A prefix is not signalled over international networks.

International Prefix—This allows a calling subscriber to obtain access to outgoing international automatic equipment. Some countries may support multiple international prefixes for different call types or to different locations. Furthermore, if several countries are using an integrated numbering plan, as is the case in North America, the international prefix need not be used among those countries. Normally, you can find the international prefixes for a country in the local telephone directory.

National Prefix—This is used within a country when the called number is outside of the caller's local calling area. This is used to distinguish a local call from a long distance call. The United States uses a "1" for this purpose, and other countries use "0" or "9." France uses "16" for this.

Escape Code—This is used to indicate that the digits following are from a different numbering plan. While this does not find common application in voice networks, it is used to connect the E.164 ISDN numbers with F.69 Telex numbers.

Country Code—This number specifies the country that the caller is calling. This is internationally standardized, and can be found in Recommendation E.163. In the United States, most telephone directories contain listings of country codes

Trunk Code—This number indicates the region in the called country. Some countries refer to these as city codes. In the North American numbering plan, these are Area Codes.

Subscriber Number—This is the number of the called telephone subscriber in the locally accepted format.

An example will help clarify these different codes. To call the International Telecommunications Union in Geneva, Switzerland, the caller must dial: 011-41-22-730-51-11. In this example, 011 is the International Prefix, 41 is the Country Code for Switzerland, 22 is the Trunk Code (or City Code) for Geneva, and 730-51-11 is the ITU's subscriber number. If this call were placed from a location in Geneva, the caller would simply dial 730-51-11. If the call were placed from elsewhere in Switzerland, the caller would dial 0-22-730-51-11. The additional "0" here is the National Prefix, defined nationally as "0." This number is omitted for international access to the Swiss telephone network.

CCITT Recommendations for the International Numbering Plan

E.160: Definitions related to national and international numbering plans

E.161: Arrangement of figures, letters, and symbols on telephones and other devices that can be used for gaining access to a telephone network

E.163: Numbering plan for the international telephone service

E.164: Numbering plan for the ISDN era

E.165: Timetable for coordinated implementation of the full capability of E.164

E.166: Numbering plan working with the ISDN era

E.167: ISDN network identification codes

In addition, Recommendation E.123 proposes standard notation for international telephone numbers. Under this recommendation, both the national and international dialing number should be provided. As you saw in the pervious example, the national number for the ITU should be given as (022) 730-51-11, whereas the international number should be listed as +41-22-730-51-11. The plus (+) indicates the international prefix for the

originating country. Using this notation, the caller need only look up the international prefix, not the country code or the trunk code. E.123 specifies four types of symbols to be used in numbers:

- Dialable symbols
- Procedural symbols
- Information symbols
- Spacing symbols

A *dialable symbol* is a digit that the caller dials. A *procedural symbol* suggests to a caller how to dial. The plus (+) is an example of this type of symbol. Parentheses () are procedural symbols used to indicate digits that are not always dialed, such as the area code in the United States or the trunk (or city) codes in other countries. Parentheses should not be used in an international number, because no part of the number is optional. Another procedural symbol, a slash (/), should be used to separate alternative numbers. A sequence of periods is recommended for direct in-dialing PBX systems, where the number of periods equals the number of digits. A tilde (~) is recommended when an additional dial tone is expected, followed by additional digits to dial. *Information symbols* can be used to inform the caller what to expect, such as an answering machine or an automated attendant. Finally, *spacing symbols* separate parts of the telephone number. A blank space or a dash (-) are the only recommended spacing digits.

Recommendation E.163 specifies the assignment of country codes. To do this, the CCITT first divided the world into nine world numbering zones. Each country code uses its zone number as the first digit. E.163 lists the country codes for all countries currently assigned numbers. Most telephone directories list the country and city codes as well.

Telephone numbering in the ISDN era intends to be more consistent across countries than the current numbering plan is. The ISDN number will consist of a Country Code, a National Destination Code, and a Subscriber Number. The ISDN number is not to exceed 15 digits. Recommendation E.165 specifies that ISDN numbering will be in force on December 31, 1996, at 23:59, Coordinated Universal Time. After this, ISDNs and PSTNs can make use of E.164 numbering to identify their user network interfaces and terminals. Escape codes will be used to facilitate the interworking of non-E.164 networks with E.164 networks.

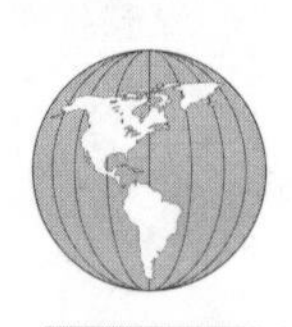

CCITT World Numbering Zones

Zone 1: The United States, Canada, and many Caribbean Islands

Zone 2: African countries

Zones 3 and 4: European Countries

Zone 5: Mexico, Central American Countries, South American Countries, and Caribbean Islands not in Zone 1

Zone 6: Southeast Asian countries, including Australia and New Zealand

Zone 7: Republics of the former Soviet Union

Zone 8: China, Hong Kong, Japan, Korea, Taiwan, other Southeast Asian countries not in Zone 6, and Maritime Mobile Service

Zone 9: India, Pakistan, Turkey, and Middle Eastern countries

International Call Routing

Although the international cable and satellite network has become much more diverse in recent years, not every country is connected to every other country. For example, suppose that no direct connection exists between Japan and Nigeria. For a caller in Japan to establish a connection with a subscriber in Nigeria, another country would need to act as a transit point. To function as a transit point, a country must have an International Switching Center (ISC) that connects to other ISCs. The national administration decides whether to provide an ISC and handle transit traffic. National administrations are under no obligation to do so, although they receive a portion of the call revenue for traffic through their ISC if they do (through the international settlements process, discussed in Chapter 3, "Economic Issues").

If an administration chooses to become a transit facility, it must operate in conformance with CCITT Recommendation E.171, which specifies an approach to international call routing that:

- Preserves the freedom of an administration to route its traffic via any transit administration it chooses and to provide transit to as many destinations as it chooses
- Provides guidance on possible international routings, subject to agreements between the relevant administrations
- Provides a nonhierarchical routing scheme, meaning that all ISCs are at the same level in a switching hierarchy

The key technical specifications of E.171 are

- No more than four international circuits should be connected in tandem to complete a call.
- Route diversity should be provided for fault tolerance.
- The inclusion of two or more satellite circuits should be avoided because of the delay inherent in satellite channels.
- The number of international trunks in tandem should be minimized.
- A country should have multiple ISCs so that transit traffic can be routed efficiently and reliably.
- The originating ISC should use a direct route if possible; otherwise, it should select a transit ISC that minimizes the length of the route and provides no more than four international circuits in tandem.

Therefore, each individual carrier or administration essentially reaches agreements with transit administrations to reach a destination. E.171 recommends that each carrier prepare a list of its international transit capabilities for distribution to other carriers or administrations in transit routing. The sample form contains the destination, the transit ISC, the route type (direct or indirect, by way of another ISC), whether a terrestrial circuit is possible to that destination, and any special restrictions that might exist on that circuit. This information is compiled for every international destination where the administration or carrier wishes to serve as a transit point.

To continue the previous example, the caller in Japan might be routed to an ISC in the United States, which, in turn, might connect to an ISC in Spain, which might connect to an ISC in Nigeria. This international call would consist of three international segments and would involve two transit ISCs (the United States and Spain). The originating, transit, and destination ISCs and administrations share in the revenues generated by the call as specified by previously agreed-to settlement terms.

Other Operational Details

In addition to the recommendations outlined previously, the CCITT also specifies other details of operational importance but of less immediate interest to the end-user. These recommendations are

E.113: Validation procedures for an automated international telephone credit card system

E.116: International telephone credit cards for use in a nonautomated environment

E.121: Pictograms and symbols to assist users of the telephone service

E.126: Harmonization of the general information pages of the telephone directories published by administrations

E.127: Pages in the telephone directory intended for foreign visitors

E.128: Leaflets for distribution to foreign visitors

An Overview of Some Common International Standards

Some standards of current importance to the telecommunications field are briefly described in the following sections. We have referred to articles focussing on these subjects in more detail if you want further information. This discussion is not exhaustive; it is intended to be representative of standards that are currently significant. Some countries, notably the United States, Canada, and Japan, do not adhere to all the international transmission standards. These countries use North American standards, which are often incompatible with international standards.

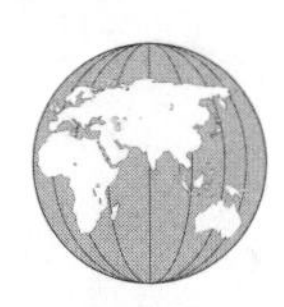

Integrated Services Digital Network

As we discussed previously, ISDN is a network capable of providing end-to-end digital transport and signaling service. One of the key concepts is that, through the widespread application of digital technology, enhanced services can be provided in a manner integrated with the transport of voice and data. This service provision is expected to be achieved using CCITT SS7. The ISDN concept emerged in the late 1970s as digital carrier and digital switching technology proliferated in the telephone network. ISDN is defined in the I-Series standards developed by the CCITT.

ISDN defines several generic equipment types and standardized interfaces. Equipment may be

- TE1 (Terminal Equipment, Type 1)
- TE2 (Terminal Equipment, Type 2)
- TA (Terminal Adapter)
- NT1 (Network Termination, Type 1)
- NT2 (Network Termination, Type 2)
- LT (Line Termination)
- ET (Exchange Termination)

Figure 5.14 illustrates the relationships among these different equipment types. TE1 is an ISDN-compatible terminal device, such as an ISDN telephone or a personal computer with an ISDN circuit card. TE2 is a non-ISDN terminal device, such as the standard analog telephone; a TE2 is attached to an ISDN by a TA. An NT2 is the network termination device belonging to the user. This may be a device such as a PBX switch, which has routing processing capability, or it may be a simple pass-through termination. An NT1 is the network termination owned by the telephone company in some countries. As its primary function, it performs the loop transmission and termination function.[9] The LT is the line termination at the telephone company's CO, and the ET is the gateway into the telephone switching exchange.

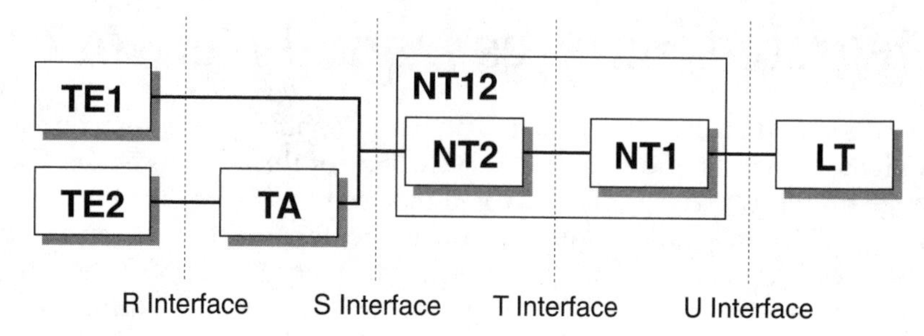

Figure 5.14. The structure of ISDN.

Therefore, the following picture emerges: The local connection begins with the ET, which is connected to the LT. The LT and the NT1 implement the local loop, which terminates on the customer's premises. The NT1 connects to an NT2, which provides additional optional functionality for the users. User devices, TE1 and TE2 (with a TA), connect to either the NT2 or the NT1, thus completing the circuit.

The standard interfaces exist between these equipment types. The "R" interface is the interface between a TE2 and a TA, and the "S" interface exists between a TE1 (or a TA) and an NT2 or an NT1. The "R" and "S" interfaces are roughly equivalent to the RJ-11 connectors currently used to connect a telephone into the telephone jack in the United States. The "T" interface exists between an NT1 and an NT2. The "U" interface defines the local loop, as it represents the interface between an NT1 and an LT. This interface was of considerable interest in the United States because of the regulatory environment, but of lesser interest elsewhere. Finally, the "V" interface defines the interface between the switching equipment (ET) and the loop termination equipment (LT).

From the user's point of view, ISDN currently offers three principle types of channels, bundled in two standard interfaces. ISDN defines a "D" channel as a *data* channel, a "B" channel as a *bearer* channel, and an "H" channel as a *high-speed* channel. The "D" channel is a 16 kbps channel normally used to carry signaling data between the network and the user device. The "B" channels are 64 kbps channels that carry the user's voice or data signals. Finally, the several categories of "H" channels include:

- An H0 channel, which is a 384 kbps channel
- An H10 channel, which is a 1.536 Mbps channel
- An H11 channel, which is a 1.920 Mbps channel

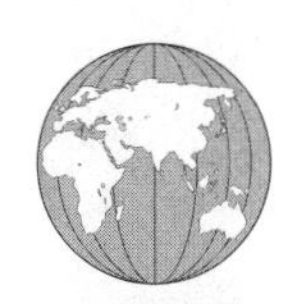

An H11 channel is equivalent to a North American DS1 circuit, and an H12 is equivalent to an E1. Table 5.2 summarizes the ISDN channel types.

Table 5.2. ISDN Channel Rates.

Channel Type	Bit Rate	Type of Switching
B	64 kbps	Circuit
D	16 or 64 kbps	Packet
H0	384 kbps	Circuit
H10	1472 kbps	Circuit
H11	1.536 Mbps	Circuit
H12	1.920 Mbps	Circuit
H21	34 Mbps	Circuit or Packet
H22	45 Mbps	Circuit or Packet
H4	140 Mbps	Circuit or Packet

These channel types are bundled in a *basic rate interface* (BRI) and a *primary rate interface* (PRI). The BRI consists of two "B" channels and a "D" channel (the BRI is referred to as a 2B+D channel), and the PRI normally consists of 23 "B" channels and a "D" channel (that is, 23B+D). The PRI can also be bundled with various combinations of "H1x" and "B" channels as well. In recent years, work has progressed on the definition of a Broadband ISDN capability, where the ISDN will be able to meet the high-speed needs of certain users (such as CAD/CAM or television).

Interconnection and Interface Standards

This category of standards refers to those that define the interfaces between equipment that must work together. The most general representative of this class is the *open systems interconnection* (OSI) reference model, discussed in the next section. Perhaps the most common representative is the EIA-232 computer interface (at the physical layer of the OSI model).

Open Systems Interconnection (OSI) Reference Model

The International Standards Organization (ISO) has developed a "reference model" for the interconnection of "open systems." This model provides a framework for organizing the various data communications functions that must occur between disparate devices that must work together. After its establishment, this reference model has been used as a guideline for developing standards that can allow the interoperation of equipment produced by various manufacturers. Because interoperability is the goal, systems that conform to these standards are referred to as "open systems" (they are open to interoperate with many other systems). Prior to this open systems paradigm, proprietary architectures and protocols dominated the industry. Such proprietary systems are "closed" and stand in contrast to the desire for ubiquitous interoperability.

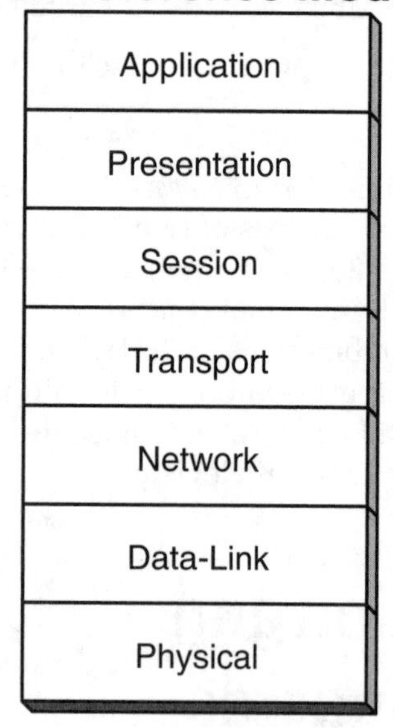

Figure 5.15. *The OSI Reference Model.*

The OSI Reference Model, illustrated in Figure 5.15, defines seven layers of functionality in data communications. We only briefly discuss each of these, because many writers have discussed this model and its components at great length.

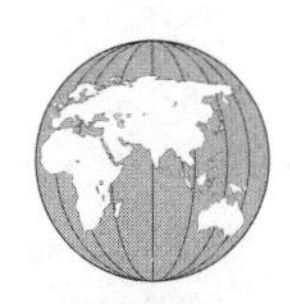

Layer 1 - Physical. The physical layer is concerned with the physical transport of bits. Therefore, physical layer standards include the V.24 interface standard, the modem standards, and certain portions of the ISDN, LAN, and MAN standards.

Layer 2 - Data-Link. The link layer assumes that bit transport occurs: it presumes the physical layer function. Using protocols, such as the high-level data-link control (HDLC) protocol, this corrects errors that might have occurred during transmission across a link. The link layer also supports a minimal amount of addressing, primarily to support multiple devices on a common link.

Layer 3 - Network. The previous two layers were concerned with getting error-free data across a link. Networks generally consist of several links and several nodes. The network layer is used to establish connections between nodes and to route data packets through the network. Network layer addresses must have global significance in the network: only one particular network address can exist in the world (or within the scope of the network).

Layer 4 - Transport. The transport layer is the first layer with an end-to-end orientation: the transport layer presumes the ability to pass through a network and provides additional services to end-users. Most notably, the transport layer enables a user to add end-to-end packet reliability.

Layer 5 - Session. This layer enables users to establish sessions across a network between machines. The session layer offers session management services.

Layer 6 - Presentation. The presentation layer is concerned with code conversions between machines and other data conversion services. Included in these services are data compression and data encryption.

Layer 7 - Application. This layer provides support for the user's network applications. Some application layer standards that have been standardized thus far are Message Handling Services for electronic mail (X.400), Directory Services (X.500), File Transfer and Management (FTAM), and Electronic Data Interchange (EDI). The user's application programs are above the application layer, not part of it.

V.24/EIA-232-C

This standard defines the interface between *data communications equipment* (DCE) and *data terminal equipment* (DTE) at an electrical and physical level. EIA-232-C is familiar to many users of terminals, personal computers, and modems and is as ubiquitous as the 25 pin D-type connector. The counterpart to EIA-232-C in the international standards arena is the CCITT standard V.24. These standards are virtually identical.

In addition to defining the connector itself, V.24 (and EIA-232-C) defines the voltage and current specifications of the signals on the pins. Each pin has a specific, well-defined function. These functions include providing ground and power voltages, data transfer, and device status indicators. The device status indicators are often used to control the flow of data across a V.24 interface.

X.25

The CCITT originally developed the X.25 standard in 1980 with a significant revision in 1984. X.25 defines an interface between DCE and DTE at the network, data-link, and physical layers. It specifies a connection-oriented network service, which requires that users establish a *virtual circuit* before data can be transmitted. Once established, a virtual circuit defines a path that all packets during that data call will follow. This stands in contrast to *connectionless network service* (CLNS). In CLNS, no virtual circuit is established; each packet may take a different path through the network. Because of this, the possibility arises that packets may arrive out of order, as each packet may follow a different path, and experience a different delay.[10]

X.25 has become synonymous with public packet-switched networks; almost all vendors support an X.25 interface. Though this interface is supported, it does not imply that the protocol between network nodes is X.25. In fact, most networks use a proprietary protocol that optimizes network performance.

At the physical layer, X.25 utilizes the V.24 DCE-DTE interface (previously described). At the link layer, X.25 specifies the use of *link access protocol-balanced* (LAP-B). This is a variation of HDLC that requires the use of a peer-to-peer link layer access method. This stands in contrast to an unbalanced mode, which is a primary/secondary access method. In this access method, one station controls the data flow on the link, usually by polling the secondary stations. X.25 currently has a maximum bit rate of 64 kbps.

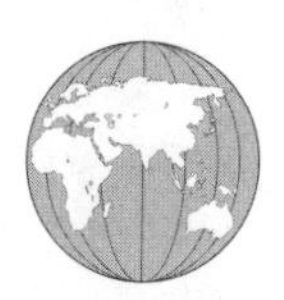

Frame Relay

Many users require higher bit rate service than X.25 can support. They also want the capability to support "bursty" traffic, such as that found in LAN-to-LAN connections. To support these requirements, the frame relay standard was developed.

Frame relay represents a significant departure from X.25 in several respects. X.25 assumes the existence of an unreliable link between nodes. With the widespread installation of fiber-optic cables in many countries, this assumption can be disregarded without significantly affecting network service. Unlike X.25, frame relay operates only at the link layer of the ISO model. In addition, no provision for error correction is made. Consequently, if errors do occur in the network, the end-user machines must detect and correct the error. In a reliable network, this mode of operation does not cause problems, but in an unreliable network, it significantly reduces throughput. Like X.25, frame relay is a connection-oriented service. Unlike X.25, it is currently specified only as a permanent virtual circuit service: the service provider establishes the circuit in advance between specified endpoints. No call setup procedure currently exists at the user interface, although standards committees are currently specifying this, and one should be available in 1993 or 1994. Table 5.3, in the following section, provides a comparison of X.25 and frame relay. Frame relay is currently defined at T1/E1 rates (1.536 to 1.920 Mbps), although this may be extended to T3/E3 (~50 Mbps) in the future.

To use frame relay service, users must have a network device that can support the frame relay service interface, which is a variant of the Link Access Procedure for the ISDN D Channel (LAPD). The user asks the carrier to provide a *committed information rate* (CIR), which is the guaranteed minimum bandwidth available on the network. Users can exceed this CIR for short bursts without violating their agreement. Framed relay currently only allows for permanent virtual circuits (PVCs), in contrast to X.25, which supports both switched virtual circuits (SVCs) and PVCs. A PVC is similar to a leased line in that it must be ordered and provisioned in advance of its use. Users of a PVC can't simply dial the destination address as needed.

Switched Multimegabit Data Service

Switched Multimegabit Data Service (SMDS) is actually a service defined in the United States by Bell Communications Research, Inc. (Bellcore), the research

arm of the Regional Bell Operating Companies. The objective of this effort was to define a service that the operating companies could use within their regions. SMDS is the service that emerged from this effort.

SMDS is a high-speed data service that can serve a regional network and can also be connected into a national network. The data rates supported begin at T1 rates and can extend into the gigabits-per-second range. Unlike frame relay and X.25, SMDS is defined as a connectionless service: users are not required to set up a connection before transmission. An analogy for a connectionless service is the postal system. You can send letters (messages) without first establishing a connection with the distant party. Like frame relay, SMDS is designed to support bursts in traffic.

Table 5.3. Comparison of public-switched data technologies.

	X.25	Frame Relay	SMDS
Link Management	Connection Oriented	Connection Oriented	Connectionless
Speed	64 kbps	T1/E1	>T1/E1
Burst Tolerance	Low	Moderate	High
Circuit Types	SVC PVC	PVC Only	N/A
Geographic Scope	Wide	Wide	Interconnected Regions

Transport Control Protocol/Internet Protocol

During development of ARPAnet, the U.S. Department of Defense developed and defined a protocol suite for this network. (A *protocol suite* is a collection of protocols used together in implementations.) This protocol suite is the *Transport Control Protocol* (TCP) and the *Internet Protocol* (IP). TCP is a transport layer protocol, and IP is a network layer protocol. TCP and IP are normally used together, thus TCP/IP.[11]

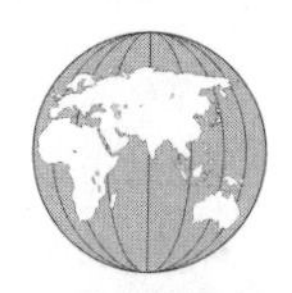

IP is a connectionless network service (CLNS) and, in contrast to X.25, an unacknowledged network service. Because the user does not have an acknowledgment of packet delivery, it is called an unreliable network protocol. Because IP is unreliable, TCP is necessary to provide the end-to-end reliability; in fact, the reliability of TCP is comparable to the highest level of transport service reliability offered by the OSI transport layer protocol.

TCP/IP is popular today in many networks, particularly among universities. Many of the TCP/IP features are included in the UNIX operating system, so it is an easy choice for organizations (such as universities) that are heavy users of that operating system. Furthermore, TCP/IP is one of the protocol suites frequently used in conjunction with LANs.

Standards related to TCP/IP and the associated application layer protocols (FTP, Telnet, SMTP, and so on) are developed by the Internet Engineering Task Force (IETF) and appear in the form of Requests for Comments (RFCs). Douglas Comer discusses the important RFCs that have been developed in *Internetworking with TCP/IP*.[12]

The IETF, a voluntary group of experts, emerged during the development of the ARPAnet project. The IETF operates under the auspices of the *Internet Architecture Board* (IAB), with the Internet Society now serving as the secretariat. These individuals continue to guide the development of this network that connects with other TCP/IP networks, called the Internet, through their participation in IETF. Draft RFCs may come from a wide variety of sources, such as system administrators or researchers. They are normally discussed and enhanced by working groups of the IETF and by other network experts before being published as RFCs. The structure and operation of the IETF is much less formal than the other organizations discussed; it strongly reflects the university research origins of the IETF.

Equipment Standards

Equipment standards are those that specify the technical characteristics of equipment that must interwork. The standards for data modems and fax machines are discussed in the following sections as representatives of this category. Also included in this category, but not discussed in detail, are multiplexers, radios, and so on.

Modem Standards

The CCITT has predominantly defined modem standards . They can be categorized by the maximum bit rate they support and their intended application: the general switched telephone network (GSTN) or leased line. The GSTN modems are

- V.22 (1,200 bps)
- V.22bis (2,400 bps)
- V.26ter (2,400 bps)
- V.32 (9,600 bps)

In some cases, leased line modems are used in half-duplex mode on the GSTN.[13] Leased line modems include

- V.21 (300 bps)
- V.26 (2,400 bps)
- V.27 (4,800 bps)
- V.29 (9,600 bps)
- V.33 (14,400 bps)

Table 5.3 contains the CCITT modem standards. Importantly, in the United States, the most common 1,200 bps dial-up modems conform to AT&T's Bell 212A standard. This standard is similar to—but does not strictly conform to—CCITT V.22. International users of these speeds may have some compatibility problems.

Additional modem standards have been developed. One is V.42, which provides error correction for modems via the *link access procedure for modems* (LAPM) protocol or microcom networking protocol (MNP), which became a de facto standard. A modem that strictly conforms to V.42 interoperates with each of these protocols in a way transparent to the user. In addition, a data compression standard, V.42bis, has been specified that can increase the effective data rate of a modem by up to a factor of four. The actual compression achieved depends heavily on the type of information being transmitted: a great deal of compression is possible for text files, whereas binary files, such as executable files, don't support as much compression. Nevertheless, a V.32bis modem supporting V.42 and V.42bis can achieve effective throughput rates up to 57.6 kbps.

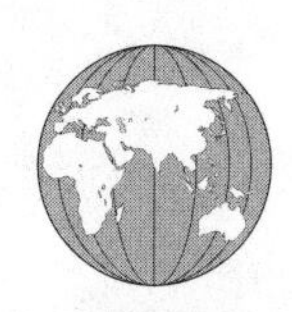

The CCITT is currently developing its last modem standard, referred to in the industry as *V.fast*. This modem standard may be complete by 1993 or 1994 and may support data rates up to 26 kbps over the GSTN. Because this is very close to the theoretical maximum capacity of a telephone line, the CCITT anticipates no need to develop additional standards in the future.

Table 5.4. CCITT Modem Standards.

Standard	Speed	Line Type
CCITT V.21	300 bps	Leased
CCITT V.22	1,200 bps	GSTN
CCITT V.22bis	2,400 bps	GSTN
CCITT V.26	2,400 bps	Leased
CCITT V.26bis	2,400 bps	GSTN (Half Duplex)
CCITT V.26ter	2,400 bps	GSTN
CCITT V.27	4,800 bps	Leased
CCITT V.27bis	4,800 bps	Leased
CCITT V.27ter	4,800 bps	GSTN (Half Duplex)
CCITT V.29	9,600 bps	Leased
CCITT V.32	9,600 bps	GSTN
CCITT V.32bis	14.4 kbps	GSTN
CCITT V.33	14.4 kbps	Leased

Facsimile (Fax) Machines

Facsimile (fax) machines have a long history in relation to the telephone network. Since early in the telephone era, people have felt a need and interest to transmit text and images over the telephone network. Until the 1970s, this process was generally slow (up to several minutes per page) with relatively poor resolution. With the advent of digital techniques, revolutionizing the transmission of text and images became a possibility. Instead of transmitting

the output of analog photodetectors that scanned the source page, the new fax machines scanned the page and encoded it into either a "0" (for white) or "1" (for black). Furthermore, because the image was converted into digital format, designers could apply advanced encoding techniques that took advantage of the redundancy in most pages to reduce the amount of data needed to be transmitted. With these advances, the transmission time can drop to six seconds per page or less. Table 5.5 summarizes the CCITT fax machine standards.

Table 5.5. Summary of facsimile standards.

Type	Technology	Transmission Speed	Maximum Vertical Resolution
Group 1	Analog	6 minutes/page	3.85 lines/mm
Group 2	Analog	3 minutes/page	3.85 lines/mm
Group 3	Digital	1 minute/page	7.7 lines/mm
Group 4	Digital	~30 sec./page	15.4 lines/mm

CCITT Group 1 (Recommendation T.2)

CCITT Recommendation T.2 was first approved in 1968. Group 1 compatible equipment can transmit a page in six minutes. These devices use an analog encoding scheme so that the signal is never converted to digital format.

CCITT Group 2 (Recommendation T.3)

Group 2 fax transmission was approved in 1976 and improves on the Group 1 standards. A Group 2 device can transmit a document in three minutes, as opposed to the six minutes required for Group 1 devices. Group 2 remains an analog technology; it is a marginal improvement in Group 1 technology. Many of the parameters are retained from Group 1. Group 2 fax machines have a scan rate of 360 lines/minute, and they also use analog encoding and transmission.

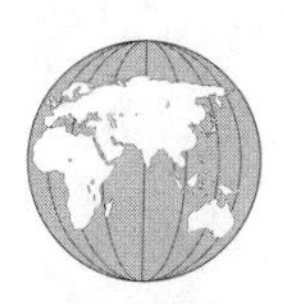

CCITT Group 3 (Recommendation T.4)

The CCITT Group 3 fax transmission system was approved in 1980, and represents a significant departure in fax technology from the previous standards in that it is a digital technology as opposed to an analog one. It offers a page transmission rate of one minute, an improvement by a factor of three over Group 2 technology. CCITT Group 3 compatible devices currently dominate the market for fax machines.

Group 3 technology offers resolution equivalent to CCITT Group 1 and 2 devices (3.85 lines/mm) with an optional mode (7.7 lines/mm) at twice the resolution. Because Group 3 is a digital technology, it specifies the number of picture elements (*pels*) per scan line (1,728) as a function of scan line length. The Group 3 standard increases the length of the scan line by 10mm (.4 in) to 215mm (8.5 in). The resolution, then, of a Group 3 fax is 203 pels/inch (horizontal) by 195 pels/inch (vertical), which approximates to 200x200 pels/inch. Being a digital technology, Group 3 makes use of a modem standardized previously by the CCITT (V.29).

The Group 3 technology uses a compression and encoding technology that takes advantage of the redundancy prevalent in most scan lines. A scan line consists of alternating sequences of black and white pels. The information on the scan line is therefore encoded into the lengths of a run of black or white. Because a scan contains 1,728 pels, you can infer the position of the change in color. This coding technique takes advantage of the horizontal redundancy that exists in many transmitted images.

Group 3 also provides the user with the option of two-dimensional coding. With this type of coding, the coding of a line depends on the previous line. So, the system takes advantage of additional redundancy: vertical in a scanned page. Most commercial machines use this two-dimensional encoding.

CCITT Group 4 (Recommendations T.5 and T.6)

Group 4 fax standards were approved in 1984 and represent a more ambitious project to further expand the scope and capabilities of fax equipment. Group 4 devices are designed for connection to public data networks as opposed to leased or GSTN telephone lines. This allows a greater transmission rate along with a reduced error rate.

Group 4 defines three classes of devices. Class I devices must be capable of transmitting and receiving fax (T.6 and T.73) information; Class II devices must be capable of transmitting fax documents and receiving fax, Teletex (T.60 and T.61), and mixed-mode documents (T.72 and T.73); and Class III devices must be capable of transmitting and receiving fax, Teletex, and mixed-mode documents. Clearly, Class I devices are practical with today's technology (products already exist that conform to this standard), whereas the Class II and Class III devices are future products.

Group 4 devices offer a resolution of 200x200 pels/in.^{2} to 400x400 pels/in.^{2} This means that a scan line contains between 1,728 and 3,456 pels. A Group 3 device offers 1,728 pels/line, with an option for increasing the vertical resolution to 400 pels/in. The nominal density is 200x200, which is identical to Group 1, 2, and 3 fax machines. The optional functions, such as transmission density, are negotiated during the handshaking sequence. Group 4 specifies a two-dimensional coding scheme for the scanned lines. The codes used are identical to those used by Group 3 devices.

The transmission of data for Group 4 devices is over a *packet data network* (PDN). The data rate for this is not specified, because this varies with the data network. The standard only specifies that the network interface will be X.21 compatible.

System Standards

In addition to the equipment and interface standards discussed in the preceding sections, a few system level standards are important as well. These include electronic messaging and directory services for electronic mail as well as electronic data interchange for business transactions.

X.400

The CCITT X.400 series of recommendations defines a system that can be used to handle messages in electronic format. The recommendations define a suite of protocols. Each of these protocols performs a specific function of the message handling system. Collectively, these protocols define the X.400 recommendation. Figure 5.16 specifies the functional model used by X.400 and illustrates the required protocols.

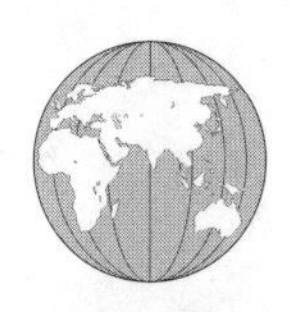

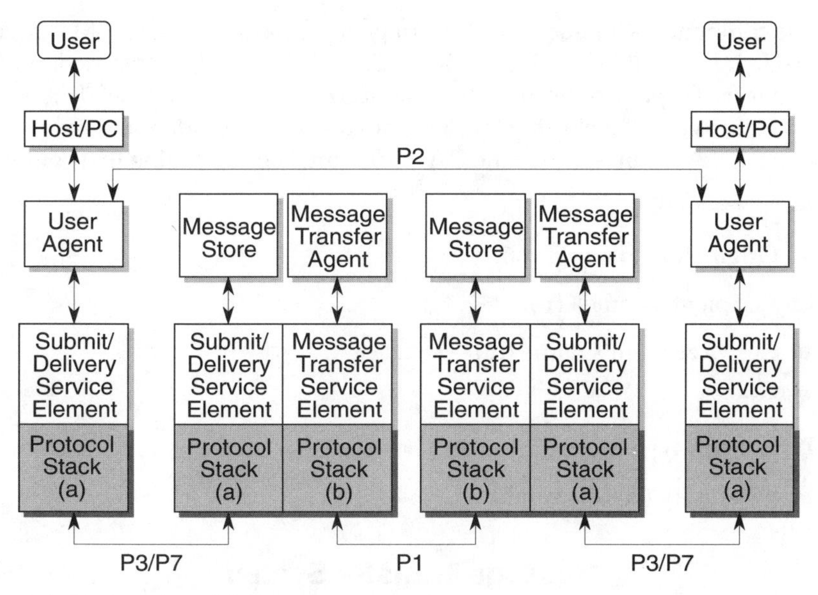

Figure 5.16. *Structure of X.400.*

In the X.400 definition, the user of the electronic mail system interacts with a mail interface under the operating system of the host computer system. This mail interface interacts with the X.400 UA on behalf of the user. X.400 defines these functional components:

- A *user agent* (UA), which is always available to the message transfer system to store messages on behalf of the user and interacts with the user's host system.
- A *message transfer agent* (MTA), which resides on intermediate mail delivery systems and transfers messages to the next MTA or end UA.
- A *message store* (MS), which exists in conjunction with the MTA on intermediate systems, and is used to store messages until the next MTA or end UA is available.
- A *submit/deliver service element* (SDSE), which is the interface of the UA to the message transfer system. The SDSE is essentially an application layer function that serves the UA or the MTA in the connection with a UA.
- A *message transfer service element* (MTSE), which is an application layer function that operates internally to the mail system (between MTAs).

These components interact using a variety of protocols: P1 (MTSE-MTSE), P2 (UA-UA), P3/P7 (SDSE/SDSE). Figure 5.17 illustrates the construction of a message as it is passed through the system. The contents of an X.400 message can be quite varied, including text, fax, videotext, voice, video, and so on. The user transfers the message to the UA, which attaches a heading that contains

- Message ID
- Originator's name (From)
- Recipient's name (To)
- Copy recipient's name (cc)
- Subject
- Content type
- Sensitivity (confidentiality)

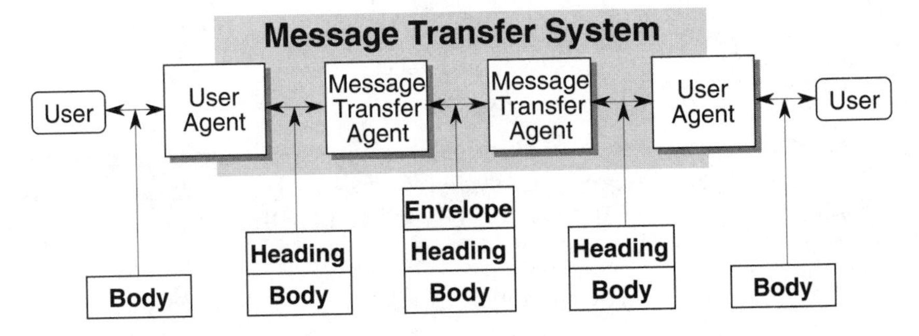

Figure 5.17. Message Construction in X.400.

The UA then transfers the message to the MTA by way of the SDSE. The MTA constructs an envelope for the message, which contains

- A message identifier
- The originator's address
- The recipient's address
- The content type
- An encryption ID

Table 5.6. X.400 Standards.

Number	Title
X.400	System and Service Overview
X.402	Overall Architecture
X.403	Conformance Testing
X.407	Abstract Service Definition
X.408	Encoded Information Type Conversion Rules
X.411	Message Transfer System: Abstract Service Definition and Procedures
X.413	Message Store: Abstract Service Definition
X.419	Protocol Specifications
X.420	Interpersonal Messaging System

X.500 Directory Services

In considering X.400-based electronic messaging in the previous section, we saw that users specify the name of the recipient(s) of the message, and the mail system takes care of delivering that message appropriately. While names and addresses are easy for users to remember, they are not always helpful to computers. Computers frequently need a binary address and a route to follow, as well as ancillary information, such as how to bill the appropriate party for the message and how to implement certain special features, such as acknowledged delivery. For these purposes, a *directory service* is required that enables the mail system to acquire necessary information about the parties in the message.

The CCITT has standardized such a directory service in the X.500 recommendation. X.500 is a hierarchical directory structure capable of providing necessary information worldwide in directory domains that are administered in a distributed fashion. As Figure 5.18 illustrates, the root of the directory service is the world. The next branch of the tree is the user's country, followed by the organization (or organization type), organization unit, department, and

so on, until the users themselves are reached. In their electronic addresses, users must specify this path. The electronic mail system then finds and queries the database for the information it needs to send a message. This resource can be managed in distributed fashion, because the organization, suborganization, and so on can build a structure consistent with the international X.500 structure without surrendering administrative or content control over the information in the X.500 database.

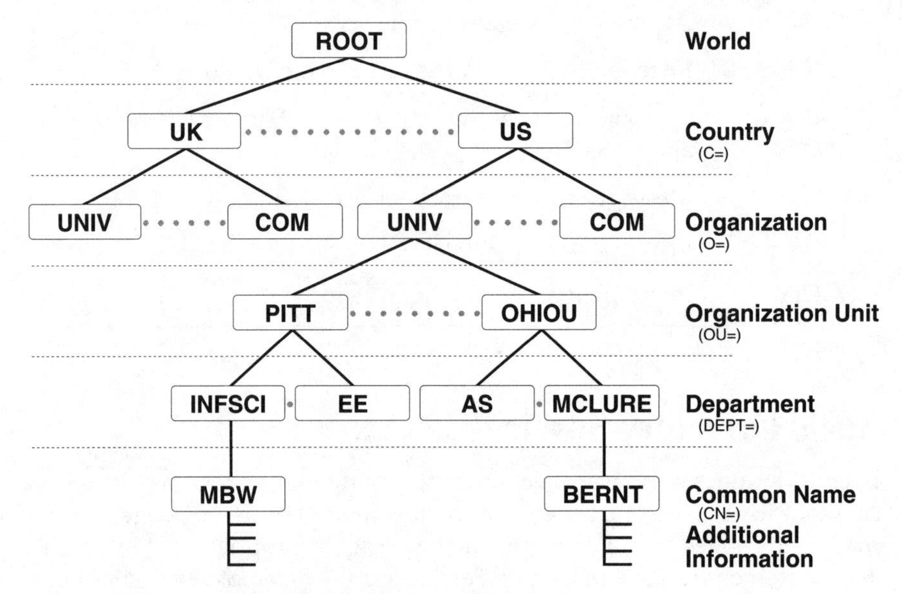

Figure 5.18. *The structure of X.500 directory information tree.*

In principle, a directory service can act as a general source of information for an organization, going beyond simple support for electronic mail. For example, include confidential information in this database if possible, because the administrators of the directory can control access to information contained there. Thus, an electronic mail system may be able to access some aspects of the information in the database, but not others.

A system structure provides directory services, just as a system structure provides mail services. Figure 5.18 illustrates this structure. As illustrated in Figure 5.19, the requesting system or user interfaces with a *directory user agent*

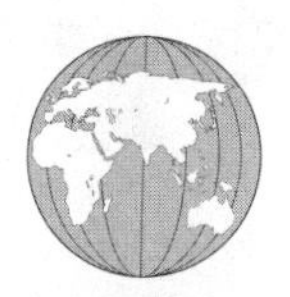

(DUA). The DUA interacts with its *directory service agent* (DSA), operating in another host, containing a portion of the Directory Information Base. If the requested information is in that portion of the database, the DSA responds to the DUA with a message containing the requested information. If not, the DSA refers the request to another DSA. This process is known as *chaining*, and it continues until the requested information is found. DSAs may also submit the request to all other DSAs of which it is aware, a process called *multicasting*.

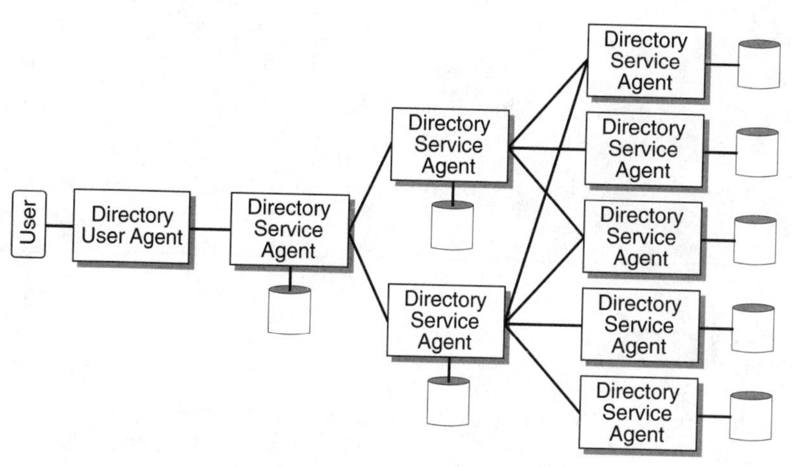

Figure 5.19. *The structure of an X.500 system.*

Summary

In this chapter, we briefly examined some of the key technical issues in telecommunications in general and international telecommunications in particular. We focussed this discussion on wide area networks, omitting the topic of local area networks because they have no significant impact on international telecommunications beyond being sources of traffic. This discussion is not intended to be complete; indeed, a complete treatment of the technical issues in telecommunications would occupy many volumes. We therefore recommend several excellent books on this topic.

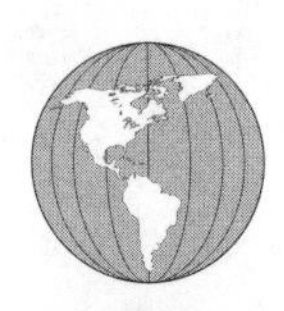

Introductory Books

William Stallings. *Business Data Communications*. New York: Macmillan, 1990.

Sanford Rowe. *Business Telecommunications*. New York: Macmillan, 1991.

Jerry Fitzgerald. *Business Data Communications*, 3rd ed. New York: J. Wiley, 1991.

Intermediate Books

Fred Halsall. *Data Communications, Computer Networks, and Open Systems*, 3rd. ed. Reading, Mass.: Addison-Wesley, 1992.

William Stallings. *Data and Computer Communications*, 3rd. ed. New York: Macmillan, 1991.

Endnotes

1. See for instance William Stallings, *Data and Computer Communications*, 3rd ed. (New York: Macmillan, 1991) or Fred Halsall, *Data Communications and Open Systems*, 3rd ed. (Reading, Mass.: Addison-Wesley, 1992).

2. It seems clear that the implicit requirement for computer use —learning to type—is also daunting to many potential users. The current effort toward graphical user interfaces, pen-based computing, and voice recognition is intended to overcome many of these obstacles.

3. Modems designed for dial-up circuits can actually be used over leased circuits, but the reverse is not true.

4. Using data over voice (DOV) technology has been demonstrated in the United States in selected local areas. It has not been implemented beyond the local area, however, and its data rates are limited compared to the digital local loop.

5. This is essentially equivalent to broadcast radio in many ways. In radio, we have a designated band for transmission (the AM and FM bands, for example), and individual radio stations are assigned to subchannels within that band. Each radio station broadcasts simultaneously and independently of one

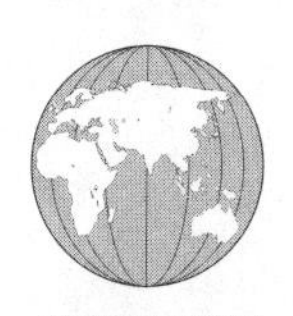

another, and each uses only a small piece of the available bandwidth. To receive a station, the user simply tunes his receiver to the designated frequency, which specifies the subchannel within the band. FDM works very much in the same way.

6. kHz is an abbreviation for kilohertz. One kHz is equivalent to 1,000 Hz, or 1,000 cycles per second. Similarly, MHz is an abbreviation for megahertz, which is equivalent to 1,000,000 Hz.

7. Channel capacities are measured in bits per second (bps) in digital communications. A bit is the fundamental unit in digital, so the capacity is a measure of how many of these fundamental units can be transmitted in one second. As mentioned before, kbps refers to 1,000 bps, and Mbps refers to 1,000,000 bps.

8. While a trunk was used to pass signaling information, it could not be used for voice conversations. Telephone companies do not collect revenue for signaling, so a substantial revenue loss occurred with in-band signaling.

9. The distinction between NT1 and NT2 is not necessary in every country. In countries where users must lease PBXs and other CPE from the telephone administration, the NT1 and NT2 may be combined into a device called an NT12.

10. X.25 originated in the CCITT and was specified only through the network layer. This is a case where the culture of the organization is reflected in the content of the standard. The connection-oriented approach is consistent with the traditional telephone approach: the establishment and termination of connections. Furthermore, telephone administrations are accustomed to providing network service to end-users. Thus, the CCITT would be reluctant to specify standards above the network layer.

11. TCP/IP emerged from the computer culture, as opposed to X.25. Therefore, an end-to-end orientation (in TCP) existed, and a connectionless network protocol allowed for greater flexibility in internetworking. Again, the culture in which the standard emerged affected its content.

12. Douglas Comer, *Internetworking with TCP/IP*, 3rd ed. (Prentice Hall, 1992).

13. For example, the Group 3 fax machine that currently enjoys market popularity includes a V.29 leased line modem, even though its application is via the GSTN.

Assessment Methodology

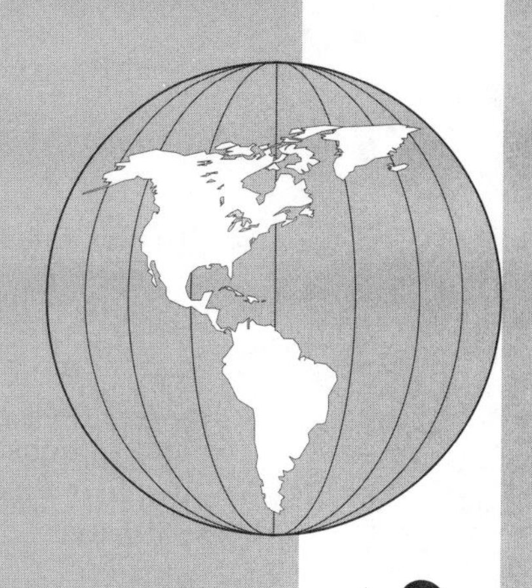

CHAPTER 6

It should be clear at this point that many factors influence international telecommunications. Developing meaningful international comparisons is difficult and requires extraordinary attention to detail.[1] Statements such as these, however, aren't particularly helpful in a practical sense for individuals who must operate an international telecommunications network, decide where to locate new facilities, or plan a network in a foreign country. For this purpose, we offer a methodology designed to support you when assessing the telecommunications industry structure, technologies, regulatory style, and level of infrastructure development in a particular country.

Objectives of the Methodology

The following methodology is designed to assist telecommunications managers as they seek and structure information about telecommunications in a foreign country. In developing this methodology, we placed a premium on simplicity and likely availability of information. As a result, a manager should quickly develop a feeling for the state of the telecommunications infrastructure in a country, even though that feeling might not be useful for a complete and fully meaningful comparison with another country.

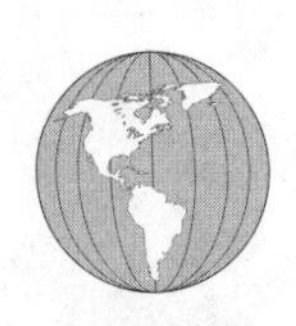

To illustrate the kinds of results that can be achieved by applying this methodology, Chapter 7, "Case Studies," contains some sample applications. We selected countries that we feel are representative of certain trends in telecommunications or certain regions of the world. Telecommunications on a global basis is highly dynamic, so a country-by-country survey of telecommunications would likely be out of date as soon as it is published. Thus, managers need to understand how to evaluate a particular country at a particular point in time.

Managers should understand the following information in order to make an informed choice:

- The state of the infrastructure
- The state of associated facilities
- The quality and quantity of international connections
- Regulations regarding privately owned facilities
- Services available through the public network

Infrastructure is important because it is an indicator of the cost and quality of telecommunications. A poor infrastructure is frequently associated with costly access and poor quality. Associated facilities, such as electricity and roads, are also important. An underdeveloped electrical system, for example, can indicate the need for backup power systems, because the local supply might be unreliable.

International connections are crucial. In developing countries, these connections often consist of satellite earth stations, although fiber-optic cables are increasing throughout the world. Many developing countries are still connected via analog cable systems or a variety of analog radio systems in addition to satellite facilities. The quality of these analog circuits can vary quite a bit and might be unsuitable for data transmission at all but the lowest speeds. These limitations often prompt managers to investigate the purchase and construction of privately owned and operated facilities. These can include the installation of cables or point-to-point microwave systems to connect local facilities or private satellite channels to connect international facilities. Because these private facilities bypass the PTT and result in a loss of revenue, they are often regulated in some way.

The regulatory environment and telecommunications industry structure in a country indicate the latitude a telecommunications manager has in meeting the needs of his or her organization. Those nations that place a premium on developing of new technologies and fostering business activity tend to be

those that turn to some degree of privatization and allow a measure of competition in telecommunications services. In those nations, telecommunications managers usually have an easier time setting up a private network or finding economically efficient services to meet specific telecommunications needs.

Sources of Information

Several sources of information are publicly available and can assist managers in assessing telecommunications in a country. You can obtain general information about countries, such as those listed in Chapter 7, from the *World Factbook*, published by the U.S. Government and available from the National Technical Information Service (NTIS) in Springfield, Virginia.

Detailed information about telecommunications in a target country is often available from the embassy of that country. Although this information can have an overly optimistic tone, it is often a useful starting point. Many telecommunications providers prepare annual reports for public relations purposes. These reports often contain useful information.

The International Telecommunications Union (ITU) in Geneva is a good source of information about worldwide telecommunications.[2] For example, the ITU publishes the *ITU Global Directory*, which provides names and addresses of telecommunications officials worldwide. The ITU also publishes the *Yearbook of Common Carrier Telecommunications Statistics*. This is a rich source of information about the telecommunications systems of countries worldwide over the preceding nine-year period. An annually updated list of publications is available from the ITU.

For information on European Telecommunications in particular, the European Community (EC)[3] puts out several publications. The EC publishes general EC-wide documents and statistics of its member countries. Several of these documents are available at no charge. The EC has many offices throughout the world where these documents are available.

The Organisation for Economic Cooperation and Development (OECD)[4] is also a useful source of information. In particular, *Communications Outlook 1990* contains detailed comparative information on member countries. It includes relative tariffs, service quality, and trade information, along with a brief summary of selected information on non-OECD members. Another document entitled *Performance Indicators for Public Telecommunications Operators* is also very useful.

Information on international telecommunications developments and telecommunications in specific countries is often published in trade magazines, such as *Telephony* and *Telephone Engineering and Management*, and academic journals, such as *IEEE Communications Magazine, Telecommunications Policy*, and *Telematics and Informatics*. All these journals are indexed by database services such as ABI/Inform and Dialog. Additionally, the ERIC database contains numerous reports and documents from conferences that might not be indexed elsewhere. These sources are more likely to be specific to an individual country than the documents from the ITU, EC, or OECD, so they are more likely to provide a basis for information about regulatory developments and history. Finally, several consulting firms specialize in international telecommunications.[5]

Methodology

The objective of this methodology is to provide a structured technique for assessing the telecommunications infrastructure of a country. Such an assessment can only be a preliminary one; detailed assessments require much more careful and thorough research or contact with independent consulting firms that specialize in international telecommunications.

We have created a series of six forms to facilitate the application of the methodology. (Appendix C contains a set of blank forms.) The methodology consists of two major components, the construction of a set of quantitative indicators (Forms 1 through 4) and some qualitative background information (Forms 5 and 6). The first set of indicators is useful in performing gross comparisons between countries. Although these indicators have some value, they don't provide a complete or thorough assessment. The qualitative background information is more useful to establish the context in which the gross quantitative indicators can be considered.

Quantitative Indicators

A telecommunications infrastructure lends itself to gross characterization by a number of high-level quantitative indicators. These indicators allow for a high-level infrastructure comparison among countries. In order to keep the analysis manageable and within the scope of available information, this first order analysis is relatively simple.

The key indicators for a telecommunications infrastructure should be broken into two major categories, direct indicators and indicators of supporting infrastructure. The direct indicators are associated specifically with telecommunications equipment and facilities. The indicators of supporting infrastructure are measures of those infrastructure components necessary to operate a viable telecommunications service.

Direct Indicators

Several statistics are useful in assessing a telecommunications infrastructure.[6] In some cases, these indicators are directly available; in others, you must compute them. Some figures aren't widely available, such as the number of international circuits. We include them in our list because, if they are available, they provide useful information. You can compute the following indicators with the help of Forms 1, 2, and 3. The direct indicators are:

- Average telephone penetration
- Average productivity per telephone
- Public telephone ratio
- Waiting time for service (per telephone)
- Average annual growth of network
- Total calls per capita and per telephone
- Number of telephones per international circuit
- Number of calls per international circuit
- Number of international calls per international circuit
- Average annual investment per telephone
- Percentage of subscriber lines served by digital switches
- Percentage of subscriber lines served by stored program control switches
- Percentage of transmission capacity that is fiber optic
- Geographic fiber-optic density ratio
- Per capita fiber-optic route miles

Average telephone penetration is the most widely used indicator in a comparative analysis of telecommunications infrastructure. This figure is normally reported as the number of telephones per 100 residents and usually varies between urban and rural areas. Typically, rural areas have a lower average

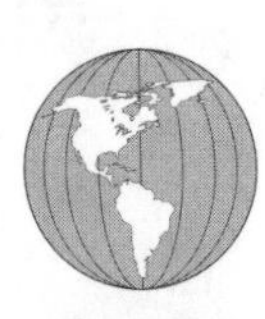

telephone density than urban areas, largely because of the cost of serving decentralized customers. The OECD's *Communications Outlook 1990* reports a low of 0.09 telephones per 100 residents in Chad to a high of 78.93 telephones per 100 residents in Monaco. Typically, an industrialized, "developed" country has in excess of 30 telephones per 1,000 residents, which implies that much of the population has easy access to telephones, and presumably, a variety of other services.

The *average productivity per telephone* is a figure that indicates how efficiently telephones are used in the economy. Although it is difficult to separate economic activity that uses a telephone from that which doesn't, this is a useful, coarse measure. An economy with a high gross domestic product (GDP) per telephone is very efficient, regardless of how much telephone instruments are actually used. Clearly, an economy with a high per capita GDP and a low telephone penetration is a high functioning economy, even though it isn't using telephones widely.

The *public telephone ratio* is an indicator of the number of private telephones. A high public telephone ratio is often associated with a low penetration rate. Public telephones are convenient mechanisms for providing access to many people for a relatively modest investment. Public telephones, however, can't serve all the roles of private telephones. Principally, it is usually difficult to contact someone by telephone when that person relies on a public telephone.[7] A high geographic public telephone ratio (the number of public telephones per square kilometer) is also an indication of a high quality infrastructure.[8] In this view, the public telephone is seen as a vehicle for universal, geographically independent access to telecommunications services.

Waiting time for service is an important indicator and is frequently reported in terms of the length of a waiting list. Waiting time is computed by dividing this by the growth of the network (in lines per month) because requests for service are presumably satisfied only with an expansion in capacity. (See, for instance, the OECD's *Communications Outlook 1990* and *Performance Indicators for Public Telecommunications Operators*.) This waiting time is therefore measured in months. A country with a long waiting time for new service frequently has an impoverished infrastructure; no equipment exists with which to provide service, inefficient bureaucracies are insensitive to the needs of business, or both. Countries with long waiting lists often have significant pent-up demand because not everyone who desires connection will apply if faced with a long waiting list.

The *average annual growth of the network* indicates the willingness of a country to make the requisite investments in the telecommunications network. Here, this indicator is stated in number of new telephones per year. This may be a

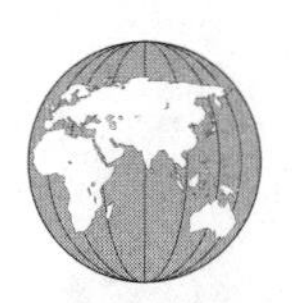

small number for developed countries with a high telephone density. These countries might continue to make substantial investments in infrastructure, but focus on modernization instead of increases in the number of telephones. Therefore, this statistic must be interpreted with some care.

The calling ratios (*calls per capita, calls per telephone,* and the *ratio of international calls to all calls*) are indicators of how comfortable the population in a country is with telephone use. The more telephony that is integrated into an economy, the larger one can expect these numbers to be. In economies with insufficient infrastructure, it might be difficult to complete calls, causing reductions in these figures. Also, economies of scale in telecommunications networks bias these figures in favor of large countries.

The *number of telephones per international circuit* is a figure designed to assess the ease with which an international call can be placed. If the ratio of telephones to international circuits is small, a caller will more likely encounter congestion when making an international call. Clearly, this depends heavily on the calling patterns of the population. Countries with few telephones per international circuit might not place many international calls. Similarly, callers in a country with many international lines per telephone might encounter frequent congestion for international circuits. You can expect a relatively high ratio of telephones to international circuits in regions, such as Europe, where many developed countries have tightly linked economies. There are two important caveats in interpreting this value:

- Due to economies of scale, you can expect more telephones per international line in countries with many telephones than in countries at the same level of service but with fewer telephones.
- If a country carries a lot of transit traffic, additional international circuits must be provided to carry it. This will bias the result toward fewer telephones per international circuit. (The converse would occur if the country carried only originating and terminating international traffic.)

The *calls per international circuit* and *international calls per international circuit* statistics are measures of the intensity of international calls and the relative use of international circuits, respectively. Countries that make fewer calls per international circuit probably provide significant transit capability to other countries; or the population places primarily domestic calls, implying that their facility with international telephony may be limited; or the international capacity is drastically underused. The second statistic attempts to measure this more precisely. Many international calls per circuit is a sign of heavy network use, and therefore a higher likelihood of congestion.[9] Again, economies of scale can result from large numbers of international circuits.

The statistics on Form 3 are attempts to assess the state of the infrastructure based on figures other than performance measures. In particular, these statistics attempt to measure the deployment of critical new telecommunications technologies, such as digital switching and fiber-optic transmission, and the investments made in telecommunications facilities. These statistics are only snapshots. You can obtain a more accurate understanding by building a history for each of these indicators.

The *average annual investment per telephone* is an indication of the amount of investment in telecommunications. Although this figure is intuitively satisfying, several important cautions must be considered when using it on a comparative basis. They are:

1. This figure doesn't differentiate between expansion and modernization, so particular care must be taken when using it to compare developed and developing countries.

2. Some countries have competitive customer premise equipment (CPE) that isn't counted in telecommunications investment, while in other countries, the telecommunications monopoly provides the CPE. If this is the case, CPE might be included in the total investment. Therefore, the investment per telephone for CPE monopolies can be overstated with respect to countries that allow competitive provision of CPE.

3. If this figure from many different countries is being converted to a common currency, such as U.S. dollars, British pounds, or French francs, additional care should be taken. Many analysts recommend the use of purchasing power parities (PPPs) instead of exchange rates to make a more equitable conversion. The reasons for this are discussed in detail in the *NTIA Infrastructure Report* and the *Performance Indicators for Public Telecommunications Operators* (OECD).

Stored program control (SPC) switches and digital switches are important technologies that facilitate modern network services. Because digital switches are always stored program control switches (the converse isn't true), we can discuss the two together. SPC switches first became available in the 1960s when they were used to control analog space division switching fabrics. (See Chapter 5, "Technical Issues," for more detail.) Although digital switches emerged in the 1980s and are now the technology of choice for developing countries, analog SPC switches are still being manufactured and installed worldwide. Thus, the deployment of these technologies is a useful indicator of a network's degree of modernization. The most widely used statistics are the percentage of subscriber lines (telephones or main lines) served by these switches.

Along with digital and SPC switches, fiber-optic cables have emerged in recent years. These cables offer many advantages over their copper- and radio-based predecessors. The deployment of fiber-optic transmission capacity in the telephone network is a useful indicator of the infrastructure's degree of modernization. Several statistics have been proposed for fiber-optic deployment.[10] The *percent of total transmission capacity that is fiber* estimates the penetration of fiber-optic technology in the overall transmission plant. The *geographic fiber density ratio* attempts a gross estimate of the fraction of landmass covered by fiber-optic transmission systems. Finally, the *population fiber density ratio* estimates the portion of the population served by fiber-optic transmission systems. These are clearly very crude measures. Their meaningfulness is somewhat limited because no attempt has been made to measure the fraction of calls carried over fiber transmission channels or where in the network the fiber is deployed. Nonetheless, as first order estimates, these are adequate.

Indirect Indicators

Form 4 suggests four indirect indicators of infrastructure that are related to telecommunications systems. These are *electricity generated per capita, paved road-km per square kilometer, railroad-km per square kilometer*, and *airports with paved runways per square kilometer*. Of these, power production per capita is perhaps most important because all telecommunications systems require power. Power production is also an important measure of the degree to which an infrastructure can support a modern economy, even when alternate sources of power, such as oil or coal, are used. An infrastructure with low power production per capita suggests the need for a backup power system to support telecommunications.

Paved roads, railroads, and airports per square kilometer are useful for assessing the transportation infrastructure in a country. A poor physical infrastructure is often associated with a weak telecommunications infrastructure because breakdowns in remote areas can't be accessed quickly and easily for repair. A poor physical infrastructure suggests the possible need to make other complementary investments before remote operations commence.

Qualitative Background

Although the quantitative indicators are helpful for comparisons, they don't completely represent telecommunications in a country. They don't, for example, indicate progress already made or changes already instituted that

might have positive long-term effects. Quantitative indicators don't represent other features, such as access to international telecommunications facilities, government objectives for foreign investment, the official policy regarding private networks, and the ease with which specific customer needs can be met through either private facilities or the public network.

Major qualitative assessments that are useful include:

- The ownership structure of the public telecommunications administration (whether a state-owned entity or a private enterprise)
- The ownership history of the telecommunications administration
- The entities that own the enterprise (foreign interests, joint ventures, and so forth)
- The existence of a regulatory body in the country
- The relationship of the regulatory body to the telecommunications administration operators
- The existence or extent of competition in CPE and VANs
- The availability of leased lines
- The ability to interconnect leased lines to the public network
- Any restrictions on resale of leased facilities
- The policy regarding private networks
- Memberships in international organizations
- Any differences in levels of service and infrastructure between rural and urban telecommunications development
- State infrastructure development goals
- Pricing relationships between international long distance and domestic long distance services
- Pricing relationships between local and long distance services
- Whether service quality reporting is required
- The availability of VPN, ISDN, and other advanced services as offerings subject to tariffs

Although this list doesn't address all the issues facing a country, it does make possible a first order assessment that complements the quantitative indicators previously discussed. Many factors included in Forms 5 and 6 were

discussed in previous chapters, so they won't be discussed in detail here. However, these factors are meant to provide assistance in assessing the policy climate for telecommunications in a country. That climate can be an indication to someone trying to assess the amount of flexibility and ease with which organizational needs can be met and the level of customer orientation. These forms are informational checklists and can be used to verify that all major areas are addressed when you research the telecommunications infrastructure and policy climate of a country.

Japan as a Sample Case

Figure 6.1. Japan.

To illustrate the application of this methodology, we will apply it to a country—we've chosen Japan. We have filled out each of the forms based on the available information. In the next chapter, we will examine several other countries. Although we won't present the completed forms as we do here, the cases were developed using this methodology.

Excerpts from the *World Factbook*

Statistical Information

Population: 124,017,137 (July 1991), growth rate 0.4 percent (1991)

Total area: 377,835 square kilometers; land area: 374,744 square kilometers; includes Bonin Islands (Ogasawara-guntō), Daitō-shotō, Minami-jima, Okinotori-shima, Ryukyu Islands (Nansei-shotō), and Volcano Islands (Kazan-rettō)

Climate: Varies from tropical in south to cool temperate in north

Terrain: Mostly rugged and mountainous

Literacy: 99 percent (male N/A, female N/A) age 15 and over can read and write (1970 estimate)

Labor force: 63,330,000; trade and services 54 percent; manufacturing, mining, and construction 33 percent; agriculture, forestry, and fishing 7 percent; government 3 percent (1988)

Organized labor: About 29 percent of employed workers; public service 76.4 percent; transportation and telecommunications 57.9 percent; mining 48.7 percent; manufacturing 33.7 percent; services 18.2 percent; wholesale, retail, and restaurant 9.3 percent

Administrative Information

Capital: Tokyo

Administrative divisions: 47 prefectures (fuken, singular and plural); Aichi, Akita, Aomori, Chiba, Ehime, Fukui, Fukuoka, Fukushima, Gifu, Gumma, Hiroshima, Hokkaidō, Hyōgo, Ibaraki, Ishikawa, Iwate, Kagawa, Kagoshima, Kanagawa, Kōchi,

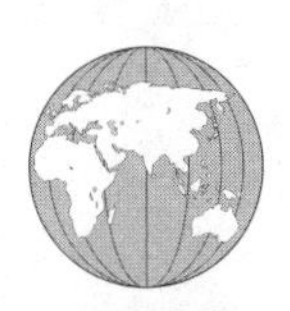

Kumamoto, Kyōto, Mie, Miyagi, Miyazaki, Nagano, Nagasaki, Nara, Niigata, Oita,Okayama,Okinawa, Ōsaka, Saga, Saitama, Shiga, Shimane, Shizuoka, Tochigi, Tokushima, Tōyō, Tottori, Toyama, Wakayama, Yamagata, Yamaguchi, Yamanashi

Legal system: Civil law system with English-American influence; judicial review of legislative acts in the Supreme Court; accepts compulsory International Court of Justice (ICJ) jurisdiction, with reservations

Economic Information

Overview: Although Japan has few natural resources, since 1971 it has become the world's third largest economy, ranking behind the United States and the former USSR. Government-industry cooperation, a strong work ethic, and a comparatively small defense allocation have helped Japan advance rapidly, notably in high-technology fields. Industry, the most important sector of the economy, depends heavily on imported raw materials and fuels. Self-sufficient in rice, Japan must import 50 percent of its requirements for other grain and fodder crops. Japan maintains one of the world's largest fishing fleets and accounts for nearly 15 percent of the global catch. Overall economic growth has been spectacular: a 10 percent average in the 1960s and a 5 percent average in the 1970s and 1980s. In 1990 strong investment and consumption spending helped maintain growth at 5.6 percent. Inflation remains low at 3.1 percent despite higher oil prices and rising wages due to a tight labor market. Japan continues to run a huge trade surplus, $52 billion in 1990, which supports extensive investment in foreign properties.

GNP: $2,115.2 billion, per capita $17,100; real growth rate 5.6 percent (1990)

Inflation rate (consumer prices): 3.1 percent (1990)

Unemployment rate: 2.1 percent (1990)

Technical Information

Electricity: 191,000,000 kw capacity; 790,000 million kwhr produced, 6,390 kwhr per capita (1989)

Railroads: 27,327 km total; 2,012 km of 1.435-meter standard gauge and 25,315 km of predominantly 1.067-meter narrow gauge; 5,724 km of double track and multitrack sections, 9,038 km of 1.067-meter narrow gauge electrified, 2,012 km of 1.435-meter standard gauge electrified (1987)

Highways: 1,098,900 km total; 718,700 km paved, 380,200 km gravel, crushed stone, or unpaved; 3,900 km national expressways, 46,544 km national highways, 43,907 km principal local roads, 86,930 km prefectural roads, and 917,619 other (1987)

Airports: 165 total, 157 usable; 129 with permanent-surface runways; 2 with runways over 3,659 meters; 29 with runways 2,440-3,659 meters; 56 with runways 1,220-2,439 meters

Telecommunications: Excellent domestic and international service; 64,000,000 telephones; broadcast stations—318 AM, 58 FM, 12,350 TV (196 major—1 kw or greater); satellite earth stations—4 Pacific Ocean Intelsat and 1 Indian Ocean Intelsat; submarine cables to the United States (via Guam), Philippines, China, and USSR

Comments on the Forms

Direct Indicators

1. If you use the data from the *World Factbook*, the telephone penetration is 52 percent. The OECD publishes a figure of 42 percent. This latter figure is more reliable. In fact, you can use it to compute the number of telephones based on the population. The number of telephones should be 124,017,137 × 0.42 = 52 million. The OECD figure is based on ITU statistics and internal OECD data, which really must be considered the reference figures. The sources for the data from the *World Factbook* is unknown.
2. Use the figure of 52 million telephones here.
3. The pay telephone figure comes from the OECD *Performance Indicators*. The number of pay telephones is probably an underestimate because it is from 1986. Therefore, the public telephone ratio estimate is probably too large.

4. The OECD *Communications Outlook 1990* indicates that the waiting time is near zero. This saves you from performing the computation.

5. This data is from the *NTIA Infrastructure Report* and is based on OECD and ITU data, so it can be considered reliable. Please note that we changed the "10" to "13" because the reported data covers an interval of 13 years.

6. The OECD *Communications Outlook 1990* provides a bar chart that contains all domestic traffic, which is the sum of local and long distance traffic. Reading the graph yields an estimate of 200,000 million call-minutes of domestic traffic in 1988. The same document includes a bar chart for international traffic, from which you can estimate 500 million call-minutes.

   The questions ask for calls, not call-minutes. For the objective of this question, the distinction is irrelevant, as long as (a), (b), and (c) are all the same. If they are different, you need more complex calculations to equalize them. You can reasonably assume that the average call lasts three minutes. Using this, convert call-minutes to calls by dividing by three, or convert calls to call-minutes by multiplying by three.

7,8. The number of international circuits is unavailable, so you can't perform this computation.

9. The *NTIA Infrastructure Report* reports the average investment per telephone between 1980 and 1989, based on ITU and OECD statistics.

10. The *NTIA Infrastructure Report* reports this figure directly.

11. The *NTIA Infrastructure Report* reports this figure directly.

12. Some of these figures are available, and others are not. The *NTIA Infra-structure Report* provides the geographic fiber density and the population fiber density. From these figures, you can compute the fiber installation as 2,108 million route-kilometers by multiplying the geographic density ratio by the land area (from the *World Factbook*). By dividing the route-kilometers of fiber by the number of telephones, you can obtain the telephone fiber density.

Form 1

Quantitative Statistics

Country: Japan
Date: 12/8/92
Completed By: mbw

A) Direct Indicators

1. Compute the Average Telephone Penetration:
   a) Total Number of Telephones 52 million (derived)

   b) Total Population 124 million

   Average Telephone Penetration = (a) divided by (b) = 42 per 100 (42%)

2. Compute the Average Productivity Per Telephone
   a) Total Gross National Product $2,115 billion
   (or Gross Domestic Product)

   b) Total Number of Telephones 52 million

   Average GNP per Telephone = (a) divided by (b) = $40,670

3. Compute the Public Telephone Ratio
   a) Total Number of Public Telephones 910,000
   (or Coin Telephones)

   b) Total Number of Telephones 52 million

   Public Telephone Ratio = (b) divided by (a) = 57.14

4. Compute Waiting Time for Service per Telephone
   a) Average Waiting Time for new service (months) ________

   b) Total Number of Telephones ________

   Waiting time per telephone = (a) divided by (b) = ~ 0

Figure 6.2. Form 1: Quantitative Statistics.

Form 2

Quantitative Statistics

Country: Japan
Date: 12/8/92
Completed By: mbw

5. Compute Average Annual Growth of Network
   a) Total Number of Telephones 10 years ago 37 million (1987)

   b) Total Number of Telephones today 52 million (1991)

   c) Total Change in Telephones = (b) - (a) = + 15 million

   Annual Average Growth = (c) divided by ~~10~~ 13 *(b) = 2.2%

6. Total Calls per Capita and per Telephone
   a) Total Local Calls } 200,000 million call-min

   b) Total Long Distance Calls }

   c) Total International Calls 500 million call-min

   d) Total Calls = (a) + (b) + (c) = 200,500 million call-min

   e) Population 124 million

   f) Number of Telephones 52 million

   Calls per Capita = (d) divided by (e) = 1,617 call-min

   Calls per Telephone = (d) divided by (f) = 3,856 call-min

   Ratio of Domestic calls to Intl. Calls = (a)+(b) divided by (c) = 400

7. Compute the Number of Telephones per International Circuit
   a) Total number of telephones

   b) Total number of satellite-based international circuits N/A

   c) Total number of cable-based international circuits N/A

   d) Total number of international circuits = (b) + (c) = N/A

   Telephones per International Circuit = (a) divided by (d) = N/A

Figure 6.3. *Form 2: Quantitative Statistics (continued).*

Form 3

Quantitative Statistics

Country: Japan
Date: 12/8/92
Completed By: mbw

8. Compute the Number of Calls per International Circuit
   a) Total number of telephones calls ____________

   b) Total number of International circuits N/A

   c) Total number of International calls N/A

   Calls per International Circuit = (a) divided by (d) = N/A

   International Calls per International Circuit = (c) divided by (b) = N/A

9. Compute Average Annual Investment per Telephone
   a) Annual Telecommunications Investment ____________

   b) Total Number of Telephones ____________

   Average Investment per Telephone = (a) divided by (b) = $243.99 (1980-1989)

10. Compute Fiber Optic Transmission Ratios
   a) Total Number of Fiber Optic Cable Route-Kilometers 2,108 million (derived)

   b) Total Number of Route-Kilometers N/A

   c) Total Land Area 374,744

   d) Total Population 124 million

   Percent of Transmission Capacity that is Fiber = (a) divided by (b) * 100 = N/A

   Geographic Fiber Density Ratio = (a) divided by (c) = 55,89

   Population Fiber Density Ratio = (a) divided by (d) = 17

11. Compute the Percentage of Subscriber Lines that are served by Digital Switches
   a) Number of Subscriber Lines Served By Digital Switches ____________

   b) Total Number of Subscriber Lines
   (Total Number of Telephones) ____________

   Percentage Served By Digital Switches = (a) divided by (b) x 100 = 31%

Figure 6.4. Form 3: Quantitative Statistics (continued).

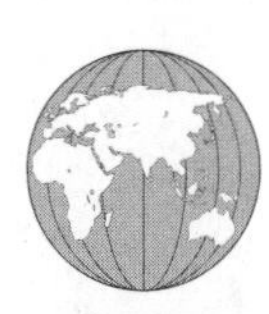

Form 4

Quantitative Statistics

Country: Japan
Date: 12/8/92
Completed By: mbw

12. Compute the Percentage of Subscriber Lines that are served by Stored Program Control Switches
    a) Number of Subscriber Lines Served By SPC Switches ______

    b) Total Number of Subscriber Lines ______
    (Total Number of Telephones)

    Percentage Served By Digital Switches = (a) divided by (b) x 100 = 44.8%

B) Indirect Indicators

1. Compute Electricity Generation per Capita
    a) Total Electricity Generated (kWh) ______

    b) Total Population ______

    Electricity Generated per Capita = (a) divided by (b) = 6,390 KWh

2. Compute Paved Road-km per Square Kilometer
    a) Total Paved Road Kilometers 718,700

    b) Total Land Area 374,744 km²

    Square km per Paved Road km = (b) divided by (a) = 0.52

3. Compute Railroad-km per Square Kilometer
    a) Total Railroad Kilometers 27,327

    b) Total Land Area 374,744 km²

    Square km per Railroad km = (b) divided by (a) = 13.71

4. Compute Airports with Paved Runways per Square Kilometer
    a) Total Airports with Paved Runways 129

    b) Total Land Area 374,744 km²

    Square km per Airport = (b) divided by (a) = 2,905

Figure 6.5. Form 4: Quantitative Statistics (continued).

Form 5

Qualitative Information

Country: Japan
Date: 12/8/92
Completed By: mbw

1. What is the current ownership of public telephone provider?
   - Public
   - Private

2. If the ownership situation has changed,

   When did the change occur? 1985

   What was the nature of the change?
   - Nationalization
   - Privatization

   If the change was privatization,
   - How was the ownership transferred? stock sale
   - Who now owns the enterprise, and in what proportions? 1/3 gov't
   - Is the ownership transfer complete? yes
     - If not, what is the schedule?

3. Is the Customer Premise Equipment (CPE) market competitive?
   Is Type Approval required? yes

4. Is there a VAN market?
   Is this market competitive? yes
   What services are available? most types

5. Are there services that are provided by a private monopoly?
   What are these services?
   - local
   - domestic long distance
   - international long distance
   - packet switching
   - other

6. What services are regulated? local, some long distance
   What is the method of regulation? N/A
   Where in the governmental structure does the regulatory agency reside? MPT
   Under what law does the agency exist? N/A

Figure 6.6. Form 5: Qualitative Information.

Form 6

Qualitative Information

Country: Japan
Date: 12/8/92
Completed By: mbw

7. Are private lines (leased lines) permitted by the carrier? Yes
   - Are restrictions placed on their use? No
   - Do usage-sensitive charges exist? No
   - Is Resale allowed? No

8. Are private networks allowed? Yes
   - Can they be connected to the public network? Data only

9. What international organizations is the government/telecommunications provider a member of?
   - (ITU) [circled]
   - ETSI
   - CEPT
   - (Intelsat) [circled]
   - Eutelsat
   - (Inmarsat) [circled]
   - Other
     OECD

10. How is network investment distributed between rural and urban areas? N/A

11. What are the stated telecommunications infrastructure goals with respect to
    - Telephone penetration? N/A
    - Conversion to digital switching and transmission? 76% by 1994
    - Installation of fiber optic transmission? N/A
    - Conversion to CCITT Signalling System 7? N/A
    - Installation of international transmission capacity? N/A
    - Privatization, deregulation and liberalization (as applicable)? N/A

12. Are service quality reports available or required? Not required
    - If so, what categories are included?

13. What are the pricing relationships among various services?
    - business vs. residential business more than residential
    - local vs. long distance flat measured local rate
    - domestic vs. international relatively inexpensive international Service
    - leased line vs. switched leased line charges reasonable

14. How do the international charges compare with those of other countries?
    Below OECD average

Figure 6.7. *Form 6: Qualitative Information (continued).*

Indirect Indicators

1. This is reported directly in the *World Factbook* excerpt in the previous section.

2-4. The source data is provided in the *World Factbook* excerpt in the previous section.

Qualitative Information

This is available from a number of documents, including *Datapro Reports on International Telecommunications* and the *APEC Report*. The source documents are cited in the next section.

Japanese Telecommunications Statistics

Telephone penetration: 42 percent (OECD statistics)
Average productivity per telephone: $40,670
Public telephone ratio: 57.14:1
Waiting time for service: Negligible
Annual growth of the network: 2 million lines per year
Total calls per capita: 1,617 call-minutes
Total calls per telephone: 3,856 call-minutes
Ratio of domestic to international calls: 400:1
Telephones per international circuit: N/A
Number of calls per international circuit: N/A
Average annual investment per telephone: $243.99 (over a 9-year period)
Percentage of lines served by digital switches: 31
Percentage of lines served by SPC switches: N/A
Fiber penetration: N/A
Geographic fiber density ratio: 55.89:1
Population fiber density ratio: 17:1
Telephone fiber density ratio: 32.94:1
Electricity per capita: 6,390 kwhr
Square kilometer per paved road-km: 0.53
Square kilometer per railroad-km: 13.7
Square kilometer per airport: 2,905

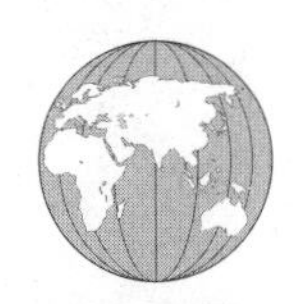

Telecommunications in Japan

Japan is one of the most powerful economies in the world today and one of the most important trading partners worldwide. Japan's telecommunications infrastructure is first-rate. As have most developed countries, Japan has been liberalizing its telecommunications markets in recent years. Japan has a tradition of government involvement in industry; this has also been the case in telecommunications.

Historically, telecommunications in Japan can be divided into four major periods:[11] before 1952, 1952 to 1980, 1980 to 1985, and after 1985. Before 1952, telecommunications services in Japan were provided by the Ministry of Posts and Telecommunications (MPT). In 1952, MPT created the Nippon Telephone and Telegraph Corporation (NTT) for all domestic services and the Kokusai Denshin Denwa Company (KDD) for international services. Until 1980, NTT and KDD were owned by the state and were the sole legal providers of telecommunications, under the regulation of MPT. In 1980, discussions began regarding the reform of telecommunications in Japan; these discussions culminated in the 1985 reforms. These reforms were codified in two legislative acts passed in 1984, the Telecommunications Business Act and the NTT Corporation Act. The former law dealt with the Japanese telecommunications market structure in general, and the latter with the privatization of NTT in particular.

The NTT Corporation Act made NTT a private, shareholder-owned company, with the proviso that shares may not be owned by foreigners and that the government would hold one-third of the shares.[12] Due to its status as the dominant carrier, NTT is obligated to provide universal service and must submit an annual business plan to MPT. MPT can then order NTT to fulfill the terms of this business plan.

Industry Structure and Regulation

Currently, the Telecommunications Business Act allows for two types of carriers, Type I carriers that own facilities and Type II carriers that don't and must lease circuits from Type I carriers. MPT regulates Type I carriers, requiring both fitness and necessity approvals to obtain operating permission. The law places several restrictions on Type I carriers, including:

- A requirement that no more than one-third of a Type I carrier may be owned by foreign interests.

- A Type I carrier may not, without reason, refuse service to anyone in its service territory.
- A Type I carrier must provide "proper service quality."
- A Type I carrier must submit tariffs to MPT for approval prior to offering or discontinuing services.
- A Type I carrier must file facilities development plans with MPT on an annual basis.[13]

Type II carriers fall into two categories, Special and General. No ownership restrictions exist for either category. Special Type II carriers are those with extensive national and international facilities. The law originally intended to regulate these strictly, but largely due to pressure from the United States, that hasn't been the case.[14] Currently, Special Type II carriers must notify MPT of changes in tariffs, but they need not receive approval for the tariffs. They also have the "proper service quality" requirement, as do Type I carriers. General Type II carriers need only notify MPT when they start their businesses; they aren't required to file tariffs.

Since the liberalization, many carriers have entered the market. As of April 1991, three new Type I long distance common carriers were competing with NTT. They began as providers of leased lines, although they later expanded into the provision of switched voice services. None of the three is a nationwide provider; rather, they operate in profitable pairings of cities. Other Type I carriers include seven regional carriers, two international carriers, and sixteen mobile service providers. The regional carriers generally provide leased services in specific cities and prefectures, while the international carriers provide nationwide telephone and leased line services to international destinations. The mobile service providers confine their services to specific metropolitan areas and prefectures.

In addition to these Type I carriers, 31 Specialized Type II common carriers were in operation as of April 1991. Of these, one is NTT International and another is NTT Data Systems Corporation. These are competitive divisions of NTT. Also included in this category are non-Japanese firms, such as AT&T JENS Corp., IBM Japan, Sprint Japan, and Nihon Unisys Ltd. Finally, 912 General Type II carriers were in operation as of April 1991.

The terminal equipment market is competitive. Devices attached to the network require type approval by the Japan Approvals Institute for Telecommunications Equipment,[15] which was designated as the examination agency in the Telecommunications Business Law.[16] Twenty-one tests are required for "the prevention of damage to or functional failure of telecommunications circuit facilities, and establishment of clear-cut lines of responsibility."[17]

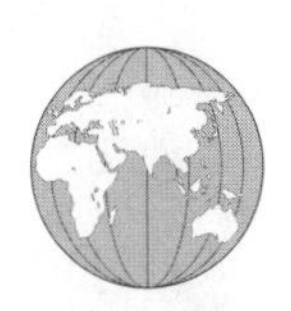

Leased lines are allowed in Japan, a fact made obvious by the large number of carriers providing leased line service. In voice communications, connections between leased lines and the public switched telephone network are prohibited, although these connections are permitted for data communications. Furthermore, two or more private voice networks may not interconnect in Japan, although data networks may. International leased circuits may be connected to switched services at one end for voice services but at both ends for data services. This effectively prevents international resale of voice services.

The State of the Infrastructure

By all measures, Japan's infrastructure is highly developed. Telephone penetration is high, and about 31 percent is served by digital switches. The target for SPC switches is 93 percent of all offices by 1994, with 76 percent of all lines served by digital offices. In addition, 76 percent of the telephone switching offices are currently capable of providing narrowband ISDN services, with a target of 100 percent by 1994. Currently, there are approximately 33 route kilometers of fiber-optics cable per telephone in the transmission plant.

Japan has direct connections to the United States via the following cables: NPC, TPC-3, TPC-4, and TPC-5. Other cable connections include those to Korea and Hong Kong via the HJK cable, and a separate cable to China. Japan also has satellite connections to four Pacific Ocean Intelsat satellites and one Indian Ocean Intelsat satellite.

Deployment of Services

A wide variety of services is available in Japan, from basic switched voice services to advanced high-speed data services, although the more advanced services are available in larger cities. Cellular services are available in all major cities.

Basic voice services are widely available throughout Japan. Local tariffs are based on the duration of a call and are uniform throughout the country. Long distance calls are both usage- and distance-sensitive, with nationwide averaging of costs. Compared with other OECD countries, Japan's business rate is slightly above average, and the residential rates are slightly below average. International tariffs for both business and residential use are well below average.

Little can be said in general about tariffs for leased lines because many Type II carriers exist, each of which is free to make whatever business arrangements it can on a case-by-case basis. Because many competitors exist, it is safe to assume that cost-effective leased line services are available in most parts of Japan.

Conclusion

This methodology was designed to assist a manager in performing a first order assessment of the telecommunications infrastructure of a country. Although this methodology is useful in preparing a briefing or a background paper, we caution the reader from using it as the basis for detailed comparative assessments between countries. Such assessments require a great deal more knowledge of the context, history, and environment of telecommunications in each country. If the reader requires more detailed information about a specific country, we recommend for further reading the *ITU Yearbook of Common Carrier Telecommunications Statistics*, *Performance Indicators for Public Telecommunications Operators*, *Critical Communications: Resources for the Future*, and the *NTIA Infrastructure Report*. Consultants who specialize in international telecommunications also can be helpful.

Endnotes

1. Chapter 5 of the *NTIA Infrastructure Report: Telecommunications in the Age of Information* (U.S. Department of Commerce, National Telecommunications and Information Administration, NTIA Special Publication 91-26, October 1991) contains a discussion of the advantages and disadvantages of methodologies and factors that can be used to perform meaningful international assessments of telecommunications infrastructure. The authors compare the infrastructure of the United States to that of other OECD members and major U.S. trading partners. The quantitative indicators used here are rooted in this discussion.

Performance Indicators for Public Telecommunications Operators, published by the OECD, is a thoughtful discussion of the difficulties involved in comparing international telecommunications infrastructures. This report constructs a set of meaningful and useful indicators. Many of these indicators require

extensive data that might not be readily available. Where this data is available, we have used the indicators recommended in this report.

2. International Telecommunications Union; General Secretariat Sales Section; Place des Nations; CH-1211 Geneva 20; Switzerland.

3. Council of the European Communities; Secretariat general du Conseil; Direction "Information et documentation"; Rue de la Loi 170; B-1048 Bruxelles, Belgium.

4. OECD Publication Service; 2 Rue Andre-Pascal; 75775 PARIS CEDEX 16; France.

5. For example, the Logica Consultancy (Ltd.) maintains a database of international tariffs, although it focuses on European tariffs. Datapro Research (Delran, N.J.) periodically publishes brief but reasonably comprehensive summaries on various countries. Larger consulting firms might have a staff that specializes in international telecommunications or have access to information from their overseas practices. Finally, some groups, such as Lynx Technologies (Little Falls, N.J.), specialize in international telecommunications.

6. In *Performance Indicators for Public Telecommunications Operators*, the following attributes are proposed for assessing service quality:

- Waiting time for a connection to the network
- Effective public telephone density
- Call failure rates
- Fault reports and clearance
- Response times for operator service

7. Institutions can emerge to facilitate this. These include a message infrastructure, so that messages can be left for people whose primary contact is through a public telephone, or "meet-me" arrangements can be made, so that parties agree on a particular time for a call.

8. This argument is made in *Performance Indicators for Public Telecommunications Operators* (OECD).

9. The *International Calls per International Circuit* statistic can be used to estimate blocking probability. If you assume an average call length of five minutes, the total international traffic carried is

(*International Calls*) (5 min)/60 = international traffic (in erlangs—a unit of telephone traffic)

The probability that an international call won't be completed can then be estimated using an Erlang-B traffic engineering table. This is a very crude estimate of blockage because the Erlang-B formula applies to trunk groups on a particular destination switch. Because we don't perform this analysis on a destination-by-destination basis, this figure is only a crude estimate. Additionally, a manager would have to account for transit traffic to obtain a more accurate estimate. Nonetheless, a manager can get a rough idea of what to expect.

10. See the *NTIA Infrastructure Report* for a discussion of these metrics.

11. Makoto Kojo and H.N. Janisch, "Japanese Telecommunications after the 1985 Regulatory Reform," in *Media and Communications Law Review*, (1991), pp. 307-340.

12. Kojo and Janisch, p. 318.

13. Asia-Pacific Economic Cooperation Telecommunications (APEC) Working Group, *The State of Telecommunications Infrastructure and Regulatory Environments of APEC Economies*, November 1991. Available through the Pacific Economic Cooperation Conference; 1755 Massachusetts Ave. NW; Washington, DC 20036.

14. Kojo and Janisch, p. 317.

15. Contact: Japan Approvals Institute for Telecommunications Equipment (JATE); 1-1-2 Toranomon, Minato-ku; Tokyo 105; International Tel: +81-03-591-4300.

16. Ministry of Posts and Telecommunications Japan, *Telecommunications Market of Japan: OPEN*, (March 1989), p. 26. Available in the United States through National Technical Information Service; Springfield, VA 22161. Order Number PB90-100140.

17. Ministry of Posts and Telecommunications Japan, p. 21.

Case Studies

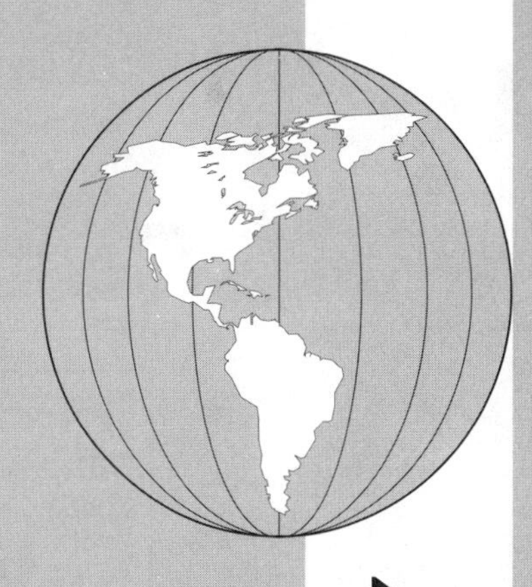

CHAPTER 7

This chapter examines telecommunications in several countries. We selected these countries because they are representative either of their region or of a "classic" situation in telecommunications regulation or privatization. Each section begins with excerpts from the *World Factbook*, published by the U.S. Government.[1]

Clearly, many other countries warranted inclusion here as a case study for their structure or for their reflection of international importance. For example:

- Singapore is the ultimate example of state-directed telecommunications.
- Argentina recently took radical steps toward privatization.
- Chile is considering alternatives for competition in local and interexchange markets.
- China is struggling to rebuild its infrastructure as it becomes an important economic player.
- Saudi Arabia has capital but a very harsh environment.

In the interest of space, we could only include a few.

Germany

Figure 7.1. The reunited Germany.

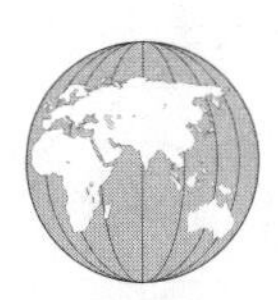

Excerpts from the *World Factbook*

General Information

Capital: Berlin

Total area: 356,910 km^2; land area: 349,520 km^2; comprises the formerly separate Federal Republic of Germany, the German Democratic Republic, and Berlin following formal unification on October 3, 1990

Population: 80,387,283 (July 1992), growth rate 0.5 percent (1992)

Literacy: 99 percent (male NA, female NA) of age 15 and over can read and write (1970 estimate)

Labor force: 36,750,000; industry 41 percent, agriculture 6 percent, other 53 percent (1987)

Organized labor: 47 percent of labor force (1986 estimate)

Administrative Information

Administrative divisions: 16 states (länder, singular—land); Baden-Württemberg, Bayern, Berlin, Brandenburg, Bremen, Hamburg, Hessen, Mecklenburg-Vorpommern, Niedersachsen, Nordrhein-Westfalen, Rheinland-Pfalz, Saarland, Sachsen, Sachsen-Anhalt, Schleswig-Holstein, Thüringen

Legal system: Civil law; judicial review of legislative acts in Federal Constitutional Court; has not accepted compulsary ICJ jurisdiction

Economic Information

Overview: The newly unified German economy presents a starkly contrasting picture. The former West Germany or the Federal Republic of Germany (FRG) has an advanced market economy and is a leading exporter. It experienced faster than projected real growth, largely because of demand in eastern Germany for western German goods. Western Germany contains a highly urbanized and skilled population that enjoys excellent living standards, abundant leisure time, and comprehensive social welfare benefits.

Western Germany is relatively poor in natural resources, with coal being its most important mineral. Western Germany's world-class companies manufacture technologically advanced goods. The region's economy is mature. Manufacturing and service industries account for the dominant share of economic activity, and raw materials and semimanufactured products constitute a large proportion of imports. In 1989, manufacturing accounted for 31 percent of GDP, with other sectors contributing lesser amounts. In recent years, gross fixed investment has accounted for about 21 percent of GDP. In 1990, GDP in the western region was an estimated $16,300 per capita.

In contrast, the obsolete command economy of the former East Germany or German Democratic Republic (GDR), once dominated by smokestack-heavy industries, has undergone a wrenching change to a market economy. Industrial production in early 1991 was down 50 percent from the same period in 1990, largely due to the slump in domestic demand for eastern German-made goods and the ongoing economic restructuring. The FRG's legal, social welfare, and economic systems have been extended to the east, but economic restructuring—privatizing industry, establishing clear property rights, clarifying responsibility for environmental cleanup, and removing Communist-era holdovers from management—is proceeding slowly, deterring outside investors. The region is one of the world's largest producers of low-grade lignite coal but has few other resources. The quality of statistics from eastern Germany remains poor; Bonn is still trying to bring statistics for the region in line with West German practices.

The most challenging economic problem of a united Germany is the reconstruction of eastern Germany's economy—specifically, finding the right mix of fiscal, regulatory, monetary, and tax policies to spur investment in the east without derailing western Germany's healthy economy or damaging relations with Western partners. The biggest danger is soaring unemployment in eastern Germany, which could climb as high as high as 30 to 40 percent. Unemployment could touch off labor disputes, renew mass relocation to western Germany, and erode investor confidence in eastern Germany. Overall economic activity grew an estimated 4.6 percent in western Germany in 1990, while dropping roughly 15 percent in eastern Germany. Per capita GDP in the eastern region was approximately $8,700 in 1990.

GDP: $1,331.4 billion, per capita $16,700; real growth rate 0.7 percent (1990)

Inflation rate (consumer prices): West—3.5 percent (1989); East—0.8 percent (1989)

Unemployment rate: West—6.8 percent (1991); East—11 percent (1991)

Technical Information

Electricity: 7,390 kwhr per capita (1991)

Railroads:

West: 31,443 km total; 27,421 km government-owned, 1.435-meter standard gauge (12,491 km double track, 11,501 km electrified); 4,022 km nongovernment-owned, including 3,598 km 1.435-meter standard gauge (214 km electrified) and 424 km 1.000-meter gauge (186 km electrified)

East: 14,025 km total; 13,750 km 1.435-meter standard gauge, 275 km 1.000-meter or other narrow gauge; 3,830 (estimated) km 1.435-meter double-track standard gauge; 3,475 km overhead electrified (1988)

Highways:

West: 466,305 km total; 169,568 km primary, includes 6,435 km autobahn, 32,460 km national highways (Bundesstrassen), 65,425 km state highways (Landesstrassen), 65,248 km county roads (Kreisstrassen); 296,737 km of secondary communal roads (Gemeindestrassen)

East: 124,604 km total; 47,203 km concrete, asphalt, stone block, of which 1,855 km are autobahn and limited access roads, 11,326 are trunk roads, and 34,022 are regional roads; 77,401 municipal roads (1988)

Inland waterways:

West: 5,222 km, of which almost 70 percent is usable by craft of 1,000-metric ton capacity or larger; major rivers include the Rhine and Elbe; Kiel Canal is an important connection between the Baltic Sea and the North Sea

East: 2,319 km (1988)

Civil air: 239 major transport aircraft

Airports: 655 total, 647 usable; 312 with permanent-surface runways; 4 with runways over 3,659 meters; 86 with runways 2,440-3,659 meters; 95 with runways 1,220-2,439 meters

Telecommunications:

West: Highly developed, modern telecommunications service to all parts of the country; fully adequate in all respects; 40,300,000 telephones; stations—80 AM, 470 FM, 225 TV; 6 submarine coaxial cables; earth stations operating in Intelsat (12 Atlantic Ocean, 2 Indian Ocean), Eutelsat, and domestic systems

East: 3,970,000 telephones; stations—23 AM, 17 FM, 21 TV (15 Soviet TV relays); 6,181,860 TVs; 6,700,000 radios; at least 1 earth station operating in Intelsat and Intersputnik Systems

Telecommunications in Germany

If the United States and the United Kingdom are examples of telecommunications liberalization and privatization, Germany serves as an example of a nation that continues to champion monopoly and state-ownership. Although pressures from the EC, large business users in Germany, and global competitors have edged Germany into adopting liberalization measures and considering the possibility of privatization, these moves haven't met with wholehearted enthusiasm.

Many consider German telecommunications a bastion of monopoly and state control. William Garrison, Jr., in discussing German telecommunications before the reforms of 1989, described Germany as presenting the best example of classic monopoly structure in the delivery of telecommunications and as being the greatest challenger to the procompetitive policies of nations like Japan, the United Kingdom, and the United States.[2] More recently, Minister of Posts and Telecommunications Christian Schwarz-Schilling, who resigned December 17,1992, was asked about the possibility of allowing utilities and railroads to serve as competitive network providers in Germany and about the efficacy of such networks in the newly liberated Eastern Europe nations. He replied:

> *For the time being, there are no plans to permit such companies as alternative network providers in Germany. The Eastern European countries cannot be recommended to open their respective telecom markets to competitive services or networks.*[3]

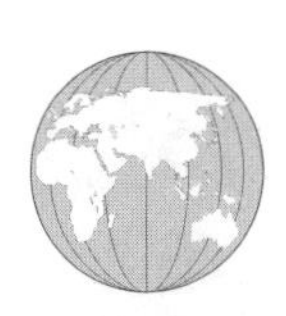

Despite movements toward liberalization, Germany is by no means willing to go as far as some other nations have.

Along with deciding exactly how much liberalization should be introduced into telecommunications, Germany faces another challenge. The former West Germany is undertaking the challenge of merging the telecommunications administrations and infrastructures of East Germany and West Germany. By 1997, the East German network will be equal to the West German network. This undertaking presents no small challenge, since at unification, East Germany had only 1.8 million telephones, or one telephone per eight inhabitants. That ratio in West Germany was one to two. To accomplish its goal, Germany plans to increase telephone installations from 1.8 million to 7 million; to add 68,000 public phones; to build 200 digital central office switches; to set up 360,000 fax connections, 300,000 mobile communications connections, and 50,000 packet-switched connections; and to lay 10 million miles of twisted pairs.[4]

The speed with which this infrastructure rebuilding takes place, the cost of the effort, and the funding mechanism used to pay for the project will have a major impact on German telecommunications. The need to raise capital for this effort might actually move Germany toward greater privatization efforts, just as competitive forces are moving the country to restructure rates and ease restrictions.

German Telecommunications Statistics

Telephone penetration: 55 percent
Average productivity per telephone: $30,074
Public telephone ratio: 272:1
Waiting time for service: No
Annual growth of the network: 6 percent
Total call-minutes per capita: 1,269
Total call-minutes per telephone: 2,304
International calls as a percentage of all calls: 2.6 percent
Average annual investment per telephone: $305.15
Percentage of lines served by digital switches: 38
Geographic fiber density ratio: 0.075
Telephone fiber density ratio: 0.056
Population fiber density ratio: 0.031
Electricity per capita: 7,390 kwhr
Square kilometer per paved road-km: 1.69
Square kilometer per railroad-km: 7.69
Square kilometer per airport: 756

Regulatory and Ownership Approaches

From the beginning of telecommunications in Germany, competition and private ownership were prohibited. The postal and telegraph authorities in Germany were brought together under the leadership of Heinrich von Stephan. The telephone was brought under this umbrella as well, and von Stephan, citing Section 48 of the Imperial Constitution of 1871, rejected all applications for the construction of private telephone networks. The first public telephone exchange was established in Berlin in 1881, and the Imperial Council in that same year established telephone service as the responsibility of the telegraph administration. In 1892, the telegraphy act was amended accordingly.[5]

Telephony has continued, for the most part, to be a state-owned and state-controlled service. This situation caused little controversy until the late 1960s. The Bundespost, which was established by the Postal Administration Law of 1954 to meet the communications needs of the nation,[6] controlled telephone service, the postal service, and a huge postal financial system. In that structure, the telephone service provided the majority of profits and, in effect, subsidized the other Bundespost services. The Bundespost was a part of the Ministry of Post and Telecommunications, so no separation of operator and regulator existed.

Discontent with this situation—with the level of infrastructure investment and the inflexibility of Bundespost procedures—led to a series of commissions that considered whether the Bundespost should be restructured and whether competition should be introduced. The list of commissions over the decades is impressive:

- **1969:** A commission to investigate the organization and role of the Bundespost
- **1974:** A commission on communications technology to discuss establishing a new telecommunications system
- **1981:** The German Monopoly Commission investigation regarding Bundespost market power and problems of monopoly
- **1981:** The Inquiry Commission on New Information and Communications Technologies exploring international aspects of these technologies; it issued report in 1984
- **1985:** The Witte Commission to formulate recommendations for the reform of telecommunications policy[7]

Although little came of the earlier efforts, on July 1, 1989, the Witte Commission recommendations resulted in the Law on Restructuring the Bundespost.

Although the 1989 law didn't fully liberalize telecommunications in Germany, it made some significant changes:

- The Bundespost was removed from the Ministry of Posts and Telecommunications.
- The three areas of the Bundespost (post office, telecommunications, and postal finance system) became three separate entities. Each entity would function independently and, though still state-owned, would relate to its customers and suppliers as if it were a private corporation.
- The telecommunications entity became DBP Telekom; the entity would contribute 10 percent of its income to the federal coffers.
- DBP Telekom retained a monopoly over voice transmission.
- No alternative physical networks could be set up. In other words, network competition wasn't allowed.
- Competition in value-added services and limited resale were allowed.
- A second cellular network was approved. Competition in radio paging was also approved.
- The CPE market was liberalized, along with the type approval process.
- Slow-speed satellite services (under 15 kbps) were allowed without regulation.
- Leased lines were made available without a usage component in the pricing.
- In addition to maintaining its voice monopoly, DBP Telekom could offer so-called "mandatory" services and unregulated, competitive services.[8]

The distinction between "mandatory" and unregulated services is instructive because it underlies the government's view of DBP Telekom's statutory role. Mandatory services are those, defined by the Federal Government and offered by Telekom on a nationwide, tariff basis. Unregulated services are those that can be offered at Telekom's option, without government involvement in the

decision. The concept of mandatory services is part of the "infrastructural obligation" held by the former Bundespost (and now passed to DBP Telekom) to make its services available to all subscribers on an equal basis. The preservation of the voice monopoly is meant to preserve that service obligation in two ways. Allowing competition in voice services could result in regional differences in service offerings; competition in voice services would erode the revenues needed to maintain the infrastructure.[9]

The restructuring law of 1989 created a regulatory environment in which the Ministry of Posts and Telecommunications is the regulator of telecommunications services. The Ministry is also responsible for type approvals in CPE. Although competitors vie for the radio paging, cellular, and CPE markets, DBP Telekom also operates in these areas and maintains a monopoly on voice transmission services. Interestingly, the restructuring law made no provision for service quality reporting.

Any discussion of the German telecommunications industry and regulatory environment would be incomplete without mentioning the impact of German Unification in 1990. Unification merged the two telecommunications infrastructures and regulatory bodies. The East German PTT merged with the West German Ministry of Posts and Telecommunications. The East German telecommunications infrastructure—and the employees who provided that infrastructure—merged into DBP Telekom.

At this writing, DBP Telekom remains a state-owned entity. Discussions are taking place, however, regarding the privatization of DBP Telekom. The former Minister of Posts and Telecommunications advocates some level of privatization in order to raise needed capital for upgrading East German facilities and to free DBP Telekom for involvement in foreign joint ventures. In referring to Post Reform II, the Minister stressed the need to get DBP Telekom out of the civil service system and to improve access to capital markets.[10] Because Article 87 of the Basic Law, or constitution, requires that posts and telecommunications be handled by a state agency, privatization requires an amendment to the Basic Law, a move that would require a two-thirds vote of both the German Bundesrat and the Bundestag. According to an official in the German Economics Ministry, privatization would be limited to 49 percent.[11] Dr. Schwarz-Schilling's recent resignation can be seen as a setback to privatization efforts.

Industry Structure

DBP Telekom dominates the telecommunications industry in Germany. It is controlled by a supervisory board comprised of seven representatives each of management, customers, and DBP Telekom staff. The entity is run by a management board comprised of four former Bundespost members and five

managers appointed from the private sector. The management board is divided into nine areas of responsibility including:

Networks
Telephone service
Broadband distribution and mobile service
Nonvoice services
Marketing
Information systems
Personnel
Finance and other support areas
The Telekom 2000 project

The *Telekom 2000 project* is the merger of East and West German telecommunications and the upgrading of East German facilities.

DBP Telekom has a monopoly on voice services and is active in all other areas of telecommunications, including cellular and value-added services. Resale and shared use of leased lines is allowed but only for nonvoice services. Leased circuits can't be connected to the public switched network (PSN), but fixed connections can be connected if they are used for data and text transmission. Value-added services (VAS) are available on a competitive basis, but VAS providers use public network facilities. CPE is provided on a competitive basis, subject to type approval. Nonvoice satellite services are unregulated; however, these facilities can't connect to the public network.

Because of the challenges presented by the eastern German infrastructure, the restriction on using satellite facilities for voice transmission has eased, at least on a temporary basis, for communications inside East Germany and connection of east and west.

The Status of the Infrastructure

The existence, at present, of the two infrastructures, east and west, complicates a discussion of German telecommunications infrastructure. The wide disparity between these two infrastructures is evident:[12]

	East	West
Population	15,000,000	60,000,000
Phone connections	1,800,000	30,000,000
Public phones	30,000	130,000
Fax connections	2,500	500,000+

continues

	East	West
Leased lines	6,000	205,000
Packet switching users	0	50,000+
Public videotext users	0	227,000
Telex connections	20,000	133,000
Mobile communication users	400	200,000

The infrastructure for both east and west is slated to improve; DBP Telekom has plans to invest DM 55 billion in the east and DM 150 billion in the west.[13]

In the west, the infrastructure includes four data transmission networks, including the public network, the public telex network, the public datex networks (both circuit-switched and packet-switched), and the public data network for fixed connections.[14] The west is extending these services into the east. For all practical purposes, the west has achieved universal service; in the east, that isn't yet the case. Before unification, the waiting period for service in the east was 10 to 15 years. The addition of new installations has accelerated quickly since unification.

Germany has 12 videoconferencing centers. Nationwide ISDN is projected for 1993 in the west, but will take longer to introduce in the east. DBP Telekom has established an intelligent network project (U 2000), projected to be operational in 1994, that will be extended to the east. DBP Telekom has also placed an emphasis on fiber deployment through several avenues. Two 20-pair, 565-megabit-per-second fiber-optic grids were undertaken to carry traffic between east and west. An overlay network from north to south is also a priority.[15]

One approach that was adopted to accelerate the process of upgrading the East German network is the creation of a series of "turnkey" projects. In these projects, DBP Telekom supplies a general plan and selects companies to install equipment and build the basic infrastructure to fulfill the plans. As an example, as part of a five-year $30 billion Fiber-in-the-Loop plan, Raynet Corporation was granted a contract to install loops to 48,500 households in several East German cities. Another example of a turnkey project is the installation of Germany's largest and most modern broadband fiber communication system in the East German state of Brandenburg. The project cost $26 million.[16]

In assessing progress made by the beginning of 1992 in the infrastructure building process, the former PTT Minister Dr. Schwarz-Schilling cited a long list of achievements and plans:

- The increase of lines between east and west from fewer than 800 at the end of 1989 to 34,000 in early 1992
- The addition of 500,000 new subscribers in 1991 and plans for another 600,000 in 1992
- The deployment of seven to eight huge digital switching systems to support a digital overlay network
- Plans to build a complete mobile communication network covering 80 percent of the population in 1992
- A pilot project in Leipzig designed to install 1.5 million subscriber lines by 1995 in a fiber-to-the-curb project[17]

Whether all the ambitious plans laid out as part of Telekom 2000 can be realized will depend on the ability to raise the needed capital to attain a modern, integrated infrastructure by 1997.

Deployment of Services

As mentioned earlier, DBP Telekom maintains the monopoly on voice transmission. Local and long distance services, including international long distance, are purchased from DBP Telekom. DBP Telekom also provides data services through either the PSN with the use of a modem, the public circuit-switched network (Datex-L), the public packet-switched network (Datex-P), the telex network, or a fixed connection data network (that is, a leased line network).

DBP Telekom also provides text communications, through telex or teletex. Fax services are available over the PSN, with the CPE available from DBP Telekom or competitors. Bildschirmtext is an interactive videotext system available through DBP Telekom. DBP Telekom also offers a mailbox service called Telebox and a telemetry service called Temex. Temex is a data-over-voice application.

DBP Telekom faces competitors to its mailbox service from other e-mail providers and from internal corporate e-mail systems. Telekom's other VAS face competition as well. In early 1991, there were over 300 international VAS alone.[18]

Cellular service has two providers and soon will have three. DBP Telekom has operated an analog cellular system (C-Netz) since 1986 and operates one

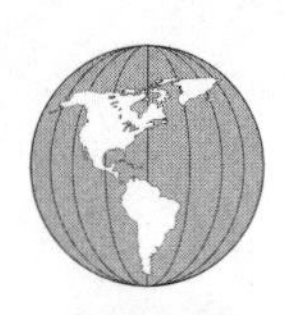

of the digital cellular systems. The second digital cellular license was granted to Mannesmann Mobilefunk, a consortium that includes Pacific Telesis. A third license was tendered for bid in late 1992.

As mentioned earlier, resale of voice transmission isn't allowed, nor are separate physical networks. Resale of nonvoice services, using network facilities, is allowed. Some accommodations are being made to expedite the rebuilding of the East German network. For example, Alpha Lyracom was granted the authority to provide voice satellite links among East German states and between east and west.[19] Otherwise, satellite circuits are limited to nonvoice and can't be connected to the PSN. Another deployment strategy involves the use of trunked mobile radio services. DBP Telekom and other licensees are expected to invest $90 million in order to reach 300,000 East German subscribers.[20]

The provision of CPE has been liberalized. Customers can use the modem, PBX, or fax machine of their choice, as long as the equipment has type approval. The liberalization of CPE was facilitated by EC pressure. EC pressure in 1985 broke the Bundespost monopoly on modems; EC action also ended the monopoly on cordless phones.

The approval process for CPE rests on the assumption that the government must protect the public network from harm. In pursuit of this goal, the Ministry for Posts and Telecommunications establishes conditions for approval that are then applied by the Central Approval Office for Telecommunications (Zentralamt für Zulassungen im Fernmeldewessen, or ZZF) in Saarbrücken. Approval by the ZZF means that equipment may be used as terminal equipment on the public network, connected to a leased line, or used as a radio installation. It doesn't mean that permission is granted to connect the equipment, merely that the equipment meets certain criteria.[21] To ensure that the approval process is an open one, the procedure must meet the requirements specified in the Ordinance on the Type Approval of Telecommunications Equipment of May 1988. Those requirements state that approval must be given if conditions are met, that fees are listed, that guidelines are established and made available, and that approvals may be transferred to another person.

One aspect of service deployment in Germany that has received substantial attention is pricing, especially the pricing of international long distance service and leased lines. OECD studies of relative pricing for these services across member nations place Germany at the high end.[22] Prior to the restructuring law of 1989, leased lines were priced with a usage component, therefore discouraging leased line use and encouraging use of the PSN. The discussion of accounting and settlement rates in Chapter 3, "Economic Issues," explains to some degree the reasons for high international long distance rates.

Although such pricing approaches can fulfill some public policy goals, they also create problems in a competitive environment. Germany is proceeding to rebalance its tariffs. Local rates are increasing while long distance and leased line rates are decreasing. In mid-1992, for example, tariff rates on calls to the United States and Canada were reduced from about $5.89 for a three-minute call to about $3.73. These decreases were part of a "fundamentally new tariff structure" in which rates are lower over longer distances and short-route rates are priced according to the market. In addition, 2-megabit leased lines are slated to be reduced 10 to 15 percent annually starting in 1993.[23] This tariff restructuring is reflective of a move toward a more competitive and cost-based approach.

Hungary

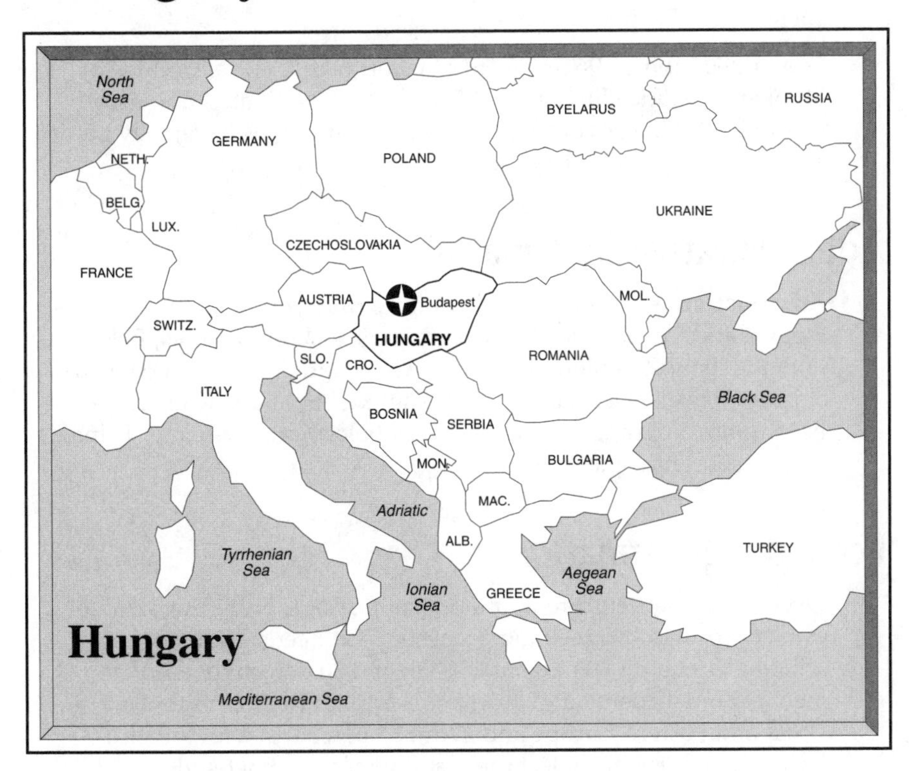

Figure 7.2. Hungary.

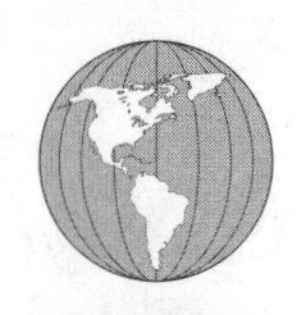

Excerpts from the *World Factbook*

General Information

Capital: Budapest

Total area: 93,030 km^2; land area: 92,340 km^2

Population: 10,333,327 (July 1992), growth rate –0.1 percent (1992)

Literacy: 99 percent (male 99 percent, female 98 percent) age 15 and over can read and write (1980)

Labor force: 5,400,000; services, trade, government, and other 43.2 percent, industry 30.9 percent, agriculture 18.8 percent, construction 7.1 percent (1988)

Organized labor: 96.5 percent of labor force; Central Council of Hungarian Trade Unions (SZOT) includes 19 affiliated unions, all controlled by the government; independent unions legal; may be as many as 12 small independent unions in operation

Administrative Information

Administrative divisions: 19 counties (megyek, singular—megye) and 1 capital city* (fovaros); Bacs-Kiskun, Baranya, Bekes, Borsod-Abauj-Zemplen, Budapest*, Csongrad, Fejer, Gyor-Moson-Sopron, Hajdu-Bihar, Heves, Jasz-Nagykun-Szolnok, Komarom-Esztergom, Nograd, Pest, Somogy, Szabolcs-Szatmar-Bereg, Tolna, Vas, Veszprem, Zala

Economic Information

Overview: Agriculture is an important sector, providing sizable export earnings and meeting domestic food needs. Industry accounts for about 40 percent of GDP and 30 percent of employment. About 40 percent of Hungary's foreign trade is with the USSR and Eastern Europe and about 33 percent is with the EC. Low rates of growth reflect the inability of the Soviet-style economy to modernize capital plant and motivate workers. GDP declined by 1 percent in 1989 and by an estimated 6 percent in 1990.

Since 1985, external debt has more than doubled, to over $20 billion. In recent years Hungary has experimented widely with decentralized and market-oriented enterprises. The newly democratic government has renounced the Soviet economic growth model and plans to open the economy to wider market forces and to much closer economic relations with Western Europe. Prime Minister Antall declared his intention to move forward on privatization of state enterprises, provision for bankruptcy, land reform, and market-based international trade, but concerns over acceptable levels of unemployment and inflation may slow the reform process.

GDP: $60.1 billion, per capita $5,800; real growth rate –7 percent (1990 estimate)

Inflation rate (consumer prices): 34 percent (1990 estimate)

Unemployment rate: 8.0 percent (1991)

External debt: $22.7 billion (1991)

Technical Information

Electricity: 6,967,000 kw capacity; 28,376 million kwhr produced, 2,750 kwhr per capita (1990)

Railroads: 7,765 km total; 7,508 km 1.435-meter standard gauge, 222 km narrow gauge (mostly 0.760-meter), 35 km 1.520-meter broad gauge; 1,147 km double track, 2,161 km electrified; all government-owned (1988)

Highways: 130,014 km total; 29,715 km national highway system—26,834 km asphalt and bitumen, 142 km concrete, 51 km stone and road brick, 2,276 km macadam, 412 km unpaved; 58,495 km country roads (66 percent unpaved), and 41,804 km (estimate) other roads (70 percent unpaved) (1988)

Inland waterways: 1,622 km (1988)

Civil air: 28 major transport aircraft

Airports: 90 total, 90 usable; 20 with permanent-surface runways; 2 with runways over 3,659 meters; 10 with runways 2,440-3,659 meters; 15 with runways 1,220-2,439 meters

Telecommunications: Telephone density is at 17 per 100 inhabitants; 49 percent of all phones are in Budapest; 12- to 15-year wait for a phone; 16,000 telex lines (June 1990); stations—32 AM, 15 FM, 41 TV (8 Soviet TV relays); 4.2 million TVs (1990)

Telecommunications in Hungary

Hungary, like most of Eastern Europe, is engaged in significant reform activities to move from a socialist regime to a market-oriented economic structure. Changes in the telecommunications environment play an important role in these reform activities.

Hungarian Telecommunications Statistics

Telephone penetration: 17 percent
Average productivity per telephone: $34,212
Waiting time for service: 12-15 years
Percentage of lines served by digital switches: 8
Electricity per capita: 2,750 kwhr
Square kilometer per paved road-km: 958
Square kilometer per railroad-km: 11.89
Square kilometer per airport: 1,026

Hungary is regarded by most outsiders as being what the American business magazine *Business Week* called "the brightest star in Eastern Europe."[24] Even before recent events in the former Soviet Union, Hungary was moving toward economic, commercial, and political liberalization. As a result, Hungary is an attractive target for foreign investment. The Market Development Manager of Ericsson Radio Systems, Christina Callmer, praised the country for having a healthy investment climate, commercial vitality, and very high customer demand.[25] This positive climate attracts substantial foreign capital from private companies and loans from development banks.

A significant amount of this capital will be applied to improving Hungary's infrastructure, which is badly in need of improvement. Hungary's telephone system has been called one of the worst in Europe, "with static-filled lines, sudden cutoffs, and phones that won't ring or won't stop ringing."[26] The city of Budapest has the lowest number of telephones of any European capital.[27]

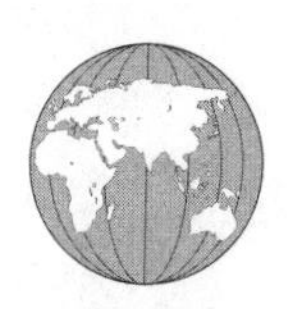

The Hungarian government has ambitious plans for infrastructure building and has taken an aggressive stance in establishing the country as a full member of the global telecommunications community. Hungary is a member of the ITU and has recently joined Eutelsat. The country has applied for membership to Intelsat, ETSI, and CEPT. Efforts to move the country's telecommunications environment closer to privatization and competition haven't moved as quickly as first anticipated because a long-awaited redrafting of the Telecommunications Law faced significant delays in the country's Parliament.

The state of the infrastructure and the vast amount of money needed to upgrade it[28] suggest that, although improvements can be expected in the 1990s, they probably won't be as extensive as the country's planning efforts envisioned.

Regulatory and Ownership Approaches

In late November 1992, the Hungarian parliament passed legislation to pave the way for the privatization of the Hungarian Telecommunications Company, which is also known as HTC or Matav.[29] Hungary's telecommunications administration was part of the Ministry of Transport, Communications, and Water Management until 1983. At that time, it was known as Magyar Posta. Before 1983, Magyar Posta faced no single regulatory body: the Price Office, the National Planning Office, and the Ministry of Finance all had some voice in determining how telecommunication services were provided.[30]

Between 1983 and 1989, Magyar Posta was both operator and regulator of the telecommunications function. In 1989, the Ministry of Transport, Communications, and Water Management took over regulatory functions. The Postal and Telecommunications Inspectorate, under the guidance of the Ministry, deals with "first level regulatory issues," while the Ministry itself is responsible for "second level" issues and for drafting new legislation governing telecommunications.[31]

In early 1990, as one step toward eventual privatization of telecommunications, Magyar Posta split into three independent companies: the Broadcasting Corporation, the Post Office, and Hungarian Telecommunications Company (HTC). While the Post Office will remain state-owned, HTC will eventually be privatized. The Hungarian government's goal is to move from state ownership of 90 percent of the nation's assets to 30 percent state ownership. To accomplish this, Hungary transferred title of state assets to a State Property Agency (SPA) and accorded the State Property Agency the right to privatize those assets. This can be done on the initiative of the state agency controlling the assets, potential investors, or the SPA itself.[32] HTC falls under the authority of the SPA.

Recently, the Hungarian Parliament established the State Asset Holding Company (AV) to manage enterprises that will be wholly or partially state-controlled. AV began operations in October 1992, and HTC is one of the agencies under its control. Also, the Parliament placed HTC on a list of companies in which it intends to keep an ownership interest. The government intends to keep a 50 percent interest for the foreseeable future.

In July 1991, HTC became a 100 percent joint stock company owned by the state and Hungarian banks.[33] After the Parliament's recent actions, a 30 percent share of HTC will probably be sold to Western telephone companies for at least $350 million in late 1993, making Hungary the first Eastern European country to privatize its state telephone monopoly.[34] Alcatel, France Telecom, and Cable and Wireless have all been mentioned as potential buyers. Privatization will provide a welcome infusion of capital.

Two laws are the basis for regulatory developments in telecommunications. The Telecommunications Bill of 1964 was amended to allow the creation of such joint ventures as Westel, the cellular company owned by HTC and US West. A new Telecommunications Law was vigorously debated in Parliament. In October 1992, Parliament drafted a new version of a Telecommunications Law that will leave privatization issues for further discussion and postpone the termination of the monopoly for local and long distance services until after privatization.[35] The eventual regulatory structure of the telecommunications industry in Hungary is expected to be a three-level approach. Monopolies will be maintained, for whatever time period, in basic network services, public telephony, and telex. A form of regulated competition (licensing of carriers) will pertain to such services as data transmission, mobile radio, and paging. Competition will be the approach for CPE and value-added services.[36]

This regulatory structure requires the ability of the state to grant licenses or concessions for various services. That power comes from the second law, which provides some basis for telecommunications regulation: the Hungarian Law of Concession allows the State to grant concessions for the operation and development of specific state assets.

Industry Structure

The final structure of the telecommunications industry is not yet certain. At the present time, the Hungarian Telecommunications Company has, for all practical purposes, a monopoly over long distance service. At the local level, the picture is a little less clear, with a degree of privatization taking place. The Company Act of 1989 made it possible to form new types of organizations in Hungary—joint ventures, subsidiaries, and so forth.[37] Local entities are

taking advantage of such new formats to create their own local telecommunications concerns. Because it is a legal requirement that HTC be involved in all telecommunications activities, HTC is involved in these ventures to some degree. For example, plans are underway for a private local telephone company to serve three districts in northeastern Hungary (Szabolcs, Szatmar, and Bereg Megye). The owners of this concern would be Helsinki Telephone Company, Kontrax (a Hungarian telecommunications firm), and HTC. These same entities provide service to the eastern portion of Budapest through the First Pest City Telephone Company.[38]

HTC provides most network services, including data switching and transmission, a packet-switched service (through a subsidiary called PLEASE), telex services, and international gateway access. Westel, which is a joint venture of HTC and US West, provides cellular service in the 450 MHz band. Two GSM licenses will soon be open for bid as well.

Significant amounts of telecommunications equipment are manufactured in Hungary through joint ventures. Ericsson Technika, a joint undertaking of Ericsson and the Hungarian company Muszertechnika, will produce digital switches and software for the network. Hungary's BGH Telecom intends to manufacture and market telephone subexchanges, transmission equipment, and CPE with a foreign partner.[39] A joint venture between Alcatel Austria and Hungarian Telecom Plc will introduce a 120-line telephone system. Northern Telecom, BHG Telecommunications of Budapest, and Austria Telecommunications are now manufacturing digital switches in Hungary.

The Status of the Infrastructure

The Hungarian telecommunications infrastructure greatly needs development. The telephone penetration rate is low, with 17 telephones per 100 people and about 10 main lines per 100 people. The waiting time for a telephone is believed to be about 12 years. Over 600,000 people are officially registered as being on a waiting list for service, though some estimates place that number close to 1.2 million.[40]

The switching situation isn't yet promising. A digital facility serves only 8 percent of the 1 million lines in the country. Crossbar switches serve 62 percent of the lines, rotary switches serve 23 percent, and manual switching still serves 7 percent.[41] Transmission facilities are either coaxial or analog microwave, though digitization of transmission facilities began in 1986. Call completion is a problem, with completion rates of 53.1 percent for local and 39.4 percent for long distance calls.[42]

The Hungarian government has ambitious plans for remedying the infrastructure situation. The country's Ten Year Development Plan calls for increasing telephone density significantly and for building three million fully automated, digital lines. On a short-term basis, the Three Year Investment Program includes a Basic Project, requiring $1.5 billion to install a digital overlay network using fiber optics and digital microwave, and a Complementary Project, which requires further monies to develop voice telephony.

Hungary has been successful in attracting loan monies from various sources to accomplish some of its infrastructure goals. In 1990, Hungary received a World Bank loan for $150 million and a European Investment Bank loan of $100 million. In mid-1992, the European Bank for Reconstruction and Development gave Hungary a loan of DM 185 million to be allocated as follows: 45 percent toward rural upgrade; 45 percent toward digitization and development of the Erzsevbet exchange in Budapest; and 10 percent toward improving connections between other Budapest exchanges.[43] The European Bank for Reconstruction and Development also granted a $10 million loan to Westel to expand its 450 MHz network.

Although Hungary has attracted significant loan monies, these loans don't provide the billions of dollars needed to get the infrastructure to the level envisioned by the Ten Year Development Plan. Capital from increased service rates and the infusion of money through privatization and joint ventures will also be necessary.

Deployment of Services

The services available in Hungary are pretty basic, and HTC offers most of them.[44] Plans for ISDN development haven't been formulated. Videoconferencing services aren't available. Analog cellular service is available through Westel and digital cellular will be forthcoming as soon as the GSM licenses are tendered.

Telex services are strong and growing, with an automated telex network capable of supporting direct dialing. HTC offers a public circuit-switched data network through a data exchange in Budapest. HTC also offers a packet-switched service. Leased lines for data use are available and can handle speeds of 2,400 bps and 9.6 kbps. HTC provides an international gateway in Budapest through an SPC/circuit-switching exchange. HTC has given NEC a $5 million contract to supply two Intelsat-A earth stations, providing more access to international traffic.

In the local service arena, HTC tariff policy is undergoing some significant changes. For social policy reasons, pricing for service was kept low until the1980s. Unfortunately, low charges didn't generate sufficient revenues for

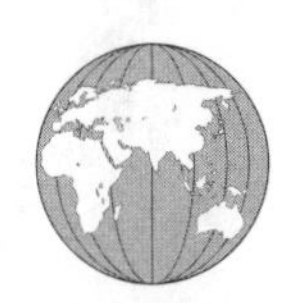

infrastructure development. No service connection charges existed until 1972 for businesses and 1978 for residential customers.[45] These connection charges eventually became very high in relation to other subscriber charges. Recent actions brought the level of connection fees and recurring charges more in line. For example, between 1986 and 1990, connection charges weren't increased, while subscriber charges were doubled for business customers and tripled for residential subscribers. Current plans call for an additional overall 50 percent increase in tariffs and a price-cap regulatory scheme that will index tariff increases to inflation.[46] Tariffs include time-of-day and usage-sensitivity components.

India

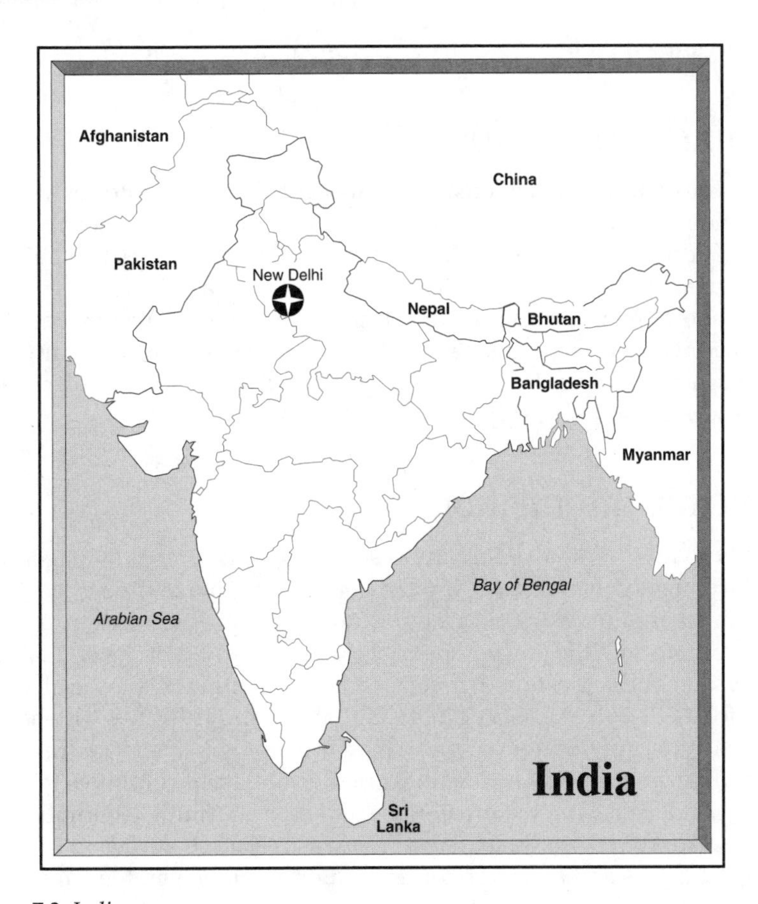

Figure 7.3. India.

Excerpts from the *World Factbook*

General Information

Capital: New Dehli

Total area: 3,287,590 km^2; land area: 2,973,190 km^2

Population: 866,362,180 (July 1992), growth rate 1.9 percent (1992)

Literacy: 48 percent (male 62 percent, female 34 percent) age 15 and over can read and write (1990 estimate)

Labor force: 284,400,000; 67 percent agriculture (FY85)

Organized labor: Less than 5 percent of the labor force

Administrative Information

Administrative divisions: 25 states and 7 union territories*; Andaman and Nicobar Islands*, Andhra Pradesh, Arunāchal Pradesh, Assam, Bihār, Chandīgarh*, Dādra and Nagar Haveli*, Daman and Diu*, Delhi*, Goa, Gujarāt, Haryāna, Himāchal Pradesh, Jammu and Kashmīr, Karnātaka, Kerala, Lakshadweep*, Madhya Pradesh, Mahārāshtra, Manipur, Meghālaya, Mizoram, Nāgāland, Orissa, Pondicherry*, Punjab, Rājasthān, Sikkim, Tamil Nādu, Tripura, Uttar Pradesh, West Bengal

Economic Information

Overview: India's economy is a mixture of traditional village farming and handicrafts, modern agriculture, old and new branches of industry, and a multitude of support services. It presents both the entrepreneurial skills and drives of the capitalist system and widespread government intervention of the socialist mold. Growth of 4 to 5 percent annually in the 1980s has softened the impact of population growth on unemployment, social tranquility, and the environment. Agricultural output continues to expand, reflecting the greater use of modern farming techniques and improved seed that helped to make India self-sufficient in food grains and a net agricultural exporter. However, tens of

millions of villagers, particularly in the south, haven't benefited from the green revolution and live in abject poverty. Industry benefits from a partial liberalization of controls. The growth rate of the service sector also is strong. India, however, has been challenged recently by much lower foreign exchange reserves, higher inflation, and a large debt service burden.

GDP: $328 billion, per capita $380; real growth rate 2.5 percent (1992 estimate)

Inflation rate (consumer prices): 12 percent (1991)

Unemployment rate: 20 percent (1992 estimate)

External debt: $72 billion (1991 estimate)

Technical Information

Electricity: 80,000,000 kw capacity; 290,000 million kwhr produced, 330 kwhr per capita (1990)

Railroads: 61,850 km total (1986); 33,553 km 1.676-meter broad gauge, 24,051 km 1.000-meter gauge, 4,246 km narrow gauge (0.762 meter and 0.610 meter); 12,617 km is double track; 6,500 km is electrified

Highways: 1,970,000 km total (1989); 960,000 km surfaced and 1,010,000 km gravel, crushed stone, or earth

Inland waterways: 16,180 km; 3,631 km navigable by large vessels

Airports: 345 total, 288 usable; 198 with permanent-surface runways; 2 with runways over 3,659 meters; 57 with runways 2,440-3,659 meters; 88 with runways 1,220-2,439 meters

Telecommunications: Poor domestic telephone service, international radio communications adequate; 4,700,000 telephones; stations—96 AM, 4 FM, 274 TV (government controlled); domestic satellite system for communications and TV; 3 Indian Ocean Intelsat earth stations; submarine cables to Malaysia and United Arab Emirates

Telecommunications in India

Unlike the other cases presented in this chapter, we didn't intend for India to be representative of a particular region. Due to its size and population, you might consider India a region of the world in itself. We present India because of its importance to the world as an emerging economy.

Indian Telecommunications Statistics

Telephone penetration: 0.57 percent
Average productivity per telephone: $64,313
Public telephone ratio: 64:1
Waiting time for service: Long (unspecified)
Annual growth of the network: 9.28 percent
Average annual investment per telephone: $1,960
Percentage of lines served by digital switches: 38
Electricity per capita: 830 kwhr
Square kilometer per paved road-km: 3.097
Square kilometer per railrod-km: 48.071
Square kilometer per airport: 15,016

Regulatory and Ownership Approaches

Although telecommunications has existed in India since the 1880s, the history of modern telecommunications begins in 1947, with the independence of India from the United Kingdom. As with Kenya, the colonial rulers established a telecommunications infrastructure suitable for their needs, but not for the needs of the country as a whole. At the time of independence, the network consisted of 86,000 connections for a population of 340 million people.

Consistent with the path that led to independence, India's new government chose to establish self-reliance as a policy goal. For telecommunications, this implied that the country would give preference to equipment designed and manufactured in India. To this end, three state-owned corporations were established: the Indian Telephone Industries (ITI), Hindustan Teleprinters Ltd. (HTL), and Hindustan Cables Ltd. (HCL). Bharat Electronics Ltd. (BEL), which

was supervised by the Ministry of Defence, and the Electronic Corporation of India (ECI) supplied these industries as much as possible.[47] ITI began manufacturing telephones in its Bangalore factory in 1948 and soon followed with the production of Strowger switches for central offices, with technology purchased from the British Associated Telephone and Electricals Co. As the next generation of crossbar-switching technology matured, ITI purchased a crossbar manufacturing technology from the Belgian Bell Telephone Manufacturing Company in 1964. After modification for the hot and dusty Indian telecommunications environment by the Telecommunications Research Centre (TRC), ITI was ready to manufacture these switches in 1982. Around this time, the Sarin Committee determined that digital technology was appropriate for the future of India. The Post and Telegraph Department (P&T) directed ITI to acquire the technology for digital switches and not manufacture the crossbar switches that had taken so long to develop. Soon thereafter, ITI signed an agreement with Alcatel to provide 200,000 lines using Alcatel's E10B system and to transfer the technology and assist in the establishment of a manufacturing facility.[48] This production process is hampered by delays so production targets are consistently missed.[49]

With the emergence of electronic switching, a significant political battle ensued between the Department of Telecommunications (DoT) and the Department of Electronics (DoE). Historically, the DoE was responsible for the production of components and systems; as electronic switching emerged, the DoE thought that the manufacture of these systems should fall under its control, while the DoT felt it was telecommunications-related equipment that, like the previous switches, should be manufactured under its control by ITI. The National Centre for Developments of Telematics (C-DoT) was established in 1984 and jointly funded by DoT and DoE. Since it was formed, C-DoT has designed a small rural switch, the Rural Automatic Exchange (RAX) with 512 ports and a Private Automatic Branch Exchange (PABX) for businesses with 128 ports. The ITI-Alcatel project provided the experience necessary to develop this switch; in fact, the RAX is a scaled-down version of the larger switch of the ITI-Alcatel project.[50]

The equipment market has experienced many changes since independence in 1947. The establishment of C-DoT in 1984 coincided with the deregulation of much of the CPE market, including the PABX market, providing an impetus for indigenously manufactured PABX equipment.[51] Although ITI is still the main producer of telephone sets (450,000 in 1990), deregulation of CPE

resulted in significant competition; production licenses were granted to 32 companies, and import permits were given to allow the sale of foreign telephone instruments. Pay telephones were also deregulated, and 11 Indian companies collaborate with foreign interests to manufacture these devices. Teleprinters were also deregulated, so HTL began facing competition for its devices as well. Also, cable manufacturing, once provided exclusively by HCL, is now open to competition. Several companies have entered the field, often with foreign interests.

After independence, India established the Post and Telegraph Department (P&T) and the Overseas Communications Service (OCS). P&T was responsible for all domestic services, including local and long distance. OCS was responsible for international communications.

Under the leadership of the late Prime Minister Rajiv Ghandi, telecommunications underwent some drastic changes in 1986. The provision of telecommunications services was separated from postal services with the formation of the DoT. The OCS was privatized and became Videsh Sanchar Nigam Ltd. (VSNL), and Mahanagar Telephone Nigam Ltd. (MTNL) took over the New Delhi and Bombay network operations. DoT was separated to end the cross-subsidization of the postal service by the telecommunications service of P&T.[52] MTNL has autonomy in the operations, but DoT handles capital purchases. Because Bombay and New Delhi represented approximately 25 percent of the country's telephone lines, a private corporation was thought to be able to raise money on the capital markets. DoT, as a government agency that obtains money from the government budget, couldn't do this. DoT regulates both MTNL and VSNL.

DoT's current five-year plan is quite ambitious. By 1995, it plans to double the number of telephones in India to 10 million and increase transmission capacity by 30 percent (in route-km). Included in the $12 billion investment target is $250 million for research and development. Other priorities for this plan include:

- The construction of a digital network to connect the 450 DoT district headquarters
- The replacement of obsolete transmission and switching equipment
- The improvement of services in rural areas[53]

Given India's other pressing problems, it's unclear whether the funds to support this ambitious plan for telecommunications will be available.

Industry Structure

Monopolies provide telecommunications services in all cases. DoT provides domestic services, with the exception of Bombay and New Delhi, where MTNL operates the network. VSNL provides international services. As mentioned previously, DoT regulates MTNL and VSNL.

CPE of most types is competitive. Joint ventures with foreign corporations are allowed, and several manufacturers produce equipment in India for the Indian market. Many large international telecommunications concerns have set up joint ventures or entered into licensing agreements with Indian companies.

The Status of the Infrastructure

India's infrastructure is very limited by most measures. The OECD estimates telephone penetration at 0.57 per 100 people. Funding for the many ambitious plans remains limited, given the shortage of foreign exchange and other pressing problems. One problem is the high staffing level per access line (104 per 1,000 lines); this compares with 7 to 12 per 1,000 for developed countries, 20 to 40 for "middle income" countries, and 60 to 90 for other "poor" countries.[54] The self-reliance strategy also limits the adoption of new technologies.

Much of the investment in new technology focuses on the more profitable urban areas. Development in the rural areas is extremely limited: only 10 percent of available access lines serve rural communities that represent 75 percent of the country.[55] The formation of VSNL and MTNL has produced positive results: line failure rates in New Delhi and Bombay are lower, and 164 countries are now reachable by direct dial, up from 19.[56] A packet-switched network exists, but it is largely inadequate.[57]

The network has grown to about 5.1 million lines supported by 14,000 exchanges. Digital microwave and fiber-optic transmission systems are replacing the transmission systems. In addition, INSAT, India's own satellite, provides domestic services. These are necessary improvements. Indian Ocean Intelsat satellites primarily provide international telecommunications.

Deployment of Services

Voice services are the easiest to obtain in India, but often these require long waits. The official waiting list contains two million subscribers. Service quality is relatively poor for those with telephones: in New Delhi, 25 percent of

the telephones break down every month.[58] Data transmission over the analog lines is possible only with slow data rates. VSNL provides international connections via satellite links and submarine cables, as well as microwave links to India's immediate neighbors. Leased line service is allowed, but leased lines might not be readily available.

Telex has become a popular service, given the low data transmission speeds and the difficulty of obtaining voice services. DoT can provide telex services on demand, and they have proven to be the most reliable means of communications in India.

Kenya

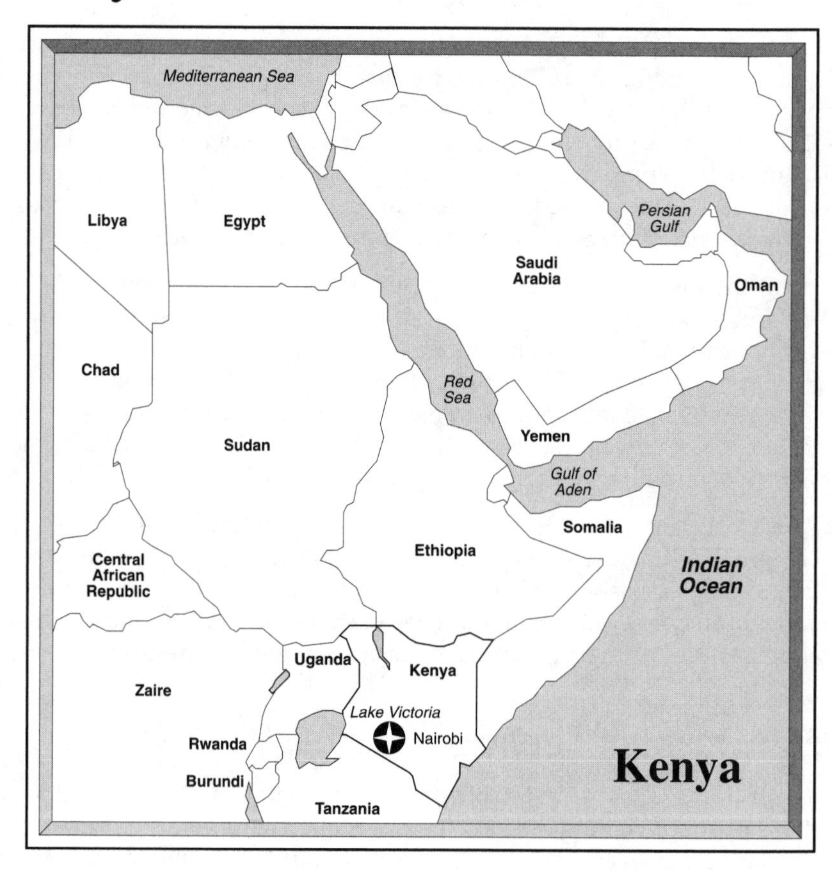

Figure 7.4. Kenya.

Excerpts from the *World Factbook*

General Information

Capital: Nairobi

Total area: 582,650 km^2; land area: 569,250 km^2

Population: 26,164,473 (July 1992), growth rate 3.6 percent (1992)

Literacy: 69 percent (male 80 percent, female 58 percent) age 15 and over can read and write (1990 estimate)

Labor force: 9.2 million (includes unemployed); the total employed is 1.37 million (14.8 percent of the labor force); services 54.8 percent, industry 26.2 percent, agriculture 19 percent (1989)

Organized labor: 390,000 (estimate)

Administrative Information

Administrative divisions: 7 provinces and 1 area*; Central, Coast, Eastern, Nairobi Area*, North-Eastern, Nyanza, Rift Valley, Western

Economic Information

Overview: A serious underlying economic problem is Kenya's 3.6 percent annual population growth rate—one of the highest in the world. In the meantime, GDP growth in the near term has kept slightly ahead of population—annually averaging 4.9 percent from 1986 through 1990. Undependable weather conditions and a shortage of arable land hamper long-term growth in agriculture, the leading economic sector.

GDP: $9.7 billion, per capita $385; real growth rate 2.3 percent (1990 estimate)

Inflation rate (consumer prices): 14.3 percent (1991 estimate)

Unemployment rate: NA, but a high level of unemployment and underemployment exists

External debt: $6.0 billion (December 1991 estimate)

Technical Information

Electricity: 730,000 kw capacity; 2,700 million kwhr produced, 110 kwhr per capita (1990)

Railroads: 2,040 km 1.000-meter gauge

Highways: 64,590 km total; 7,000 km paved, 4,150 km gravel, remainder improved earth

Airports: 249 total, 213 usable; 22 with permanent-surface runways; 2 with runways over 3,659 meters; 2 with runways 2,440-3,659 meters; 47 with runways 1,220-2,439 meters

Telecommunications: In top group of African systems; consists of radio relay links, open-wire lines, and radiocommunication stations; 260,000 telephones; stations—11 AM, 4 FM, 4 TV; satellite earth stations—1 Atlantic Ocean Intelsat and 1 Indian Ocean Intelsat

Telecommunications in Kenya

In many respects, telecommunications in Kenya is representative of telecommunications in many African countries. Some countries, such as Chad and Somalia, have infrastructures that are much worse than Kenya's; other countries, such as South Africa, have better infrastructures. Taken as a whole, the telecommunications problems of Africa are substantial. Only 3.5 million telephones serve Africa's total population of approximately 570 million.[59] As a comparison, Tokyo has more telephones than all Africa.[60] Of the approximately 151,000 villages in Africa, 80 percent have no telephone. Local call completion rates are less than 30 percent, compared with the OECD average of 70 percent.[61] The current ITU estimate is that $50 billion is required to bring the average telephone penetration on the continent to 1 telephone per 100 people.[62] Because many African nations are impoverished, it is unclear where

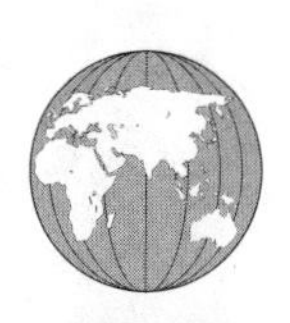

this investment will come from if it doesn't occur in the form of aid from developed countries.

Nevertheless, a number of activities are underway to improve telecommunications in Africa. The Pan African Telecommunications Network (Panaftel) is a microwave system designed to provide connections among the major cities in African countries, but not off the continent. Panaftel currently provides access to most countries. The ITU coordinates the Regional African Satellite Communications System (Rascom) to provide a second approach for internal African telecommunications.

Privatization is taking hold in Africa, as it is in other parts of the world. Nigeria, Senegal, and Zambia have all privatized, and liberalization is taking place in other countries, including Kenya.[63] In many cases, this will be advantageous because the privatized systems will reinvest profits in the telecommunications infrastructure rather than use profits to cover postal deficits.

Kenyan Telecommunications Statistics

Telephone penetration: 1.51 percent
Average productivity per telephone: $48,500
Public telephone ratio: 23.5:1
Annual growth of the network: 64 percent
Telephones per international cricuit: 216 (est.)
Average annual investment per telephone: $3,217
Electricity per capita: 110 kwhr
Square kilometer per paved road-km: 81.3
Square kilometer per railrod-km: 279
Square kilometer per airport: 25,875

Regulatory and Ownership Approaches

Telecommunications in Kenya traces its roots to the British colonial administrators, who developed a network to govern the country. As might be expected with such a network:

- It was concentrated in major urban areas.

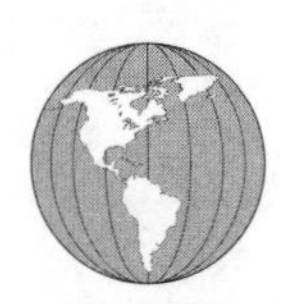

- International communications occurred via transit points in Europe.
- The system depended heavily on expertise "imported" from the United Kingdom.[64]

When Kenya gained independence in 1963, the telephone network was operated by a regional "customs union arrangement" that provided telecommunications to Kenya, Tanzania, and Uganda. This arrangement collapsed in 1977, causing the development of the Kenya Posts and Telecommunications Corporation (KPTC) and The Kenya External Telecommunications Company (Kenextel). KPTC provided domestic services, and Kenextel provided international services. In 1982, Kenextel merged with KPTC, which now holds a monopoly over all telecommunications and postal services, including spectrum allocation.

KPTC is supervised by the Ministry of Transport and Communications, which communicates the government's policy goals to KPTC. Although KPTC is a separate corporation, it is wholly owned by the government. Since 1991, competition has been introduced, forcing KPTC to compete with the private sector for Customer Premise Equipment (CPE). Plans are in place to introduce competition in more sectors of the telecommunications industry.[65] Although competition is permitted for CPE, KPTC dominates all aspects of the telecommunications market. Recent government initiatives require corporations like KPTC (so-called parastatal corporations) to either show a profit or be sold to the private sector.

KPTC took the somewhat unusual step of entering the manufacturing business with the establishment of a manufacturing plant in Gilgil. This plant is designed to manufacture telephones, switchboards, power supplies, and other subassemblies for Kenya, other African countries, and the world.[66]

The Status of the Infrastructure

Although KPTC has made significant improvements in the infrastructure since its incorporation in 1977, much more remains to be done. At independence in 1963, 20,000 telephone lines were available in Kenya; today, this number has grown to 259,626. Although this is still meager, telephone penetration has increased from 0.5 per hundred people at KPTC's founding in 1977 to 1.4 per hundred today.[67] The percentage of the GDP devoted to telecommunications investments rose from 3.28 in 1978 to 8.61 in 1987 and reflects this growth in access lines. With what is reportedly the world's fastest growing population, significant investments will be required to maintain this level.

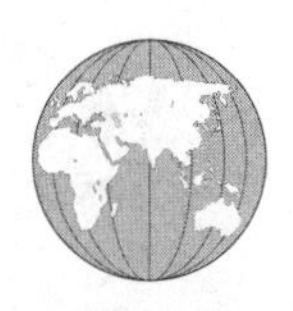

Much of the new infrastructure has been through electronic digital exchanges (System X) manufactured by GEC Plessey Telecommunications (UK). Currently, more than 92 percent of the access lines are connected to these digital switches. KPTC is also investing in digital microwave facilities to build a nationwide backbone network to interconnect these switches. Some of these switches, such as the one in Nakuru, are programmed to provide features, such as abbreviated dialing, three party calling, and so forth.

In an attempt to slow the migration of the population to cities, the government has directed KPTC to devote significant resources to the rural districts. Much of this rural development is done through public pay telephones, which currently number approximately 8,500. This network is being expanded using solar-powered pay telephones to eliminate the dependence of pay telephones on the local power grid, which may be inadequate. These pay telephones, manufactured in Kenya under an agreement with GEC Plessey Telecommunications, are connected to exchanges via radio links, eliminating the need for cable.[68]

Intelsat earth stations currently provide international connections. Kenya is a part owner of the Southeast Asia, Middle East, and Western Europe (SEA-ME-WE) cable. When the system is completed, a link between Mombasa and Djibouti will provide access to the cable. Panaftel provides microwave transmission to Uganda and Tanzania.

Deployment of Services

Several services other than traditional voice services are available in Kenya. Telex is a rapidly growing part of the telecommunications network. Telex capacity grew from 920 lines in 1977 to 2,800 in 1988. The number of connections to the telex network grew from 848 in 1977 to 2,416 in 1988. Datapro estimates that by mid-1992, the number of telex customers will have exceeded 4,000. KPTC has allocated an additional 3,500 telex lines for future expansion. Telex centers are located in Mombasa and Nairobi; customers must purchase leased lines to these centers, although multiple customers are permitted to share lines to reduce the cost of telex service.[69] Cellular services don't exist, although mobile services via VHF radio are available in Nairobi. Public packet-switched services are available from KPTC to supplement leased lines for private networks.

Mexico

Figure 7.5. Mexico.

Excerpts from the *World Factbook*

General Information

Capital: Mexico City

Total area: 1,972,550 km^2; land area: 1,923,040 km^2

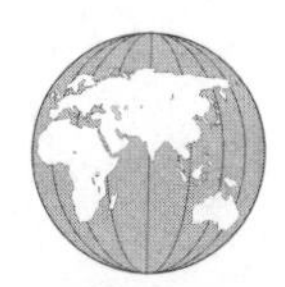

Climate: Varies from tropical to desert

Terrain: High, rugged mountains, low coastal plains, high plateaus, and desert

Population: 92,380,721 (July 1992), growth rate 2.3 percent (1992)

Literacy: 87 percent (male 90 percent, female 85 percent) age 15 and over can read and write (1985 estimate)

Labor force: 26,100,000 (1988); services 31.4 percent, agriculture, forestry, hunting, and fishing 26 percent, commerce 13.9 percent, manufacturing 12.8 percent, construction 9.5 percent, transportation 4.8 percent, mining and quarrying 1.3 percent, electricity 0.3 percent (1986)

Organized labor: 35 percent of labor force

Administrative Information

Administrative divisions: 31 states (estados, singular—estado) and 1 federal district* (distrito federal); Aguascalientes, Baja California, Baja California Sur, Campeche, Chiapas, Chihuahua, Coahuila, Colima, Distrito Federal*, Durango, Guanajuato, Guerrero, Hidalgo, Jalisco, México, Michoacán, Morelos, Nayarit, Nuevo León, Oaxaca, Puebla, Queretaro, Quintana Roo, San Luis Potosí, Sinaloa, Sonora, Tabasco, Tamaulipas, Tlaxcala, Veracruz, Yucatán, Zacatecas

Economic Information

Overview: Mexico's economy is a mixture of state-owned industrial plants (notably oil), private manufacturing and services, and both large-scale and traditional agriculture. In the 1980s, Mexico experienced severe economic difficulties: the nation accumulated large external debts as world petroleum prices fell; rapid population growth outstripped the domestic food supply; inflation, unemployment, and pressures to emigrate became more acute. Growth in national output, however, appears to be recovering, rising from 1.4 percent in 1988 to 3.9 percent in 1990. The United States is Mexico's major trading partner, accounting for two-thirds of its exports and imports. After petroleum, border assembly plants and tourism are the largest means of foreign exchange. The government, in consultation with international economic

agencies, is implementing programs to stabilize the economy and foster growth. In 1992, the government completed negotiations with the United States and Canada on a free trade agreement.

GDP: $289 billion, per capita $3,200; real growth rate 4 percent (1991)

Inflation rate (consumer prices): 18.8 percent (1990)

Unemployment rate: 14 to 17 percent (1991 estimate)

Technical Information

Electricity: 26,150,000 kw capacity; 144,277 million kwhr produced, 1,270 kwhr per capita (1991)

Highways: 212,000 km total; 65,000 km paved, 30,000 km semipaved or cobblestone, 62,000 km rural roads (improved earth) or roads under construction, 55,000 km unimproved earth roads

Airports: 1,815 total, 1,505 usable; 200 with permanent-surface runways; 2 with runways over 3,659 meters; 33 with runways 2,440-3,659 meters; 284 with runways 1,220-2,439 meters

Telecommunications: A highly developed system with extensive radio relay links; connection to Central American Microwave System; 8.5 million telephones; stations—679 AM, no FM, 238 TV, 22 shortwave; 120 domestic satellite terminals; earth stations—4 Atlantic Ocean Intelsat and 1 Pacific Ocean Intelsat

Telecommunications in Mexico

Many aspects of telecommunications in Mexico mirror the developments throughout Latin America. With telephone penetration ranging from 2.3 per 100 people (Paraguay) to 11.7 (Uruguay),[70] the availability of telephones is better here than in most of sub-Saharan Africa and the Indian subcontinent. Still, many Latin American countries face aging infrastructures as they attempt to modernize.

In Latin America, several countries recently have engaged in privatization (Chile, Venezuela, and Argentina, for example). Other countries, such as Brazil and Uruguay, are just beginning the privatization process. The prospect of a hemispheric free trade zone has driven much of this.[71] Some countries such as Bolivia, Ecuador, and Peru are characterized by rugged terrain that hinders the installation and operation of a telecommunications system. Estimates are that $27 billion must be invested annually throughout Latin America to meet user demand by the year 2000.[72] Additionally, many countries are characterized by political instability, which, as in Africa, limits the development of the infrastructure.

Although Mexico differs from the rest of Latin America due to its long border and close economic ties with the United States, it is still representative of Latin America from a telecommunications viewpoint. Mexico is also interesting from this standpoint because it often stands at the forefront of developments in Latin America.

Mexican Telecommunications Statistics

Telephone penetration: 9.2 percent
Average productivity per telephone: $34,000
Public telephone ratio: 184:1
Waiting time for service: 17.6 months
Annual growth of the network: 8.8 percent
Average annual investment per telephone: $1,352
Goegraphic fiber density ratio: 142
Population fiber density ratio: 6,843
Electricity per capita: 1,270 kwhr
Square kilometer per paved road-km: 29.6
Square kilometer per airport: 9,615.2

Regulatory and Ownership Approaches

Public telephone service in Mexico began in 1882 when the International Telephone and Telegraph (ITT) subsidiary, Compañía Telefonica y Telegrafica

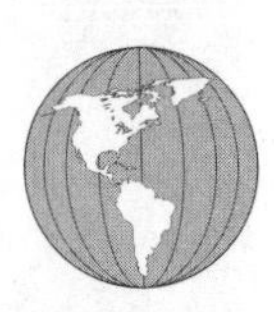

Mexicana, began operations. In 1907, Ericsson also began operations, later merging with ITT to form Teléfonos de Mexico S.A., also known as TelMex, in 1947. TelMex was purchased in 1958 by some Mexican interests, who subsequently, in 1972, sold 51 percent of the company to the Mexican government. Services were provided by the Secretariat for Communications and Transportation (SCT) via the Director General for Telecommunications (DGT) and TelMex.

By 1989, in addition to the 51 percent share of TelMex, the government owned over 120 subsidiaries that provided local services to small and rural communities. In 1989, SCT and TelMex operated separate national long distance networks, although TelMex had 70 percent of the network. The TelMex concession for voice telephone services is due to expire in 2006, with the monopoly concession on long distance service expiring in 1997. In 1990, a controlling interest of 20.4 percent of TelMex was sold to a consortium of Southwestern Bell, France Telecom, and Grupo Caruso (a Mexican conglomerate) for $1.76 billion. The remainder of the shares are traded on the New York Stock Exchange.[73]

In 1988, TelMex was organized into seven directorates with separate responsibilities for Finance, Planning, Long Distance, Large Accounts, the Metro/Central Region, the North Area Region, and the South Area Region. In 1990, SCT established a new carrier, SCT Telecomm, by merging DGT and Telégrafos Nacionales (Telenales). SCT Telecomm acts as an autonomous company within SCT, its owner.

Telecommunications manufacturing is an important component of the infrastructure here, as in several of the other cases. Ericsson de México operates a factory outside Mexico City, which produces switches for both public and private networks, and one in San Luis Potosí for the production of copper and fiber-optic cables. TelMex purchases 70 percent of what these facilities produce.[74]

Industry Structure

With the exception of cellular services, TelMex and SCT Telecomm currently provide telecommunications services. Although these are two distinct entities, they are more likely to cooperate with each other than to compete. TelMex currently operates 70 percent of the long distance network, and SCT Telecomm provides long distance and local services to rural communities. SCT Telecomm also operates specialized services such as the public switched data network, the fax network, and the Telex network.[75]

The Status of the Infrastructure

The infrastructure is highly variable in Mexico, partly due to damage from the 1985 earthquake. In Mexico City, telephone penetration is 19 per 100 inhabitants, for about 40 percent of the country's telephones. As illustrated by the 50,000 service complaints received in 1991, much work remains. TelMex operated 5.5 million lines at the beginning of 1992, with plans to add 750,000 lines per year through 1994. Some of these lines will be dedicated to public telephones, which numbered 0.5 per 1,000 people in 1992. With the planned addition of 96,000 pay telephones, this ratio will increase to 2 per 1,000 people in 1994. The 1995 service level target is to satisfy network service requests within six months, down from the current wait of one year. The current backlog is approximately 1.1 million lines. By the year 2000, the TelMex telephone penetration target is 30 per 100, a substantial increase from today's levels.

Deployment of Services

Most telephone services are available through TelMex, although SCT Telecomm also provides long distance and local voice services. Leased lines are available from these carriers, although the quality of the lines is highly variable and suited only for low speed data transmission or voice transmission. As the digitization of the Mexican network proceeds, quality will likely improve dramatically. SCT Telecomm has been operating the national data network, TELEPAC, since 1981, although this network has significant capacity limitations. SCT Telecomm operates a fax service in 27 cities in Mexico through its former network. Telex is also available through SCT Telecomm in conjunction with TelMex.

International services are available through Intelsat satellites and numerous cables operating across the United States border. Five of the border crossings were digitized in 1991. The Mexican Morelos-1 and Morelos-2 satellites are also available for national telecommunications.

The cellular market was completely liberalized in 1990, with licenses awarded to eight separate foreign consortia, most of which include foreign interests. Currently the cellular systems are to operate in Mexico City, resulting in perhaps one of the most highly competitive cellular markets in the world. TelMex's cellular operation, Telcel, will also cover Tijuana, Guadalajara, and Monterrey in the near future.

New Zealand

Figure 7.6. *New Zealand.*

Excerpts from the *World Factbook*

General Information

Capital: Wellington

Total area: 268,680 km^2; land area: 268,670 km^2; includes Antipodes Islands, Auckland Islands, Bounty Islands, Campbell Island, Chatham Islands, and Kermadec Islands

Population: 3,308,973 (July 1991), growth rate 0.4 percent (1991)

Literacy: 99 percent (male NA, female NA) age 15 and over can read and write (1970)

Labor force: 1,603,500 (1991); services 67.4 percent, manufacturing 19.8 percent, primary production 9.3 percent (1987)

Organized labor: 681,000 members; 43 percent of labor force (1986)

Administrative Information

Administrative divisions: 93 counties, 9 districts*, and 3 town districts**; Akaroa, Amuri, Ashburton, Bay of Islands, Bruce, Buller, Chatham Islands, Cheviot, Clifton, Clutha, Cook, Dannevirke, Egmont, Eketahuna, Ellesmere, Eltham, Eyre, Featherston, Franklin, Golden Bay, Great Barrier Island, Grey, Hauraki Plains, Hawera*, Hawke's Bay, Heathcote, Hikurangi**, Hobson, Hokianga, Horowhenua, Hurunui, Hutt, Inangahua, Inglewood, Kaikoura, Kairanga, Kiwitea, Lake, Mackenzie, Malvern, Manaia**, Manawatu, Mangonui, Maniototo, Marlborough, Masterton, Matamata, Mount Herbert, Ohinemuri, Opotiki, Oroua, Otamatea, Otorohanga*, Oxford, Pahiatua, Paparua, Patea, Piako, Pohangina, Raglan, Rangiora*, Rangitikei, Rodney, Rotorua*, Runanga, Saint Kilda, Silverpeaks, Southland, Stewart Island, Stratford, Strathallan, Taranaki, Taumarunui, Taupo, Tauranga, Thames-Coromandel*, Tuapeka, Vincent, Waiapu, Waiheke, Waihemo, Waikato, Waikohu, Waimairi, Waimarino, Waimate, Waimate West, Waimea, Waipa, Waipawa*, Waipukurau*, Wairarapa South, Wairewa, Wairoa, Waitaki, Waitomo*, Waitotara, Wallace, Wanganui, Waverley**, Westland, Whakatane*, Whangarei, Whangaroa, Woodville

Dependent areas: Cook Islands, Niue, Tokelau

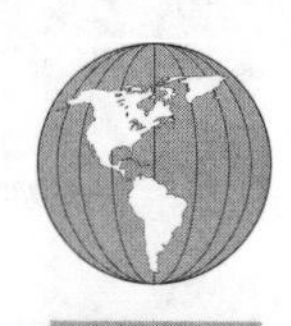

Economic Information

Overview: Since 1984 the government has been reorienting an agrarian economy dependent on a guaranteed British market to an open free market economy that can compete on the global scene. The government hoped that dynamic growth would boost real incomes, reduce inflationary pressures, and permit the expansion of welfare benefits. The results are mixed: inflation is down from double-digit levels, but growth is sluggish, and unemployment—always a highly sensitive issue—is at a record high 7.4 percent. In 1988, GDP fell by 1 percent. In 1989, it grew by a moderate 2.4 percent, and was unchanged in 1990.

GDP: $46.2 billion, per capita $14,000; real growth rate –0.4 percent (1990)

Inflation rate (consumer prices): 1.0 percent (1991)

Unemployment rate: 10.7 percent (1991)

External debt: $17.4 billion (1989)

Technical Information

Electricity: 7,800,000 kw capacity; 28,000 million kwhr produced, 8,500 kwhr per capita (1990)

Railroads: 4,716 km total; all 1.067-meter gauge; 274 km double track; 113 km electrified; over 99 percent government-owned

Highways: 92,648 km total; 49,547 km paved, 43,101 km gravel or crushed stone

Inland waterways: 1,609 km; of little importance to transportation

Airports: 157 total, 157 usable; 33 with permanent-surface runways; none with runways over 3,659 meters; 2 with runways 2,440-3,659 meters; 46 with runways 1,220-2,439 meters

Telecommunications: Excellent international and domestic systems; 2,110,000 telephones; stations 64 AM, 2 FM, 14 TV; submarine cables extend to Australia and Fiji; 2 Pacific Ocean Intelsat earth stations

Telecommunications in New Zealand

New Zealand has a traditional agricultural base that it is attempting to diversify. To develop its infrastructure, the New Zealand government has taken some dramatic steps with respect to its telecommunications policies. As with the case of many countries, New Zealand has chosen to separate the provision of telecommunications services from the provision of postal services and then privatize it. This approach is dramatic compared to other countries: New Zealand has placed few restrictions on foreign ownership and operations, and it has gone to auction-based radio frequency spectrum deployment for significant portions of its spectrum. As with the cases of Japan and the United Kingdom, the former step is a significant development.

New Zealand's ownership and regulatory outlook is so liberal that it is viewed as a laboratory for substantial deregulation; the extent to which the New Zealand experience will be applicable to other countries is a topic of some debate because the country is small in population and geography. Although we haven't discussed radio spectrum allocation, New Zealand has become a laboratory for this approach to spectrum allocation; auction-based spectrum management has historically been the object of policy debates among engineers and economists. Consequently, in some senses, New Zealand represents one extreme of privatization approaches, whereas Japan and the United Kingdom represent the other, more cautious, extreme.

New Zealand Telecommunications Statistics

Telephone penetration: 43.65 percent
Average productivity per telephone: $21,000
Waiting time for service: No
Annual growth of the network: 2.9 percent
Average annual investment per telephone: $195
Percentage of lines served by digital switches: 87
Electricity per capita: 8,500 kwhr
Square kilometer per paved road-km: 5.423
Square kilometer per railrod-km: 56.97
Square kilometer per airport: 8,141.5

Regulatory and Ownership Approaches

New Zealand today allows competition in all aspects of telecommunications service provision. Furthermore, no ownership restrictions encumber firms that want to provide services. The current arrangement traces its roots to the Telecommunications Act of 1987, which formed the Telecommunications Corporation of New Zealand (TCNZ) as a publicly held corporation to assume the responsibility of the telecommunications operation of the Post Office.[76] In addition, this Act also allowed competition in the Customer Premise Equipment (CPE) market.

The initiative taken in the 1987 law was solidified and refined in subsequent legislation. The Telecommunications Amendment Act of 1988 removed the restrictions for entry into the telecommunications marketplace, effectively deregulating the market. The Telecommunications (International Services) Regulations Act of 1989 and the Telecommunications (Disclosure) Amendment Act of 1990 were designed to ensure a fair and competitive market. In particular, the Disclosure act:

> *... requires Telecom New Zealand to release certain information which would be available in a normal commercial environment, [and]*
>
> *... ensure[s] that it does not cross-subsidize its competitive areas from areas that do not yet face effective competition.*[77]

With these laws, New Zealand provided the structure of its telecommunications marketplace.

In September 1990, the government sold Telecom New Zealand to Bell Atlantic Corp (U.S.), Ameritech Corp (U.S.), Fay Ricwhite Holdings (a New Zealand investment firm), and Freightways Holdings (a New Zealand transportation firm) for NZ $4.25 billion (U.S. $2.6 billion). The target investment percentages of these partners are: Bell Atlantic and Ameritech, each 24.95 percent; Fay Richwhite and Freightways, each 5 percent. The remaining shares must be held by public investors.[78] Current law stipulates that no single foreign entity may own more than 49.9 percent of Telecom New Zealand. The Articles of Association for Telecom New Zealand require that free local calls remain available, that access rates don't rise faster than inflation, that the prices be geographically averaged, and that residential services remain widely available. Beyond these agreements, no regulations exist governing tariff changes.

The context for telecommunications liberalization exists within the context of overall market liberalization in New Zealand, initiated by the Commerce Act of 1986 and governed by the Fair Trading Act. While no single government agency is responsible for regulating telecommunications providers in particular, the agencies that have authority and interest in this area are the Ministry of Commerce and the Commerce Commission.[79] The Commerce Commission was created under the Fair Trading Act and is responsible for supervising the provisions of the Commerce Act of 1986. The Commerce Commission has responsibility for all markets, not just telecommunications.

Industry Structure

Competition in telecommunications came to New Zealand in 1991 in the form of Clear Communications, Inc. Clear was formed as a result of the merger of two consortia. One consortium consisted of MCI Communications (U.S.), Todd Corporation (NZ), and New Zealand Rail; the other consisted of Television New Zealand and Bell Canada International. Thus far, Clear provides toll, business access, leased line, and international services, while Telecom New Zealand provides toll as well as business and residential access services. Another service provider, Netways Communications Ltd. is a joint venture of Telecom New Zealand and Freightways Services Ltd. that provides messaging, voice and data communications, and network management services.

The Status of the Infrastructure

Telecom New Zealand has engaged in an ambitious modernization program since privatization. It has invested $1.3 billion between 1989 and 1991, with an additional $800 million planned for 1992 and 1993. As a result of this investment, by 1993 all access lines should terminate in digital switches, and the interoffice network should consist almost exclusively of fiber-optic cable. Clear Communications has the benefit of New Zealand Rail's fiber-optic transmission system in addition to other microwave and fiber capacity. At present, Clear Communications depends on New Zealand Telecom for access to international circuits. In addition, because equal access currently doesn't exist, Clear Communications' customers must dial access codes on New Zealand Telecom's network, unless they are attached via leased lines.[80]

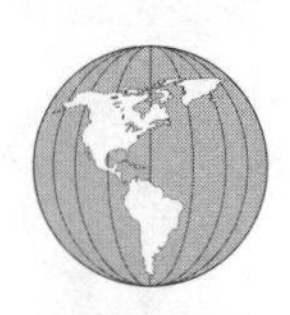

Cable and satellite earth stations provide international capacity. The current cables are TASMAN 1 connecting New Zealand to Australia and ANZCAN connecting New Zealand to Australia and Canada. The TASMAN 2 system is being completed and will connect several islands in the South Pacific. Also under construction is the PACRIM system, which will connect New Zealand to North America and Asia. Pacific Intelsat satellites, as well as the AUSSAT provide satellite capacity. Telecom New Zealand is the Intelsat signatory.

Deployment of Services

Many services are available from both Telecom New Zealand and Clear Communications. While Telecom New Zealand provides most residential service, both carriers provide switched and leased business services (both national and international). Because neither carrier has set significant regulations, no restrictions exist on the use or resale of leased circuits or the connection of private networks and the PSN. All networks must negotiate connection to Telecom New Zealand's network separately. Clearly, this isn't sustainable as a market structure over the long term.

Telecom New Zealand furnishes numerous special calling arrangements, such as 0800 service (where the called party is charged for the call) and 0900 service for information services (where the caller is charged a premium for access to the information service). Leased line and data switched services are also available.

Clear Communications also provides a variety of services. Clear 050 is a toll calling service for business and residential customers and Clear Gateway is a network-based local switching service for businesses, similar to the Centrex service available in the United States. Finally, Clear offers a variety of leased line services to business customers.

Cellular services are available in most cities on the Telecom Cellular Ltd. system. BellSouth (U.S.) is planning a competing network based on the Digital European Cordless Telephone (DECT) standard. Cellular services are growing rapidly.

Tariffs have decreased since deregulation and the introduction of competition. Residential toll charges are below the OECD average, and business toll rates rest at the OECD average. Residential local tariffs are characterized by fixed charges that are above the OECD average but with no usage-sensitive charges. International tariffs rank below the OECD average for both residential and business users.

The United Kingdom

The term *United Kingdom* used in this book refers to the "United Kingdom of Great Britain and Northern Ireland." This includes England, Northern Ireland, Scotland, and Wales. The term *United Kingdom* once included all of Ireland prior to Irish independence, but here the term reflects the current situation.

Figure 7.7. *The United Kingdom, including England, Northern Ireland, Scotland, and Wales.*

Excerpts from the *World Factbook*

General Information

Capital: London

Total area: 244,820 km^2; land area: 241,590 km^2; includes Rockall and Shetland Islands

Population: 57,797,514 (July 1992), growth rate 0.3 percent (1992)

Literacy: 99 percent (male NA, female NA) age 15 and over can read and write (1978 estimate)

Labor force: 26,177,000; services 60.6 percent, manufacturing and construction 27.2 percent, government 8.9 percent, energy 2.1 percent, agriculture 1.2 percent (June 1991)

Organized labor: 40 percent of labor force (1991)

Administrative Information

Administrative divisions: 47 counties, 7 metropolitan counties, 26 districts, 9 regions, and 3 islands areas

England: 39 counties, 7 metropolitan counties* (capital: London); Avon, Bedford, Berkshire, Buckingham, Cambridge, Cheshire, Cleveland, Cornwall, Cumbria, Derby, Devon, Dorset, Durham, East Sussex, Essex, Gloucester, Greater London*, Greater Manchester*, Hampshire, Hereford and Worcester, Hertford, Humberside, Isle of Wight, Kent, Lancashire, Leicester, Lincoln, Merseyside*, Norfolk, Northampton, Northumberland, North Yorkshire, Nottingham, Oxford, Shropshire, Somerset, South Yorkshire*, Stafford, Suffolk, Surrey, Tyne and Wear*, Warwick, West Midlands*, West Sussex, West Yorkshire*, Wiltshire

Northern Ireland: 26 districts (capital: Belfast); Antrim, Ards, Armagh, Ballymena, Ballymoney, Banbridge, Belfast, Carrickfergus, Castlereagh, Coleraine, Cookstown, Craigavon, Down, Dungannon, Fermanagh, Larne, Limavady, Lisburn, Londonderry, Magherafelt, Moyle, Newry and Mourne, Newtownabbey, North Down, Omagh, Strabane

Scotland: 9 regions, 3 islands areas* (capital: Edinburgh); Borders, Central, Dumfries and Galloway, Fife, Grampian, Highland, Lothian, Orkney*, Shetland*, Strathclyde, Tayside, Western Isles*

Wales: 8 counties (capital: Cardiff); Clwyd, Dyfed, Gwent, Gwynedd, Mid Glamorgan, Powys, South Glamorgan, West Glamorgan

Economic Information

Overview: The United Kingdom is one of the world's great trading powers and financial centers, and its economy ranks among the four largest in Europe. The economy is essentially capitalistic with a generous mixture of social welfare programs and government ownership. Over the last decade, the Thatcher government halted the expansion of welfare measures and promoted extensive reprivatization of the government economic sector.

Agriculture is intensive, highly mechanized, and efficient by European standards, producing about 60 percent of food needs with only 1 percent of the labor force. Industry is a combination of public and private enterprises, employing about 27 percent of the work force and generating 22 percent of GDP. The United Kingdom is an energy-rich nation with large coal, natural gas, and oil reserves. Primary energy production accounts for 12 percent of GDP, one of the highest shares of any industrial nation.

In mid-1990, the economy fell into recession after eight years of strong economic expansion that raised national output by one quarter. Britain's inflation rate, which has been consistently well above those of her major trading partners, declined in 1991. Between 1986 and 1990 unemployment fell from 11 percent to about 6 percent, but it rose 8 percent in 1991 because of the economic slowdown. As a major trading nation, it will continue to be greatly affected by factors such as world boom or recession, swings in the international oil market, productivity trends in domestic industry, and the terms on which the economic integration of Europe proceeds will continue to greatly affect the United Kingdom's status as a major trading nation.

GDP: $915.5 billion, per capita $15,900; real growth rate –1.9 percent (1991)

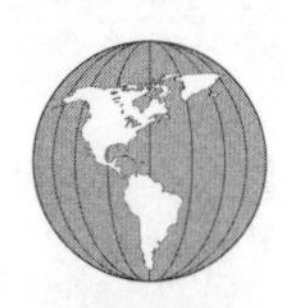

Inflation rate (consumer prices): 5.8 percent (1991)

Unemployment rate: 8.1 percent (1991)

External debt: $10.5 billion (1990)

Technical Information

Electricity: 98,000,000 kw capacity; 316,500 million kwhr produced, 5,520 kwhr per capita (1990)

Railroads: Great Britain—16,629 km total; British Railways (BR) operates 16,629 km 1.435-meter standard gauge (4,205 km electrified and 12,591 km double or multiple track); several additional small standard-gauge and narrow-gauge lines are privately owned and operated; Northern Ireland Railways (NIR) operates 332 km 1.600-meter gauge, 190 km double track

Highways: UK, 362,982 km total; Great Britain, 339,483 km paved (including 2,573 km limited-access divided highway); Northern Ireland, 23,499 km (22,907 paved, 592 km gravel)

Airports: 498 total, 385 usable; 249 with permanent-surface runways; 1 with runways over 3,659 meters; 37 with runways 2,440-3,659 meters; 133 with runways 1,220-2,439 meters

Telecommunications: Modern, efficient domestic and international system; 30,200,000 telephones; excellent countrywide broadcast systems; stations—223 AM, 165 (401 relays) FM, 207 (3,210 relays) TV; 40 coaxial submarine cables; satellite communication ground stations operating in Intelsat (7 Atlantic Ocean and 3 Indian Ocean), Marisat, and Eutelsat systems

Telecommunications in the United Kingdom

A significant feature of the telecommunications environment in the United Kingdom is its primacy in introducing competition and privatization to Europe. The United Kingdom was the first European nation to adopt a competitive regulatory policy toward telecommunications network services and CPE and the first nation in Europe to privatize its PTT. UK policymakers have the distinction of adopting an incentive regulatory pricing scheme, price cap regulation, which has become a model for other nations, including the United States.

In its approach to liberalization, the United Kingdom has proceeded cautiously. Rather than adopting a full-blown scheme of total competition, as the United States has done, policymakers adopted a more moderate approach by first opting for a duopoly. In its "fixed links" duopoly approach, the United Kingdom selected one competitor for the former PTT and legislated terms to facilitate connection and competition between the two entities. When the duopoly approach didn't produce the desired level of network modernization and efficiency, UK policymakers recently decided to pursue a broader-based competitive approach.

British Telecommunications Statistics

Telephone penetration: 41.4 percent
Average productivity per telephone: $40,740
Public telephone ratio: 294:1
Waiting time for service: No
Annual growth of the network: 5.1 percent
Total call-minutes per capita: 3,468.3 million
Total call-minutes per telephone: 8,966.8 million
International calls as a percentage of all calls: 1.5 percent
Average annual investment per telephone: $160.58
Percentage of lines served by digital switches: 65
Geographic fiber density ratio: 100.82 route-km per 100 km^2
Population fiber density ratio: 0.043 route-km per 100 pop.
Electricity per capita: 5,250 kwhr
Square kilometer per paved road-km: 0.712
Square kilometer per railrod-km: 14.53
Square kilometer per airport: 970

Regulatory and Ownership Approach

The early history of telecommunications in the United Kingdom was one of private ownership. Competition existed in local telephony among privately owned companies. The inefficiencies inherent in this early competitive scheme resulted in a series of mergers. The Telephone Company merged with Edison Telephone Company to form the United Telephone Company in 1879. Mergers continued with the National Telephone Company emerging from a merger of United, Lancashire, and Cheshire Telephone companies.[81]

While telephony grew, the postal authority sought to protect its revenue stream, especially its telegraph revenues. The postal authority had the monopoly over electric telegraph, and in 1880 was granted the monopoly for telephony. The post office granted United Telephone a 31-year license to operate in return for 10 percent royalties on receipts. When the post office lost its battle to inhibit intercity provision of telephony, it levied a 10 percent royalty on intercity revenues as well. In 1896, the post office nationalized the trunk (intercity) facilities of the National Telephone Company and in 1905 changed the licensing provisions for telephone companies to include an option for the government to buy the telephone company assets at the lowest possible price. These actions inhibited expansion and investment in telephone companies.[82] Finally, on the basis of a committee recommendation, the telephone industry was nationalized on December 31, 1911 at a cost to the government of $12.5 million.[83]

The performance of the newly nationalized telephone service was significantly short of stellar. Complaints about the lack of investment in the infrastructure and about poor quality of service were frequent and constant, resulting in a series of studies and White Papers and in a growing momentum for reform. Finally, a White Paper resulted in the Post Office Act of 1969. This act changed the post office from a government department to a public corporation, separated the postal and telecommunications functions at the regional level, and transferred responsibility for personnel from the civil service commission to the post office itself.[84] However, the situation didn't substantially improve. During the 1960s, 70s, and 80s, the PTT's performance remained poor. A comparison to the situation in the United States, Germany, and France shows that Britain was less productive than its counterparts, as demonstrated by the significantly higher number of British Telecom employees per subscriber (55 percent higher than Germany). The British PTT also earned less per employee (90 percent less than AT&T for example), and invested substantially less per subscriber line than the other three nations did.[85]

Poor service and underinvestment encouraged more discontent among policymakers, resulting in a 1977 report that advocated the separation of the post and telecommunications functions. The British Telecommunications Act of 1981 implemented this separation. A second bill introduced in 1982 initiated the privatization of British Telecom, which occurred in 1984 through a public stock offering in which 51 percent of the company was sold for about £4 billion. As a result of this legislation, British Telecom split from the PTT and was granted a license to operate. The license required BT to meet the following criteria:

- Meet universal service obligations
- Subject itself to price regulation

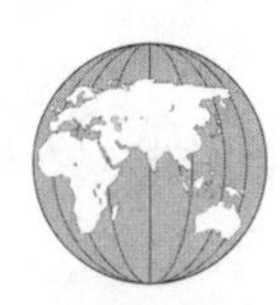

- Practice accounting methods that would prevent cross-subsidization between monopoly and competitive services
- Avoid discrimination among customers
- Publish tariffs
- Provide interconnection to customers and other network providers, notably Mercury (which was a private company, originally owned by Cable & Wireless, Barclay's Bank, Merchant Bank and British Petroleum, now wholly owned by Cable & Wireless)

The British Telecommunications Act of 1981 liberalized the CPE market, ending the monopoly on all but the first telephone set. Customers were still required to get their initial telephone set from BT. Additional sets could be purchased from other sources. The bill also provided for one competitive network, and the policy adopted was one of "fixed links duopoly."[86] BT and Mercury would provide the "fixed links," or facilities connecting the customer to the PSN. Competitors for nonvoice services (VANs) weren't allowed to resell capacity on BT and Mercury facilities until 1989; however, shared use and third party use were allowed. The government wanted to prevent any further switched voice services beyond the created duopoly. Mercury received the exclusive license to compete with BT until 1990 in the area of trunk traffic and international services. Mercury's license entailed no universal service obligation, no obligation to create a nationwide network, and no price regulation.

In separating the operation from the regulation of the telecommunications network, the government created Oftel, the Office of Telecommunications. Oftel is a nonministerial level government department that, while it doesn't set rates, has the authority to make recommendations regarding rates. Oftel also ensures that licensees meet their obligations and works to promote competition and consumer interests. Oftel is headed by a Director General of Telecommunications. The DGT consults with the Secretary of State for Trade and Industry in the granting of licenses. The Secretary has the actual license-granting authority. The Department of Trade and Industry (DTI) has authority over telecommunications.

The duopoly policy was slated for review in 1990. That review took place through an advisory document called *Competition for Choice: Telecommunications Policy for the 1990s*. That document called for discussion of broad-based changes to the regulatory environment. It also called for the end of the duopoly policy and replacement of that approach with extensive liberalization, which would introduce competition into all areas of the network: local, long distance, and international.

As a result of the duopoly review, the duopoly policy ended and was replaced with extensive competition. Any entity could be considered as a fixed links provider, not just BT and Mercury. Mobile operators were allowed to offer fixed services, though fixed service providers weren't allowed to offer mobile services under their main licenses. Cable operators were allowed to provide voice telephony; however, BT and Mercury were barred from providing entertainment services until 2001.[87]

As part of the 1984 restructuring of British telecommunications, British policymakers adopted an incentive regulatory scheme known as Price Caps. This scheme resulted from extensive discussion regarding the best approach for establishing a separate regulatory body and regulatory regime after the separation of BT from the former PTT. The ultimate result of deliberations (led by Professor Stephen Littlechild)[88] was a price cap regime, which established a basket of BT principal services.[89] The price movement of this basket couldn't rise above the increase in the British Retail Price Index (RPI) minus three percent. Additionally, BT pledged that residential and single line business installation and recurring rental charges wouldn't increase by more than RPI plus two percent in any year.

Though Oftel doesn't set rates, it does influence the process through oversight of the price cap regime. Oftel's indirect influence is also seen in the area of service quality reporting. BT stopped publishing service quality reports after the 1984 reforms. However, because service quality continued to be an important issue, Oftel pressured BT to resume publishing semiannual reports, starting in 1987. BT reports:

- Waiting time for connection
- Rate of fault reports and fault clearance
- Waiting times for operator and directory inquiry
- Public pay telephone maintenance
- Call failure rates

Mercury also collects statistics, including:

- Provision of service
- Fault reports
- Operator services including directory inquiries, billing inquiries, and sales services
- Call success rate

- Post-dial delay
- Customer satisfaction indicators
- Pay telephones[90]

Industry Structure

After the duopoly review, the current British telecommunications industry structure is positioned to become an industry of burgeoning competition. Three firms now provide service: BT, Mercury, and Kingston Communications in Hull. Kingston Communications is a public telephone network, owned by local authorities. The enterprise is publicly owned by the Hull City Council and offers voice and data services locally, domestically, and internationally. It is the only company that wasn't bought by the national government in the 1911 nationalization process.

Both BT and Mercury hold 25-year licenses. The BT license includes a universal service obligation; a requirement to provide emergency (999), directory, and public pay telephone service; a requirement to allow interconnection to licensed providers; and an obligation to connect CPE that has met type approval. BT is still the overwhelmingly dominant provider of service, with more than a 90 percent market share.[91] BT offers local, domestic long distance, and international voice and data services. BT also provides VANs, telex, mobile, and videotext services.

Mercury's license grants it the authority to provide domestic and international services, but not land mobile, maritime, or cable TV program services. Mercury's customers are connected to the Mercury network either directly (through a fiber-optic or microwave connection) or indirectly (through a BT exchange line). Customers who indirectly access the Mercury network must have a phone with a dedicated button to gain manual access to the network or a "Smartbox," which provides automatic access. Equal access dialing to connect to the Mercury network—a procedure worked out in the United States—hasn't yet been introduced in the United Kingdom.

Interconnection will undoubtedly become a major issue as a result of the newly liberalized competition policy. An acceptable interconnection arrangement was worked out between BT and Mercury with some difficulty. With no specific rules in place governing interconnection terms and conditions, the newly competitive environment could require extensive negotiations.

The telecommunications industry includes two cellular providers, Cellnet and Vodafone. These cellular network operators are precluded, until 1993, from

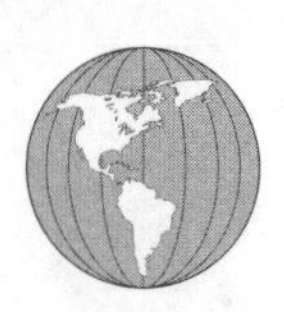

selling directly to customers; their services are sold to customers through service providers or their agents. There are also holders of Telepoint (CT-2) licenses and PCN licenses. There are also radio paging providers, including BT and Mercury, and several videoconferencing service providers, in addition to BT.

BT can provide VANs, but it must do so on a nationwide basis. Large VAN providers are subject to licensing and fair trading conditions (no undue discrimination, protection against cross-subsidization, and so forth); however, smaller VAN providers don't need to register with Oftel as licensed providers.

New entrants in the fixed links area are emerging. At the beginning of 1992, a total of 13 cable operators provided local telephone services over their cable networks. Several entities have applied for licenses to compete with BT and Mercury for national and international services. They include BR Telecommunications, a subsidiary of British Rail; Sprint International (UK); National Network Ltd., a part of the Post Office; and several others.[92]

Sprint has also applied to provide facilities-based international service between the United Kingdom and the United States However, the status of the Sprint application was still not clear late in 1992.[93] The Sprint application to provide international service raised an interesting point. The United Kingdom, in its duopoly review, decided to liberalize conditions for international service, including simple resale, between the United Kingdom and nations with sufficiently "liberal competitive regimes." British policymakers don't regard the U.S. environment as sufficiently liberal in this regard, probably because of U.S. restrictions on foreign ownership.

Additional types of companies that might seek to offer fixed link service include PCN providers seeking to replace the last linkage to the customer with a wireless connection. The CPE market in the United Kingdom is liberalized. CPE does need to meet type approval to be connected to the network. Approvals are handled by the British Approvals Board for Telecommunications (BABT). The BABT follows guidelines from the British Standards Institute. The BABT will also accept ETSI standards.

State of the Infrastructure

The United Kingdom, with a population of about 57 million, had 44.58 main lines per 200 people at the beginning of 1990. With the introduction of new fixed link service providers, the issue of infrastructure status will become more difficult to assess. However, given the dominant position of BT, a look at the BT network provides a good guide to the state of the infrastructure.

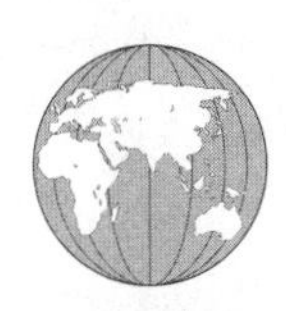

BT is on a network modernization program that, by early 1992 had spent £15 billion.[94] The three-tiered BT structure includes a top tier of 56 digital switches (most of them GPT's System X switches). The next tier comprises 300 principal exchanges. Over 6,000 local exchange switches exist. In addition to GPT's System X switches, Ericsson AXE 10 switches are also employed. BT plans to have all digital local exchange switches by 1994. BT's trunk network and backbone are fully digital. The company is also in the process of installing fiber-optic cable.

BT provides five international switching centers. Customers can use international direct dialing to over 200 nations. BT has close to 100,000 pay telephones deployed and has introduced pay telephones that can handle both coin and card payment methods. BT offers a foreign exchange type service (called Linkline 0345) and also a toll-free 800-type service called Linkline 0800. ISDN service, both ISDN 30 (or primary access) and ISDN 2 (or basic rate access) are available. BT also offers international ISDN connections.

Although Mercury doesn't represent a large percentage of the telecommunications industry, it has added to the level of digitization in the nation. Using Northern Telecom DMS switches and GPT's System X, Mercury had the first fully digital network in Europe. Mercury has a fully digital network comprised of fiber-optic cable or microwave links. Mercury also offers ISDN services and Centrex services.

Both BT and Mercury offer public packet-switched data networks and high-speed leased lines. Both carriers have satellite facilities (earth stations) for the provision of international voice and private line services.

The telecommunications infrastructure in the United Kingdom will undoubtedly continue to expand and improve. The facilities of the newly emerging cable TV providers and the facilities which will be built by prospective network providers will add to the already existing infrastructure provided by BT, Mercury, and the cellular service providers.

Deployment of Services

Under the new competitive regime, voice and data services at the local, national, and international level will be available from a variety of providers other than BT and Mercury. These providers will include cable TV operators and PCN operators. The cable TV providers will supply the connection to the customer through existing systems. The mobile providers will do so through a wireless connection. A license was granted in early 1992 to a company that will provide customer connection via radio waves.

Newly liberalized satellite services will be available. Alpha Lyracom and British Aerospace were granted licenses to connect their satellite circuits to the PSN for the provision of voice and data services. Licenses were also granted to four companies proposing to offer data services over radio-based fixed and mobile services.[95] Mobile services are particularly rich and diverse in the United Kingdom. At the end of 1991, there were 22 national mobile licenses. Two were for cellular telephony; four were Telepoint licenses; three were for PCN; four were for mobile data; four were for trunked private mobile radio; and seven were for paging services.[96] The two cellular licenses belong to Cellnet and Vodafone. As network operators, these two companies haven't been able to directly market to users. They have gone through service providers. This situation changed in 1992 allowing both network providers to market directly to users. Both Cellnet and Vodafone began operation of analog systems in 1985. They are both upgrading their systems to digital and to meet the GSM standard.

Telepoint or CT-2 service is available through four providers (Zonephone, Phonepoint, Rabbit, and Mercury Callpoint). The CT-2 service hasn't been highly successful. It was used as an alternative to public pay telephones, rather than as a promising service. On the other hand, PCN services, which are perceived to offer truly wireless capability, are more likely to succeed. PCN is slated for introduction in 1993. The three licenses granted are for Microtel, Mercury PCN, and Unitel.

Wide area paging services are offered by BT Radiopaging, Air Call (which belongs to BellSouth), Mercury Radiopaging, Vodapage, Inter-City Paging, and Hutchison Paging.[97] BT, Mercury, and Comtext International offer telex services. As is the case all over the world, fax services are cutting into the telex market.

BT is active in both videotext and videoconferencing and faces competition from VAN providers in both arenas. BT's interactive videotext service uses the Prestel network, which dedicated terminals or PCs equipped with modems can access.

BT is facing increased competition from VAN providers in general. VAN providers can operate privately or through the facilities of licensed network providers. While the majority of VANs use network facilities, private VANs are increasing.[98] BT's electronic mailbox and e-mail services face increased competition.

A basic question that arises in discussions regarding the deployment of services is the issue of pricing. The United Kingdom appears to have a rate structure that favors competition and global traffic. In an OECD study, which

examined the relationship between local charges and long distance charges by expressing the local charge as a percentage of the long distance charge, the United Kingdom's local charges were by far the largest percentage.[99] This suggests that very little subsidy can be drawn from long distance to local rates. In like manner, the United Kingdom's break-even point between leased lines and usage-based charges on the public network is one of the lowest of the OECD nations. British rates for international leased lines are also low. Pressure has been applied to lower international usage rates even further by decreasing those rates 10 percent and then subjecting them to the price cap regime.

The United States of America

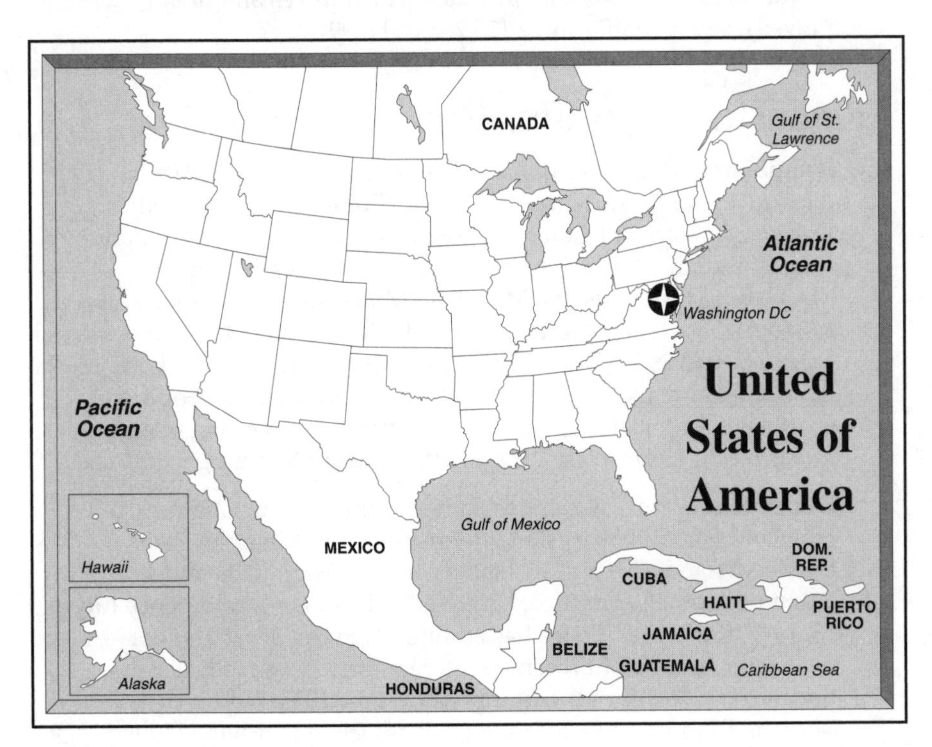

Figure 7.8. The United States.

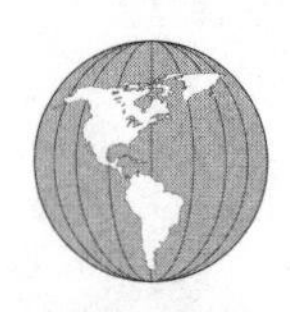

Excerpts from the *World Factbook*

General Information

Capital: Washington D.C.

Total area: 9,372,610 km^2; land area: 9,166,600 km^2; includes only the 50 states and District of Colombia

Population: 254,521,000 (July 1992), growth rate 0.8 percent (1992)

Literacy: 98 percent (male 97 percent, female 98 percent)

Labor force: 126,867,000 (includes armed forces and unemployed); civilian labor force 125,786,000 (1991)

Administrative Information

Administrative divisions: 50 states and 1 district*; Alabama, Alaska, Arizona, Arkansas, California, Colorado, Connecticut, Delaware, District of Columbia*, Florida, Georgia, Hawaii, Idaho, Illinois, Indiana, Iowa, Kansas, Kentucky, Louisiana, Maine, Maryland, Massachusetts, Michigan, Minnesota, Mississippi, Missouri, Montana, Nebraska, Nevada, New Hampshire, New Jersey, New Mexico, New York, North Carolina, North Dakota, Ohio, Oklahoma, Oregon, Pennsylvania, Rhode Island, South Carolina, South Dakota, Tennessee, Texas, Utah, Vermont, Virginia, Washington, West Virginia, Wisconsin, Wyoming

Dependent areas: American Samoa, Baker Island, Guam, Howland Island; Jarvis Island, Johnston Atoll, Kingman Reef, Midway Islands, Navassa Island, Northern Mariana Islands, Palmyra Atoll, Puerto Rico, Virgin Islands, Wake Island. Since July 18, 1947, the United States has administered the Trust Territory of the Pacific Islands but recently entered into a new political relationship with three of the four political units. The Northern Mariana Islands is a Commonwealth associated with the United States (effective November 3, 1986). Palau concluded a Compact of Free Association with the United States that was approved by the U.S. Congress, but to date the Compact process hasn't been completed in Palau, which the United States continues to

administer as the Trust Territory of the Pacific Islands. The Federated States of Micronesia signed a Compact of Free Association with the United States (effective November 3, 1986). The Republic of the Marshall Islands signed a Compact of Free Association with the United States (effective October 21, 1986).

Economic Information

Overview: The United States has the most powerful, diverse, and technologically advanced economy in the world, with a per capita GDP of $22,470, the largest among major industrial nations. In 1989, the economy enjoyed its seventh successive year of substantial growth, the longest in peacetime history. The expansion featured moderation in wage, consumer price increases, and a steady reduction in unemployment to 5.2 percent of the labor force. In 1990, however, growth slowed to 1 percent due to a combination of factors, such as the worldwide increase in interest rates, Iraq's invasion of Kuwait in August, the subsequent spurt in oil prices, and a general decline in business and consumer confidence. Ongoing problems for the 1990s include inadequate investment in education and other economic infrastructures, rapidly rising medical costs, and sizable budget and trade deficits.

GDP: $5,673 billion, per capita $22,470; real growth rate –0.7 percent (1991)

Inflation rate (consumer prices): 4.2 percent (1991)

Unemployment rate: 6.6 percent (1991)

External debt: $581 billion (December 1989)

Technical Information

Electricity: 776,550,000 kw capacity; 3,020,000 million kwhr produced, 12,080 kwhr per capita (1990)

Railroads: 270,312 km

Highways: 6,365,590 km, including 88,641 km expressways

Inland waterways: 41,009 km of navigable inland channels, exclusive of the Great Lakes (estimate)

Civil air: 3,297 commercial multiengine transport aircraft, including 2,989 jet, 231 turboprop, 77 piston (1985)

Airports: 14,177 total, 12,417 usable; 4,820 with permanent surface runways; 63 with runways over 3,659 meters; 325 with runways 2,440-3,659 meters; 2,524 with runways 1,220-2,439 meters

Telecommunications: 182,558,000 telephones; stations—4,892 AM, 5,200 FM (including 3,915 commercial and 1,285 public broadcasting), 7,296 TV (including 796 commercial, 300 public broadcasting, and 6,200 commercial cable); 495,000,000 radio receivers (1982); 150,000,000 TV sets (1982); satellite communications ground stations—45 Atlantic Ocean Intelsat and 16 Pacific Ocean Intelsat

Telecommunications in the United States

The U.S. experience in telecommunications is different from that of most other nations. While other nations offer services through a state-owned monopoly, telecommunications in the United States is a privately owned undertaking. During the 1980s, the U.S. government adopted a policy that endorsed deregulation and the replacement of regulated monopolies with market competition wherever possible. Airlines, bus service, and interstate long distance services were deregulated.

Policymakers in the United States and in other nations identify the U.S. telecommunications infrastructure as one element contributing to the country's financial development and ability to build an information-based economy. The U.S. approach to telecommunications—liberalization and privatization—has, therefore, become a model for other nations seeking to expand and develop their own infrastructure and services.

United States Telecommunications Statistics

Telephone penetration: 49 percent
Average productivity per telephone: $45,487
Public telephone ratio: 72.7:1
Waiting time for service: About 0
Annual growth of the network: 3.8 percent

Total call-minutes per capita: 7,087 million
Total call-minutes per telephone: 14,464
International calls as a percentage of all calls: 0.3 percent
Average annual investment per telephone: $217.9
Percentage of lines served by digital switches: 80
Percentage of lines served by SPC switches: 96
Geographic fiber density ratio: 34.5 route-km per 100 km^2
Population fiber density ratio: 0.127 route-km per 100 pop.
Electricity per capita: 12,080 kwhr
Square kilometer per paved road-km: 1.44
Square kilometer per railroad-km: 33.9
Square kilometer per airport: 1,901.8

Regulatory and Ownership Approach

The U.S. regulatory structure is complicated by the dual regulation of the PSN based on geographical jurisdiction. Traffic and services that cross state boundaries fall under the authority of the Federal Communications Commission (FCC); traffic and services offered inside state boundaries come under the jurisdiction of state public service commissions. Broadcast services that use the airwaves (broadcast radio and television) are deemed interstate and regulated by the FCC. Cellular and microwave facilities must be licensed by the FCC; any interstate traffic conveyed through those facilities also falls under the FCC's authority.

Both the FCC and state commissions are creatures of legislative activity. State commissions are founded by actions of state legislatures, and they derive their functions and authority from such legislation. The FCC was created by the Communications Act of 1934, a broad-based law that defined the status and basis for most forms of interstate communications. The Communications Act of 1934 gives the FCC its universal service mandate, which requires the creation and maintenance of a nationwide network of affordable national and international telephone service.

As expected, tensions often arise in this system of dual regulation because both state commissions and the FCC regulate services that, for the most part, are provided over the same facilities. The FCC can preempt state authority in those instances in which state action would inhibit interstate activity.

The U.S. courts also have jurisdiction over telecommunications within the context of the antitrust laws. The former Bell system was divested as a result of the settlement of an antitrust action. Oversight of the divestiture process rested in the hands of the courts. The courts also serve as avenues of appeal when regulated entities disagree with actions taken by the state or federal commissions.

Telecommunications companies in the United States are—and have always have been—privately owned. In order to provide service, those companies must be designated as common carriers. At the state level, local telephone companies, which provide local service and state long distance service, are granted the exclusive right to serve a specific local service area. In return for this franchise, the local telephone company must meet certain quality of service standards and is limited to offering services authorized by the state commission at prices that the commission approves.

At the federal level, the FCC no longer grants AT&T an exclusive franchise to provide interstate long distance services. The FCC's policy introduces competition into the interstate long distance arena. The FCC still maintains regulatory oversight over AT&T, but the Commission has chosen not to regulate AT&T's competitors. Indeed, as competition increases, the Commission is increasingly lessening its regulation of AT&T, with the ultimate goal of allowing competitive forces to replace regulation.

A variety of regulatory approaches are evident in U.S. commissions. Traditional rate base regulation was the norm at both the federal and state levels. Increasingly, commissions are experimenting with incentive regulatory schemes designed to encourage innovation and greater efficiencies in the PSN. At the federal level, the FCC has adopted a price cap regime for the Bell Operating Companies (BOCs) and for AT&T. At both the federal and state levels, competition is being introduced into all areas of the PSN.

A Brief History of U.S. Telecommunications

The telephone industry grew fairly rapidly in the United States following the introduction of commercial telephone service in Boston in 1877. When the Bell System lost its patents in 1894, thousands of non-Bell (independent) phone companies developed across the country. This industry approach presented problems of inefficiencies and duplication of facilities. Consolidations began with the Bell system buying independent telephone companies.

The growth of the Bell System attracted the attention of Congress, which placed it under the control of the Interstate Commerce Commission and the attention of the Department of Justice, which persuaded the company not to acquire more companies without governmental permission. Due to the lack

of oversight by the Interstate Commerce Commission, the FCC was formed in 1934. The FCC approached the interstate telephone network as one that should be regulated as a unified, natural monopoly.

This monopoly was soon breached. Microwave technologies (developed during World War II) offered a way to bypass the public network. In 1959, the FCC allowed private use of microwave frequencies above 890 megahertz. This private use was soon extended to commercial use, as Microwave Communications, Inc. (MCI) received permission to offer point-to-point, non-switched service between Chicago and St. Louis. MCI soon began the push into providing switched services and, through a series of lawsuits and FCC actions, eventually was granted interconnection to the public network for competition with AT&T.

Comparable interconnection to the public network was the result of a series of actions by the courts and the FCC. As part of the Modification of Final Judgment (the end of the Department of Justice's long antitrust suit against AT&T), the duties of the BOCs to provide equal treatment to AT&T and its competitors were defined. With FCC actions in several proceedings, a new relationship was defined between local telephone companies and long distance carriers, and methods to ensure customer choice in selecting a long distance carrier were developed.

An equal access regime was created in which long distance carriers (interexchange carriers, or IXCs) would gain equal access to customers through use of local telephone company facilities. Local telephone companies would levy access charges on all IXCs for use of their services; these access charges would be subject to tariff and applied in a nondiscriminatory manner. Customers would have access to the long distance company of their choice through a uniform dialing pattern. Interconnection, dialing patterns, and access charging would all be equivalent. The stage was set for successful competition.

While competitive inroads were made in long distance, competition came to the CPE market as well. As a result of FCC and judicial actions, non-Bell-provided CPE could be connected to the public network. The CPE would first need to meet the technical requirements outlined in Part 68 of the FCC's Rules and Regulations.

Another challenge to U.S. policymakers was the proper approach to growing computerization. In its Computer Inquiries, the Commission sought to determine where to draw the line between basic voice service and "enhanced" services. The FCC determined that enhanced services shouldn't be regulated, but it struggled with the problem of whether regulated telephone companies should be allowed to offer these enhanced services. If these services were provided, how would the regulators ensure that no cross-subsidization of

costs and price took place between regulated and enhanced services? The Commission at first required that enhanced services be offered through a separate subsidiary but then determined that accounting safeguards were adequate to protect against cross-subsidization.

An additional outgrowth of Computer Inquiry III (the last of the three Computer Inquiry proceedings) was the creation of an Open Network Architecture (ONA) process and series of tariffs. Because the BOCs wanted to provide enhanced services in competition with enhanced service providers, who were also BOC customers, for the interconnection necessary to reach the enhanced services customer, the possibility existed that the BOCs could enjoy a distinct competitive advantage. The ONA process was designed to insure that the BOCs would treat their enhanced service competitors equitably and would not take advantage of their position as network providers.

The outcome of the Justice Department action in its antitrust suit against AT&T placed the BOCs' capability of offering enhanced services (or value-added services) in question. In an effort to curb the growth and range of services controlled by AT&T, the Justice Department, in 1949, filed an unsuccessful antitrust suit against AT&T, which was settled in 1956. The settlement limited AT&T to providing regulated services only, not unregulated computer services. The Justice Department filed a second suit in 1974, which eventually resulted in the Modification of Final Judgment (MFJ). The MFJ required AT&T to divest itself of its local services, which were given to seven newly created Regional Bell Operating Companies (RBOCs). In return, AT&T was allowed to move into nonregulated services and to retain its equipment manufacturing activities.

The RBOCs were prohibited from providing interexchange long distance and information services and from manufacturing equipment. The information services prohibition was recently lifted. Efforts are underway to lessen the manufacturing restriction as well.

The seven newly created Regional Bell Operating Companies (RBOCs) include Ameritech, Bell Atlantic, BellSouth, NYNEX, Pacific Telesis, Southwestern Bell, and U.S. West. The RBOCs are holding companies; in other words, they are comprised of individual companies (called BOCs). For example, NYNEX is comprised of New York Telephone and New England Telephone. The RBOCs are "regional" in the sense that they each cover a specific region of the United States.

Industry Structure

The U.S. telecommunications industry structure is complex and filled with many players. Local telephone companies (the seven RBOCs and over 1,300 independent telephone companies) provide service in their franchised service areas. The country was divided into Local Access Transport Areas (LATAs) by the MFJ. Local telephone companies can provide long distance service in those LATAs, which encompass several local service areas. The BOCs cannot provide service across LATA boundaries; non-BOC local telephone companies don't face these same restrictions. Local telephone companies also provide local service on a franchise basis.

Long distance companies such as AT&T and its over 500 competitors provide long distance service between LATAs. These long distance companies, or IXCs, either provide service through their own facilities (facilities-based carriers) or by reselling services usually purchased from AT&T.

Local telephone companies and mobile telephone competitors (Radio Common Carriers) provide mobile telephone service. Cellular service is provided as a duopoly. The nation is divided into metropolitan service areas and rural service areas. Two licenses are granted for each area, one of which is reserved for a wireline company. The FCC grants cellular licenses. The state commissions can regulate intrastate traffic carried over the cellular network.

Until recently, the MFJ precluded the BOCs from offering information services. That provision of the MFJ was overturned, and the BOCs are now free to offer videotext, audiotext, and other value-added services. BOCs face competition from a growing number of enhanced services and information services providers.

Telephone companies cannot provide cable television service in their own serving area. The telephone industry is lobbying to overturn those restrictions. The cable companies have expressed an interest in providing voice and data services. The cable companies' interest is strengthened by growing alliances with alternative local access providers, companies that offer fiber-optic facilities as alternatives to the PSN. Both cable companies and alternative local access providers are actively involved in PCN/PCS experiments, as are local telephone companies. No decisions have yet been made about how PCN/PCS will be deployed, though a series of special licenses was granted to PCN experimenters.

A deviation from this structure is found in Puerto Rico, where the telephone company, Puerto Rico Telephone Co. (PRT), is owned by the government. An effort to privatize PRT in 1991 failed because the offers of prospective buyers weren't high enough and because trade unions resisted. In December 1992, the FCC approved the sale of the Puerto Rican long distance company, Telefonica Larga Distincia de Puerto Rico (TLD), to Telephonica of Spain. The U.S. carriers protested this sale because the Spanish telephone market remains closed to them.

Status of the Infrastructure

In a report to the FCC regarding the status of telephone penetration in the United States, the percentage of households with telephones was 93.3 in 1990.[100] In the view of many policymakers, this constitutes the achievement of universal service.

The U.S. infrastructure is regarded by many as the most developed, most modern in the world. A comparison of the U.S. infrastructure to those of other developed nations can be useful, particularly, with a nation that is engaged in network modernization and liberalization efforts. The U.S. Department of Commerce took such a look in its recent *NTIA Infrastructure Report: Telecommunications in the Age of Information.*[101] Examining the results of that report may be instructive to assess not only the current state of the U.S. infrastructure but also opinions about its future state.

The NTIA report looked at results from Canada, France, Italy, Japan, the United Kingdom, the United States, and West Germany.

Data used tended to lie within the period from 1980-89. In the areas of deployment and investment, the results for the United States were as follows:[102]

Average annual capital investment per main line for modernization	(2nd of 7)	$183.03
Average percentage of total investment in modernization	(1st of 7)	86.6
Average total calls per line	(1st of 7)	3,516
Lines per 100 population	(2nd of 7)	49.0

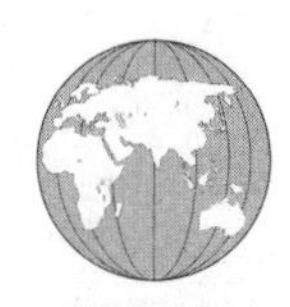

Percentage of digital switches by subscriber line	(3rd of 7)	42.5
Percentage of electronic switches by subscriber line	(1st of 7)	96.6
Fiber route km per 1,000 population	(3rd of 4 reported)	1.27
Percentage of ISDN coverage by access	(3rd of 3 reported)	0.1
Percentage of offices with SS7 capability	(3rd of 4 reported)	6.3

The projected figures for 1994 are perhaps even more enlightening.[103]

Percentage of digital switches by subscriber line	(5th of 6 reported)	68.2
Percentage of electronic switches by central office	(1st of 6 reported)	98.2
Percentage of ISDN coverage by access line	(4th of 4 reported)	49.8
Percentage of offices with SS7 capability	(4th of 4 reported)	57.0

While all comparative numbers are subject to question, the NTIA study does suggest that the U.S. infrastructure is strong in such areas as telephone penetration and usage levels and that companies are investing heavily in modernization efforts. The report also suggests, however, that the United States is trailing in some areas (ISDN and digital switching, for example) and that, given existing planning information, might continue to do so.

The picture presented by the NTIA figures isn't a total one—a fact that underlines the difficulty in amassing a complete picture of the U.S. infrastructure. Most of the NTIA numbers represent results at the local telephone company level, usually results of RBOC operations. NTIA numbers don't reflect results of long distance carriers such as AT&T and MCI. These IXCs have many miles of fiber-optic cable, increasing numbers of digital switches, and aggressive deployment schedules for SS7 capability. The assets of the top three IXCs alone totalled over $60 billion in 1990. Neither do the NTIA numbers reflect the assets of alternative access providers. The assets of these companies are almost totally fiber-optic and digital.

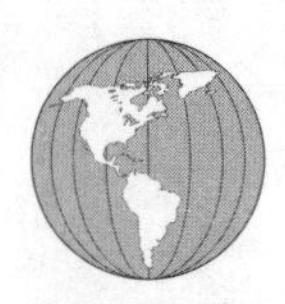

Underlying a study of the state of U.S. infrastructure is the question of what constitutes infrastructure. Should IXC networks be considered infrastructure if they aren't owned by regulated entities? This is one question that should be considered as more nations liberalize and allow more network providers to enter the telecommunications market.

Deployment of Services

Local service is offered as a monopoly regulated at the state level. The usual rate structure is flat rate unlimited calling, though Local Measured Service is an option in many areas. Long distance service is usually provided with time-of-day and distance-sensitive rate components. Depending on the jurisdiction of the call, the regulatory entity and the type of company offering the service vary.

If the call is intraLATA, it is usually offered by a BOC or an independent telephone company and is regulated by a state commission. In some states, competition for intraLATA service is allowed, but equal dialing and equal access aren't available. If the call is interLATA, it is offered by an IXC. If it is an intrastate call, the regulatory entity is the state commission; if the call crosses state boundaries, the regulatory entity is the FCC. International calls are handled by IXCs under FCC jurisdiction. The local telephone company charges the IXC some form of access charge for interLATA and international calls.

Specialized services like toll-free 800 calls and bulk-billed long distance services (Wide Area Telecommunications Service or WATS) are offered by BOCs within LATAs and by IXCs across LATAs. This often leads to confusion. The majority of such services are interstate in jurisdiction. BOCs and IXCs offer operator services. Directory inquiry is a local telephone company function, as is the provision of a telephone directory.

Dedicated lines are offered for all jurisdictions—local, intraLATA, interLATA, and international. Both analog and digital lines are available. T1 or DS1 (24 voice-grade equivalent 1.5 megabit) services tend to be readily available from both local telephone companies and IXCs. T3 or DS3 facilities (674 voice-grade equivalents) are a more prevalent offering in recent years.

Private networks are allowed with no licensing requirement. IXCs and some local telephone companies have begun to offer Virtual Private Networks (VPNs) to keep customers on the network. Through software, the public network functions as if it were a private network for a specific customer. Resale

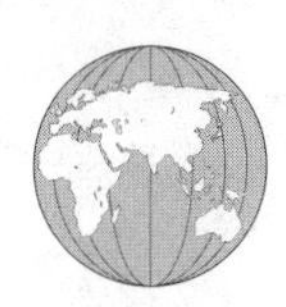

and interconnection of private networks to the public network are allowed. Enhanced, or value-added, services are provided by a host of companies, including the network providers.

CPE is deregulated. The customer owns all telephone sets, including the first one at a location. The wiring in the customer property has also been deregulated and belongs to the property owner. The PBX market is active. Centrex services offer competition to that market. The battle between PBXs and Centrex is one indication of the tension between CPE and the network. The question revolves around intelligence in the network. Should it be in the CPE (hence the PBX) or should it be in the network (Centrex)? With the development of Signaling System 7, the creation of an intelligent network was reinforced. (See Chapter 5, "Technical Issues," for a discussion of SS7.)

U.S. telecommunications during the past decade has been marked with the steady incursion of competition into all areas of the public network. At the peripheries of the network, CPE and location wiring have been deregulated. In the long distance arena, competition exists at the international and interstate levels. In most states, interLATA competition is also allowed. In some states, intraLATA is being opened to competitive carriage as well. Alternative access providers have emerged to offer methods for bypassing the public network or at least the local telephone company portion of the network. Customers can use the alternative access provider's facilities, which are usually fiber-optic, to get to a long distance carrier or to another location without going through the local telephone company. It appears that the only areas of the network not yet open to competition are local service and the local loop. These areas are also being opened to competition. In New York state, alternative providers may provide loop connections. In some ways, cellular service and, when it is developed, PCN service will provide a local loop alternative.

As in other discussions of service deployment, the issue of pricing emerges. In the pursuit of universal service, the FCC and state commissions kept local rates low by allowing long distance services to provide a subsidy. With the introduction of competition in the 1980s, that process began to be reversed. Long distance rates are dropping, and more of the cost of the local loop has shifted out of the long distance arena and onto the subscriber's bill. Competitive pressures have brought down the prices of leased lines and the cost of international long distance calling. Indeed, as we explained in Chapter 3, the U.S.'s relatively low rates for international calls might be a contributing factor to accounting and settlement deficits.

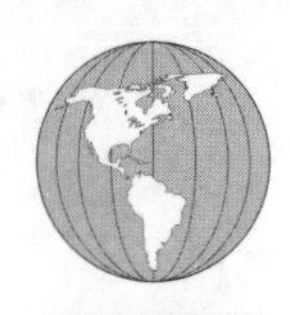

Conclusion

We have presented what we believe to be important models of telecommunications ownership, regulation, and infrastructure throughout the world. The United States is the leader in private ownership and competition. New Zealand has taken this strategy and applied it in the most extreme way, providing few regulations and barriers to entry in services of all kinds. In the developed world, Japan (discussed in Chapter 6, "Assessment Methodology") and the United Kingdom provide examples of more cautious and measured approaches to privatization and competition.

Several of our cases address key countries with telecommunications infrastructures that require development. For these, we identified examples that we feel are representative of selected regions in the world. Because it is always dangerous to make general conclusions about a region on the basis of scant data, we have avoided doing so in favor of presenting several countries. Our survey of the world's countries gives us some confidence that these are reasonably representative.

Telecommunications in Kenya is indicative of the problems faced by many, although not all, African countries. Mexico, a North American country, is in many ways indicative of telecommunications in Latin America, although its close relationship with the United States diminishes its value somewhat in this regard. Nonetheless, the infrastructure and problems of Mexican telecommunications are similar to those faced by Latin American countries. Finally, India is representative of the Indian subcontinent. Telecommunications in neighboring states is equivalent to or less developed than in India. The Japanese experience is similar to that of many Pacific Rim countries, all of which are important trading partners. Finally, telecommunications in Eastern Europe and the Commonwealth of Independent States is in a state of great flux today. We have presented the case of Hungary as a leading example of what users can expect to find there.

Of these cases, all but New Zealand have a significant manufacturing infrastructure for telecommunications equipment. Although we didn't deliberately select these cases because of their infrastructures, this is noteworthy. Many countries don't have a significant infrastructure. Further research is needed to determine if this is an important indicator.

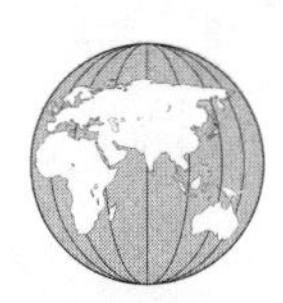

Endnotes

1. United States Central Intelligence Agency *1991 World Factbook.* Available through the U.S. National Technical Information Administration (NTIA), Springfield, VA 22161

2. William B. Garrison, Jr., *Four Case Studies of Structural Alterations of the Telecommunications Industry* (D.C.: Annenberg Washington Program, January 1988), p. 64.

3. *Eastern European and Former Soviet Telecom Report* (EESTR), vol. 3, no. 8 (August 1, 1992), p. 9.

4. See *Doing Business with DBP Telekom in the Center of Europe* (Bonn: DBP Directorate, April 1991).

5. Raymond M. Duch, *Privatizing the Economy: Telecommunications Policy in Comparative Perspective* (Ann Arbor: University of Michigan Press, 1991), p. 125.

6. Eli Noam, *Telecommunications in Europe* (New York: Oxford University Press, 1992), p. 80.

7. Noam, pp. 91-95.

8. See *Reform of the Postal and Telecommunications System in the Federal Republic of Germany: Concept of the Federal Government for the Restructuring of the Telecommunications Market* (Heidelberg: Federal Minister of Posts and Telecommunications, May 1988), for a discussion of these provisions.

9. *Reform of the Postal*, pp. 5-6.

10. EESTR, vol. 3, no. 8, p. 8.

11. *Telecommunication Reports International* (August 7, 1992), pp. 1-2.

12. *Doing Business with DBP Telekom*, p. 33. Though not specified, it is assumed that these figures are 1990 data.

13. "Germany: The Commercial and Regulatory Environment," *Datapro Reports on International Telecommunications* (September 1991), p. 327.

14. "Germany: The Commercial," p. 330.

15. "Germany: The Commercial," p. 328.

16. EESTR, vol. 3, no. 9 (September 1, 1992), p. 14.

17. *Telecommunication Reports International* (February 7, 1992), p. 14.

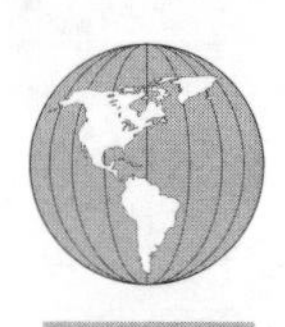

18. *Doing Business with DBP Telekom*, p. 32.

19. EESTR, vol 3, no. 3 (March 1, 1992), p. 13.

20. EESTR, vol 3, no. 6 (June 3, 1992), p. 14.

21. The details of the approval process are taken from *Doing Business with DBP Telekom*, pp. 37-46.

22. See *Performance Indicators for Public Telecommunications Operators* (Paris: OECD, 1991), p. 66, for one example.

23. *Telecommunication Reports International* (March 6, 1992), p. 7.

24. "Hungary: A Giant Step Ahead," *Business Week* (April 15, 1991), p. 58.

25. John Williamson, Steve Titch, and Peter Purton, "The Curtain Rises on Telecommunications in Eastern Europe," *Telephony* (July 6, 1992), pp. 32-33.

26. "What the Brochures Don't Tell You," *Business Week* (April 15, 1991), p. 52.

27. Clare McCarthy, "Hungary: The Commercial and Regulatory Environment," *Datapro Reports on International Telecommunications* (March 1992), p. 361.

28. One estimate is U.S. $7 to $8 billion. McCarthy, p. 361.

29. Dow Jones News Retrieval Service, December 1, 1992.

30. Erik Whitlock and Emilia Nyevrikel, "The Evolution of Hungarian Telecommunications Policy," in *Telecommunications Policy* (April 1992), pp. 253-4.

31. McCarthy, p. 362.

32. *Eastern Europe Business Bulletin* (March 1991), pp. 1-2.

33. McCarthy, p. 362.

34. Dow Jones News Retrieval Service, November 30, 1992.

35. *Eastern European and Former Soviet Telecommunications Report* (October 1, 1992), p. 4. Some developments in Hungary occurred too late to be fully addressed at this writing. According to *Telecommunications Reports International* (December 11, 1992), p. 22, the Hungarian parliament passed legislation that would eventually allow foreign investment in local telephone, mobile, and paging services; and which would limit the length of time to two years during which the Hungarian Telephone Company could remain a monopoly provider.

36. McCarthy, p. 362.

37. Whitlock and Nyevrikel, p. 254.

38. *Telephony*, pp. 32-3.

39. EESTR, vol. 3, no. 4 (April 1, 1992), p. 14.

40. McCarthy, p. 361.

41. McCarthy, p. 365.

42. Whitlock and Nyevrikel, p. 249.

43. EESTR, vol. 3, no. 5 (May 1, 1992), p. 6.

44. The source for much of the following material is *McCarthy*, pp. 363-366.

45. Whitlock and Nyevrikel, p. 251.

46. Whitlock and Nyevrikel, p. 251.

47. Steven van Slageren and Pit Gooskens, *Connected with the Future: Telecommunications and Economic Development, India as an Example*, Stichting Onderzoek Bedrijfstak Elektrotechniek, Eindhoven (Netherlands), 1990. Available in the U.S. through the National Technical Information Service (NTIS), Springfield, VA 22161, as report number PB91-164848.

48. van Slageren and Gooskens, p. 47.

49. Datapro Reports on International Telecommunications, *India: The Commercial and Regulatory Environment* (September 1991), p. IT10-040-405.

50. Datapro.

51. Jean-Pierre Vercruysse, "Telecommunications in India: 'Deregulation' vs. Self Reliance," *Telematics and Informatics*, vol. 7, no. 2, pp. 109-121.

52. Vercruysse.

53. Vercruysse.

54. *Datapro*, pp. IT10-040-403.

55. Datapro, *India*, p. 114.

56. Datapro *India*, pp. IT10-040-405.

57. Vercruysse, p. 113.

58. T.H. Chowdary, "Telecommunications Restructuring in Developing Countries," *Telecommunications Policy* (September/October 1992), p. 599.

59. *Datapro*, pp. IT10-040-402.

60. Heather E. Hudson, "Telecommunications in Africa: The Role of the ITU," *Telecommunications Policy* (August 1991), pp. 343-350.

61. Meheroo Jusawalla, "Is the Communications Link Still Missing?" *Telecommunications Policy* (August 1992), pp. 485-503.

62. Hudson, p. 343.

63. Jusawalla, p. 500.

64. Raymond U. , "Telecommunications in Kenya: Development and Policy Issues," *Telecommunications Policy* (September/October 1992), pp. 603-611.

65. Akwule, p. 608.

66. In fact, the Gilgil complex supplies cable forms for AT&T's Northern Ireland assembly plant.

67. Datapro Reports on International Telecommunications, *Kenya: The Commercial and Regulatory Environment* (September 1992), pp. IT10-011-352.

68. Datapro *Kenya*, pp. IT10-011-353.

69. Akwule, p. 606.

70. *OECD Communications Outlook 1990,* Paris: OECD, 1990.

71. Bruce Willey, "A Latin America Telecommunications Primer," *Telecommunications (North American Edition)* (March 1992), pp. 45-50.

72. Peter Heywood and Stephen Saunders, "¡Competition, Sí!" *Data Communications* (September 1991), pp. 70-74.

73. Larry Luxner, "Mexico Reaches for New Telecom Heights," *Telephony* (February 3, 1992), pp. 22-28.

74. Luxner, p. 23.

75. Datapro Reports on International Telecommunications, *Mexico: The Commercial and Regulatory Environment* (February 1991), pp. IT10-056-255.

76. Contact address: Telecommunications Corporation of New Zealand, Ltd.; PO Box 1471; Wellington, New Zealand.

77. APEC Telecommunications Working Group, *The State of Telecommunications Infrastructure and Regulatory Environment of APEC Economies* (November 1991), p. 187. Available through: Pacific Economic Cooperation Conference; 1755 Massachusetts Ave. NW; Washington, D.C. 20036.

78. Datapro Reports on International Telecommunications, *New Zealand: The Commercial and Regulatory Environment*, Report No. IT10-060-601 (September 1991).

79. Crawshay, Grant R. "Recent Commercial Developments in the New Zealand Marketplace," *Proceedings of the 1992 Pacific Telecommunications Conference* Jan. 1992, p. 546. Available through: Pacific Telecommunications Council; 2454 South Beretania Street, Suite 302; Honolulu, HI 96822

80. Crawshay, p. 547.

81. Raymond M. Duch, *Privatizating the Economy: Telecommunications Policy in Comparative Perspective* (Ann Arbor: University of Michigan Press, 1991), pp. 219-221.

82. Duch, pp. 219-221.

83. Noam, p. 22.

84. Duch, pp. 228.

85. Duch, pp. 218-219.

86. See William Garrison, *Four Case Studies of Structural Alterations of the Telecommunications Industry* (Washington D.C.: Annenberg Washington Program, January 1988), for an excellent discussion of this policy.

87. Alan Bell and Martin Cave in "Lessons from the UK Duopoly Review," *Telecommunication: New Signposts to Old Roads*, ed. Franca Klaver and Paul Slaa (Amsterdam: IOS Press, 1991), pp. 83-88, discuss the models used in assessing the effects of approaches to competition in these newly opened areas.

88. These deliberations are well presented in what has come to be called the Littlechild Report. See Stephen C. Littlechild, *Regulation of British Telecommunications' Profitability: Report to the Secretary of State, February 1983*, Department of Industry, 1983.

89. The concept of Price Caps is discussed further in Chapter 3.

90. See *Performance Indicators for Public Telecommunications Operators* (Paris: OECD, 1990), pp. 123-124.

91. Sarah Griffiths and Clare McCarthy, "The United Kingdom: The Commercial and Regulatory Environment," *Datapro Reports on International Telecommunications*, (May 1992), p. 803.

92. Griffiths and McCarthy, p. 808.

93. See *Telecommunication Reports International*, (August 7, 1992), pp. 2-3.

94. Griffiths and McCarthy, p. 805. Much of this discussion is based on information from this source.

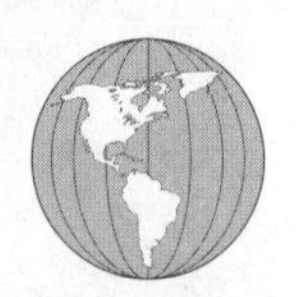

95. *Telecommunication Reports International* (March 6, 1992), pp. 6-7.

96. Griffiths and McCarthy, p. 809.

97. Griffiths and McCarthy, p. 811.

98. Griffiths and McCarthy, p. 813.

99. *Performance Indicators*, p. 84, Figure 4.2.

100. Federal Communications Commission, CC Docket No. 80-286, January 1991.

101. Released in October 1991 as NTIA Special Publication 91-26.

102. *NTIA Infrastructure Report*, pp. 8-9.

103. *NTIA Infrastructure Report*, p. 10.

A Look to the Future

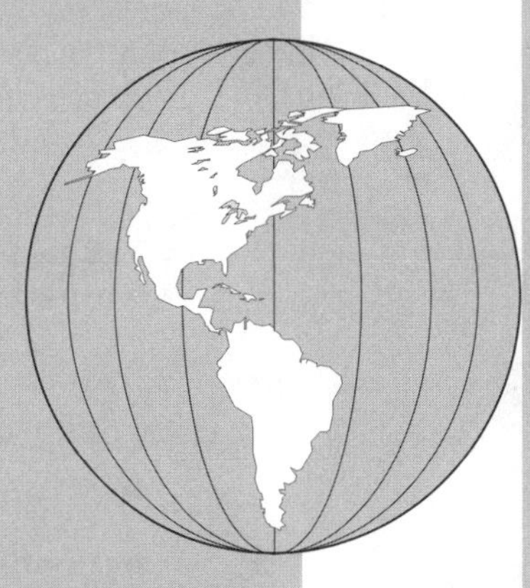

CHAPTER 8

This book focuses on historical issues and current developments in global telecommunications. Before discussing the situation in specific nations, we reviewed some underlying forces that led to the present state of international telecommunications. We've discussed the growth of privatization, the continuing push for increased competition, the pressure for cost-based pricing, the continued development of new technologies and services, the creation of new organizations and the evolution of old ones, and the challenges facing standards bodies.

Many forces and developments that brought us to this point will continue to shape telecommunications during the next decade. Technologies will continue to develop at a fast pace, countries will persist in seeking the best method for building and financing infrastructure given their history and current situation, and users will be ever more insistent that services be provided with the maximum efficiency and flexibility. While any view of the future is speculative at best, we can make the safe assumption that many of the forces that have dominated the recent past will continue to play a strong role in the future.

In this chapter, we consider what the near future might bring to international telecommunications. While prognostications such as these are always risky, we will attempt to make predictions not about specific countries but about international telecommunications in general.

Regulatory Trends

Four interwoven trends will continue to affect the development and provision of telecommunications services:

- Continued globalization of commercial enterprises
- Continued technological advances that "raise the bar" for what customers consider acceptable and expected telecommunications service
- Continued competition among nations to attract capital and economic activity
- Continued emphasis on user needs and improved customer service

Telecommunications is one vehicle that has facilitated the globalization of enterprises. Organizations can overcome limitations of space and time through satellites, transoceanic cables, digital switches, and other telecommunications equipment. Through these conduits, organizations can send voice and data messages, facsimiles, and images to virtually any location. Globalization will continue as a dominant force in the world economy as organizations search for new markets and new opportunities. As a result, organizations will continue to pressure telecommunications administrations to provide more and better services.

As telecommunications technologies and services continue to grow in sophistication and capability, the definition of "more and better" will also continue to grow in scope and complexity. The capability to place a simple phone call on the first try and with a minimum of interference will not suffice as acceptable service. Organizations will want the capability to videoconference, to move large amounts of data efficiently and economically, to combine voice and data for greater economy, and to have a larger available capacity at declining cost.

These increased telecommunications requirements present a significant challenge to telecommunications administrators. Providing more sophisticated services may be an expensive undertaking. However, nations will have to provide a robust telecommunications infrastructure to attract business because business will not settle for less.

All these trends will encourage the continuation of regulatory activities discussed earlier in this study.

- Privatization
- Competition
- Cost-based pricing
- Flexibility in tariff structures
- Greater emphasis on service quality standards

Privatization will continue to provide an attractive mechanism for the capital needed to finance expensive infrastructure building or rebuilding. In nations where the telecommunications enterprise is state-owned and state-run, funds for infrastructure building usually come from increased rates or increased taxes. Neither are attractive options. To avoid either alternative, Britain turned to privatization and the infusion of capital through the sale of equity shares. For developing nations that don't have well-developed capital markets, foreign investment has provided needed capital. Signals indicate that the privatization trend continues. In Eastern Europe, for example, countries such as Hungary are considering the privatization of their telecommunications function. In Western Europe, countries taking a conservative view of infrastructure ownership, such as Germany, are also looking at privatization as a valuable step. Developing countries such as Chile, Argentina, and Venezuela are also relying on external investment through privatization to help build their infrastructures to international standards.

As the technology becomes more complex and the demands for customer service become clearer, those nations lacking a work force trained to provide the demanded services will continue to look for concessionaire arrangements and management contracts to fulfill service provision needs. While these nations may not go so far as privatization of ownership, they will privatize their operations in order to encourage economic activity.

Competition, a dominant theme during the past decade, will continue to play a major role in the international telecommunications arena. As the case in the U.S. experience shows, after competition is unleashed, it is difficult to contain. In the United States, competition began in the customer premise equipment (CPE) market, was introduced into interLATA long distance, and has made inroads into intraLATA long distance and now local transport. (See the United States case study in Chapter 7, "Case Studies," for an explanation of interLATA and intraLATA.)

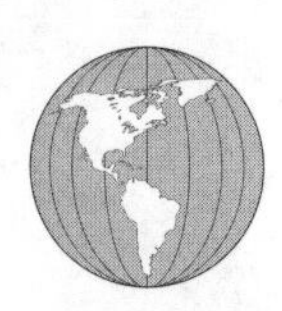

In the United States and other large telecommunications markets, competition continues to grow. The United Kindom is building on its earlier experience with the duopoly configuration that provided limited competition by opening telecommunications services to an expanded list of competitors. Japan continues to explore broader competitive initiatives as well. The European Community (EC) continues to foster competition in telecommunications services through a series of directives seeking to liberalize all services short of local service.

The prospects for competition in developing countries are not so clear. For many developing nations, the major thrust of telecommunications policy may still be the attainment of some measure of universal service. In pursuit of more broadly available, affordable service, these nations may continue to seek subsidies from long distance services and relatively high accounting rates to finance infrastructure development and to keep local rates low. Developing nations that have sought foreign investment sometimes must offer monopoly protection to foreign service providers in order to attract foreign network operators. Such activities preclude competition, at least in the short term.

As developing nations build and expand their infrastructure and as they welcome global enterprises, they undoubtedly will need to open some services to competition. Competition at the peripheries of the public network (value added services, for example) will become more prevalent in developing nations if the network provider cannot provide enhanced services and if increasingly sophisticated users exert pressure for such services.

Indications that the forces of competition are hard to contain after they are released are the failed efforts at the World Administrative Telephone and Telegraph Conference (WATTC) held in Melbourne to form anticompetitive recommendations (see the section on the ITU in Chapter 4, "Organizational Issues," for details). Another factor that suggests competitive forces will be difficult to contain is the growing number of new players who have entered the marketplace ready to provide service. Competitors and potential competitors have developed in all areas of telecommunications: be it competitive satellite system providers, consortia placing transoceanic fiber-optic cables, CPE manufacturers, or providers of VANs. Curtailing the activities of these new players, especially because they meet user needs and demands, is not a likely priority of telecommunications policymakers in most nations, especially in light of the continuing pressure to treat telecommunications as a commodity subject to the provisions of the free trade agreements. (See the discussion of the General Agreement on Trade in Services (GATS) in Chapter 3, "Economic Issues.")

The growing demands of users will continue to fuel the expansion of competition, as users seek new services and efficient, customer-oriented service. The growing demands of users will also continue to impact tariffs and service quality standards.

In the past, the ability to form private networks, to order leased lines, to connect leased lines to the public network, and to resell capacity on leased lines were far from universally available. Today, private networks, leased lines, and resale are becoming common features of the telecommunications landscape. They will be more prevalent in the future as network providers seek to meet user demands.

The demand for leased lines provides only one indication of user demands for more flexible arrangements. As users seek greater efficiency and more responsive services, telecommunications administrators must respond by bringing new services (such as VPN and ISDN) under tariff and allowing for customer-specific arrangements. The alternative is to risk losing the customers, perhaps permanently, to other nations in the global marketplace. These tariffs will also reflect a continuing trend to reduce both domestic and international long distance rates, as administrations seek to provide an attractive environment for business enterprises.

The globalization of companies will continue to exert pressure on all nations to provide similar services at comparable rates. Telecommunications managers responsible for assembling global networks will increasingly demand continuity in pricing and the type of available services. These telecommunications managers will also continue to push for continuity in the quality of services offered, regardless of national boundary. Such pressures will lead to more reporting of service quality standards; indeed, such pressures will lead to the development of standards and quality indicators.

These pressures on tariffs, quality standards, prices, and the type of services offered will bring some uniformity to the telecommunications landscape in all countries. As an example, a recent EC directive sought to provide users with the right to access leased lines all across the EC and to have those lines meet a common technical standard. Efforts such as this will become more prevalent in the years ahead.

Economic Trends

Although we discussed "regulatory issues" and "economic issues" separately in this book, separating these two areas is difficult. No discussion involving competition can occur without the consideration of pricing issues. Tariff issues include the rate relationships between local and long distance services. The trend toward privatization and competition profoundly affects accounting and settlement rates and procedures. Just as the various regulatory topics we discussed are intertwined, they are also intertwined with topics we categorized as "economic."

As is the case with regulatory trends, current economic trends will continue into the future. These economic trends will be fueled by the continuing movement toward competition and the replacement of "business as usual" with a market approach. The area of accounting and settlements is a good example of this process.

The present method of negotiating accounting and settlement rates was formulated at a time when most telecommunications administrations were state owned and operated as monopolies. A monopoly PTT might sit at one end of the negotiations; AT&T might sit at the other end (it was a privately owned entity but a monopoly nonetheless). Negotiations were handled on a bilateral basis, with agreements binding on only the two entities involved. The whole process was, and still is, time consuming and cumbersome. With the introduction of privatization and competition, the process is becoming unworkable.

Because of privatization, PTTs increasingly do not own and operate the telecommunications system. The question then becomes who should do the negotiating and on whose behalf. In the United States, AT&T has traditionally negotiated on its own with minimal FCC involvement. With the introduction of competitive carriers (such as MCI and Sprint), the process has become more involved, with the FCC playing a greater role, but with the individual companies still negotiating on their own. As other nations privatize and open their markets to competitive carriers, the picture becomes murkier at both ends of the negotiating table. Should PTTs negotiate on behalf of the new competitive carriers, or should the carriers negotiate directly with one another? Should the rates that are negotiated be binding for all carriers from a specific country? These are questions that governments all over the globe will need to address.

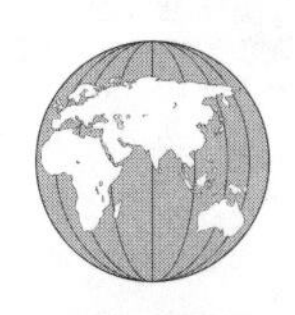

The introduction of competitive carriers ushers new players into the accounting and settlement arena. The growing number of joint ventures and consortia introduces the same carriers to a variety of telecommunications markets. The easing of rules regarding leased lines and resale creates networks that are controlled at both ends by the same entity. All these factors suggest that, rather than conducting accounting and settlement agreements between national entities, we may see the rise of single-company controlled international networks that need no international settlement agreements, intercompany agreements, and other mechanisms inherent to a multicarrier environment.

Because of increasing access to leased lines and the liberalization of rules regarding resale (two more signs of increased competition), a growing amount of international traffic will not fall under the accounting and settlement process at all. Indeed, leased lines and resale provide a mechanism for bypassing national public switched networks and the issue of accounting and settlements.

The threat of bypass has traditionally driven down prices for public switched network services. The best example of this is long distance services in the United States. Bypass in the international arena will be yet another force for bringing accounting and settlement rates closer to cost. The dominant force for such a decrease has, of course, been complaints from the United States, which faces a large deficit in accounting and settlement payments (see Chapter 3). Another factor driving down accounting rates will be the increased competition among nations to attract telecommunications traffic and business. Nations seeking a positive business climate will want to keep incoming and outgoing calling as economical as possible.

Complaints regarding above-cost accounting rates appear as one manifestation of a movement away from subsidies to a more cost-based approach. As countries strive to be globally competitive, they will continue to lower their international long distance charges. They will also lower the subsidies from domestic long distance to local service and will reduce any business to residential subsidies inherent in their rate structures.

International policy-making bodies seem to be encouraging this move toward cost-based pricing. CCITT recommendations, for example, speak to the advantages of cost-based pricing. The EC's telecommunications policies espouse cost-based pricing as a laudable goal. The continuing move toward competition will encourage cost-based pricing. In theory at least, an effective competitive marketplace drives prices closer to cost.

Another underlying trend that will continue to encourage competition and cost-based pricing is the inclusion of telecommunications as a service regarded

as an item of trade. When policymakers considered telecommunications purely as a utility, the major policy considerations involved the provision of access to a nation's citizenry. A utility functions as a facilitator, a key ingredient needed in order for other activities to occur, such as some business functions. A utility is usually provided to the public on a monopoly basis.

Because of technological developments, telecommunications has become more than a utility. It is a commodity and a means by which entities communicate. Competitive manufacturers can produce telecommunications equipment. Competing carriers can provide VANs, long distance services, and cellular and mobile communications. Privately owned companies now seek entry into newly competitive telecommunications markets all over the globe. For many nations, telecommunications has become a good to be traded, a fact that is recognized in the current efforts to include telecommunications in the General Agreement on Trade in Services (GATS, see Chapter 3).

Trade issues surrounding telecommunications, as well as other goods and services, include market entry, fair pricing, and perhaps of greatest significance, multilateral treatment. Arrangements concerning international telecommunications have traditionally been bilateral. That approach is not reflective of what has happened to international telecommunications during the past decade. Most global entities seek network configurations that span a number of countries and continents. Building a network through bilateral negotiations and arrangements is difficult, complex, and inefficient. International telecommunications has become a multilateral undertaking. Trade negotiations, with their basic tenet of "most favored nation" treatment, reflect that multilateral configuration.

As we pointed out in Chapter 3, some uncertainty arises about the relationship between GATS provisions and ITU recommendations regarding telecommunications. Some question even exists about whether the GATS provisions will actually supersede the ITU's procedures and recommendations in establishing policies for the provision of telecommunications services. Reliance on the GATS process would certainly strengthen the current move toward competition. Even if the GATS does not supersede the ITU's CCITT recommendations, the GATS process has strengthened the move toward a competitive telecommunications environment.

As the number of players in the telecommunications marketplace increases, the role of technical standards becomes more important. In the past, standards were essentially a ratification of accepted practice, developed over years of

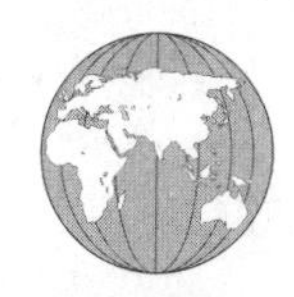

experience. This suggested a deliberate and conservative process. As competition emerged in the marketplace and as the technological change accelerated, this way of developing standards became unacceptable. Until recently, for example, CCITT standards were approved every four years. In today's marketplace, an entire generation of technology can evolve and reach maturity in that time frame. Clearly, this process was no longer acceptable, so a set of accelerated procedures was adopted. The very makeup of the CCITT voting membership complicated the standard setting process; the membership was dominated by administrations, not manufacturers.

The process has been somewhat better in the International Standards Organization (ISO), but it is still time consuming to the point where the effectiveness of the process is seriously in question. Excessive complexity and substantial delays in ISDN and OSI standards development have effectively hampered the emergence of those standards as the dominant marketplace standards that they were meant to be. Standards organizations responded by developing standards in advance of the market, something referred to as *anticipatory standards*. The concept follows that the standard be ready by the time the technology matures to make the standard feasible and cost effective to implement. While this has succeeded in a number of cases, particularly in modem standards, this approach resulted in the development of standards that have never been implemented. Inevitably, this has led to standards that are overly complex and difficult to implement.

As a result of all these trends, consortia (such as the Open Software Foundation and X/Open) have begun to form, and this trend will likely continue. A consortium of firms effectively bypasses the formal standards process and attempts to develop a de facto standard on the strength of the market power of its members. These efforts have produced mixed success in practice.[1] The standards development organizations will need to adopt more changes to their procedures for further streamlining the standards development process and for including the consortia in the process.

Organizational Trends

In Chapter 4, we surveyed the major organizations involved in international telecommunications. These organizations were formed to solve a set of specific problems or to address specific issues that were pressing at the time of their formation. There has never been a comprehensive review of international organizations from a systems standpoint. As a result, we have some functions

that are overrepresented, such as the tariff and service quality areas (CCITT and GATS) and the standards development areas (CCITT and ISO), and other areas that are not served at all (accounting and settlements). This lack of efficient organizational coverage of international issues will present increasing problems to a growing global marketplace.

Furthermore, some organizations, such as Intelsat, are caught by changing technology and competition. Undersea fiber-optic cables have weakened Intelsat in the developed nations, but it remains an important organization for developing nations. This results in a stratification of capability and options that is clearly not feasible in the long run. Some countries, such as the United States, allow competition with Intelsat, further reducing its influence. As with all firms in a competitive global marketplace, organizations like Intelsat must continue redefining their objectives and providing customer service to remain in existence. Despite some of their problems, the international community requires their services.

Plans for a unified European market substantially strengthen regional organizations, such as the Commission of the European Communities (CEC), the European Telecommunications Standards Institute (ETSI), and the Comité Europé en de Normalisation (CEN). These organizations are developing standards and policies that will have an international influence as nations wish to conduct trade with what would be the world's largest market. Despite the recent setbacks that this plan has endured in Denmark, and the weak popular support of France and the United Kingdom, the influence of these regional bodies will continue to be important. Nations of other regions have noted the success and influence of the EC and are likely to form similar regional bodies. It is easy to imagine regional organizations in the Pacific Rim, Africa, the Organization of Independent States, North America (perhaps under the auspices of the newly signed North American Free Trade Agreement), and South America. Clearly, the ITU, Intelsat, the ISO, and other international organizations will need to recognize these emerging regional organizations and develop strategies to incorporate them into the international process.

User organizations such as the International Telecommunications Users Group (INTUG) will likely gain strength in the future because of the trend toward viewing telecommunications as a user-oriented service. This will cause service providers to meet the requirements of the users, who are represented through these organizations. There is also a trend toward stimulating user involvement in technical standards committees. In the past, standards committees included representatives from telecommunications administrations,

service providers, and product manufacturers. Users were absent, and perhaps even discouraged, from attending and participating. Most standards committees are making significant efforts to increase the role of users in the standards process; again, this may occur through user organizations.

Technological and economic changes will change the roles of the traditional telecommunications organizations, such as the ITU and Intelsat. As we discussed previously, substantial overlap in authority exists between organizations. This will not be feasible in the long term, and the boundaries between these organizations will need to be defined more clearly. Economic and regulatory trends, such as privatization and competition, will force the existing organizations to reconsider their traditional roles. A privatized, competitive telecommunications market will cause the ITU to focus primarily on its standards development and data gathering activities, both of which are important for the efficient operation of international networks. The tariff and economic development activities may become more relevant to the GATS framework than to the ITU activities, and may come to reside there. Intelsat must also adjust to increasing competition from other satellite organizations and from competing technologies, primarily fiber optics. Its role as a provider of mainline telephone circuits is likely to diminish, and its role as a provider of backup telephone circuits and television broadcast channels is likely to increase. Intelsat will need to adjust its organizational strategy appropriately.

As in business, organizations that don't adjust to changing realities will eventually cease to exist. As a result, we can expect to observe substantial changes in the functions and operations of the current organizations. Indeed, many of these changes are already taking place in the ITU in general and the CCITT in particular. Changes will have to occur in the other organizations as well. New organizations may also emerge, with their roots in consortia, ad hoc committees, or regional organizations to address issues not handled adequately by international organizations today, such as accounting and settlements. The need for an international organization will become more apparent as the range of provided international telecommunications services increases. The number of bilateral agreements that will need to be reached will become too large to be practical; each agreement is costly to achieve, so service providers will feel compelled to create a more efficient exchange mechanism than exists today.

Technical Trends

Many of the changes discussed previously in this chapter have been stimulated by advancements in technology. As the cost of providing service decreased, pressure to permit competitive carriers increased because the cost of entry into the market increased. As fiber-optic transmission systems became feasible, the role of satellites in international telecommunications diminished. We can expect only technology to continue changing at the rate we have witnessed in recent years.

The advent of fiber-optic transmission brought about high capacity transmission systems (several tens of gigabits per second are possible today), an increase in the transmission quality, and a reduction in the cost of providing capacity. This has spawned fast packet-switching techniques, where all kinds of traffic (voice, data, and video) are collected into very small packets (typically 48 bytes) and transmitted over very high-speed networks with low delays and low error rates. Therefore, the entire bandwidth of the transmission channel can be dynamically assigned based on demand. This stands in contrast to today's approach, where the multiplexing of low speed channels creates a high speed channel. If one of these subchannels is idle, a user who requires more bandwidth cannot use that capacity. This traditional approach is often called *synchronous transfer mode* (STM), and the emerging approach is called *asynchronous transfer mode* (ATM). In many ways, ATM is the ultimate incarnation of voice and data integration because integration occurs in all network components, to some extent even in the user's terminal equipment. While product manufacturers around the world are implementing the ATM today, many questions remain before this technology can emerge as an important mainstream technology.

The emergence of fiber-optic transmission, coupled with modern digital switching technology, has also stimulated the development of many new public network services. Most often, these public services can be provided at lower cost than a private network can provide similar services. As a result, we will see a trend away from private networks to custom public services, such as virtual networks and frame relay. The widespread implementation of Signaling System 7 (SS7), which was standardized by the CCITT, makes many of these services possible. SS7, which is being widely implemented in developed countries, can provide a wide range of network-based calling features not available before. Widespread availability and interoperability of SS7 will result in innovative services for users across national boundaries, further strengthening the role of public networks in modern business operations.

While these new technologies will likely cause an increase in public network usage, the emergence of open network management standards will stimulate interest from users to manage their virtual networks privately. International business users are constantly seeking cost-effective approaches for their operations that will stimulate interest in public networks. On the other hand, they will require control of the network resources to the extent that the telecommunications system represents a strategic technology or capability. This will require appropriate system architectures and changes in business practices for carriers so that changes that one user makes do not affect other users.

Although considering ISDN an emerging technology is unusual because it has been under development since the 1970s, widespread use of ISDN may yet occur. The problems that have existed pertain to incompatibilities among different manufacturers' implementations. Many of these difficulties have been solved. Furthermore, countries such as Germany and France have made widespread ISDN capability a national priority. While basic rate ISDN may not be completely satisfactory for all applications, it significantly improves upon currently available technologies. As ISDN becomes available across national boundaries, it may yet become a viable technology for international telecommunications users.

The principal technology trends in the information technology industries in general and telecommunications in particular have been "smaller, faster, cheaper." This applies to nearly all components of a telecommunications system: transmission technology, computer equipment, memory, storage, and so on. This drives the development of new architectures, new services, and new applications. With the emergence of technical standards as a critical infrastructure for the information technology industries, these new architectures, services, and applications have become the fundamental basis for competition among rival providers.

Conclusions

The trends we've discussed are tightly interwoven. As you have seen, technological changes cause organizational and economic changes. Similarly, changes in regulatory structure can stimulate technological developments. These four threads, then, are tightly interwoven. This strong relationship exists in all countries.

Developed and developing countries face different kinds of problems as they move toward modern network technologies. Developed countries tend to be at the forefront of developments, but they have a large installed base of existing equipment, not yet depreciated, using older technologies. These countries seek changes in regulations and tariffs so they can proceed with the conversion to digital and fiber-optic networks as soon as possible. Developing countries frequently do not have a large base of installed equipment to depreciate, so they could build widespread modern networks, if they only possessed the capital to do so. Therefore, each of these countries faces unique problems.

As you have seen in the case studies, different countries have proposed a variety of solutions to these problems. In many senses, the countries adopting riskier strategies act as laboratories for other countries; techniques adopted in one country frequently emerge in other countries. For example, the United Kingdom introduced price cap regulation, and the U.S. FCC subsequently adopted it. The United States led the world by introducing deregulation and competition; these approaches have been adopted by Japan and others (and by New Zealand in a more radical form). Similarly, Germany and France have made significant technological commitments to the ISDN as a strategic technology; whereas other countries are waiting for the demand for ISDN services to develop. Undoubtedly, the success or failure of the German and French experience will serve as a lesson for the less aggressive countries.

International telecommunications will continue to be an interesting and dynamic field. We hope that we have provided you with the background needed to interpret developments as they occur and to use them in your business or enterprise. As industries continue to globalize, the ability to adjust to a changing climate in telecommunications will become increasingly important.

Endnote

1. See Martin B.H. Weiss and Carl Cargill, "Consortia in the Standards Setting Process," *Journal of the American Society for Information Science (JASIS)* (September 1992).

Country Codes

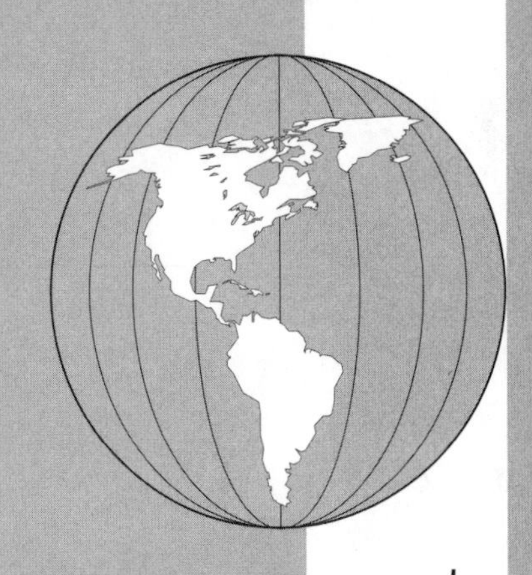

APPENDIX A

This appendix contains a table of country codes, taken from the CCITT Recommendation E.163 (1988). The world numbering zone of each country is the first digit of its country code. Some zones (notably Zone 1 and Zone 7) have a jointly administered numbering plan. Further, the countries of Algeria, Libya, Morocco, and Tunisia have several country codes. You should contact the embassies of these countries to find out which are currently active. In addition to the country codes, many countries may use city codes, trunk codes, or area codes within their countries. These are nationally administered and therefore not a matter of international standardization. Contact the telephone administration of the country in question to obtain a list of these codes within that country.

This list of countries does not reflect all the recent political changes in Eastern Europe, Yugoslavia, and the former Soviet Union. For the most part, you can assume that the current country codes are still active. If you need further details on one of these countries, contact the country's embassy or the CCITT for updated information.

Table A.1. International Country Codes.

Country	Country Code
Afghanistan	93
Albania	355
Algeria	213, 214, 215
American Samoa	684
Angola	244
Anguilla	1

continues

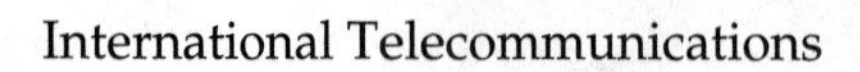

Table A.1. continued

Country	Country Code
Antigua and Barbuda	1
Argentina	54
Aruba	297
Ascension	247
Australia	61
Australian External Territories	672
Austria	43
Bahamas	1
Bahrain	973
Bangladesh	880
Barbados	1
Belgium	32
Belize	501
Benin	229
Bermuda	1
Bhutan	975
Bolivia	591
Botswana	267
Brazil	55
British Virgin Islands	1
Brunei Darussalam	673
Bulgaria	359
Burkina Faso	226
Burma	95
Burundi	257
Cameroon	237

Country	Country Code
Canada	1
Cape Verde	238
Cayman Islands	1
Central African Republic	236
Chad	235
Chile	56
China (People's Republic)	86
China (Taiwan)	866
Colombia	57
Comoros	269
Congo	242
Cook Islands	682
Costa Rica	506
Cote d'Ivoire (Ivory Coast)	225
Cuba	53
Cyprus	357
Czechoslovakia	42
Denmark	45
Diego Garcia	246
Djibouti	253
Dominican Republic	1
Ecuador	593
Egypt	20
El Salvador	503
Equatorial Guinea	240
Ethiopia	251
Falkland Islands (Malvinas)	500

continues

Table A.1. continued

Country	Country Code
Faroe Islands	298
Fiji	679
Finland	358
France	33
French Polynesia	689
Gabonese Republic	241
Gambia	220
Germany	(Former West) 49 (Former East) 37
Ghana	233
Gibraltar	350
Greece	30
Greenland	299
Grenada	1
Guadeloupe	590
Guam	671
Guatemala	502
Guiana	594
Guinea	224
Guinea-Bissau	245
Guyana	592
Haiti	509
Honduras	504
Hong Kong	852
Hungary	36
Iceland	354

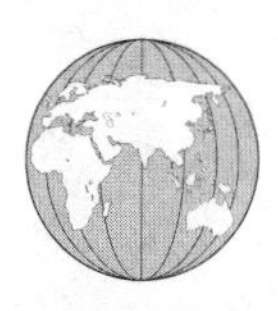

Country	Country Code
India	91
Indonesia	62
Iran	98
Iraq	964
Ireland	353
Israel	972
Italy	39
Jamaica	1
Japan	81
Jordan	962
Kampuchea	855
Kenya	254
Kiribati	686
Korea (Democratic People's Republic of)	850
Korea (Republic of)	82
Kuwait	965
Laos	856
Lebanon	961
Lesotho	266
Liberia	231
Libya	218, 219
Liechtenstein	41
Luxembourg	352
Macao	853
Madagascar	261
Malawi	265
Malaysia	60

continues

Table A.1. continued

Country	Country Code
Maldives	960
Mali	223
Malta	356
Mariana Islands	670
Maritime Mobile Service	87
Marshall Islands	692
Martinique	596
Maruitius	230
Mauritania	222
Mexico	52
Micronesia	691
Monaco	33
Mongolia	976
Montserrat	1
Morocco	210, 211, 212
Mozambique	258
Namibia	264
Nauru	674
Nepal	977
Netherlands	31
Netherlands Antilles	599
New Caledonia	687
New Zealand	64
Nicaragua	505
Niger	227
Nigeria	234

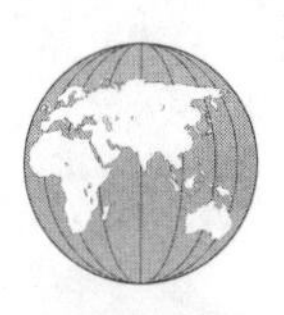

Country	Country Code
Niue Island	683
Norway	47
Oman	968
Pakistan	92
Palau	680
Panama	507
Papua New Guinea	675
Paraguay	595
Peru	51
Philippines	63
Poland	48
Portugal	351
Qatar	974
Reunion	262
Romania	40
Rwandese Republic	250
San Marino	295
Sao Tome and Principe	239
Saudi Arabia	966
Senegal	221
Seychelles	248
Sierra Leone	232
Singapore	65
Solomon Islands	677
Somalia	252
South Africa	27
Spain	34

continues

Table A.1. continued

Country	Country Code
Sri Lanka	94
St. Kitts and Nevis	1
St. Lucia	1
St. Pierre and Miquelon	508
St. Vincent and the Grenadines	1
Sudan	249
Suriname	597
Swaziland	268
Sweden	46
Switzerland	41
Syria	963
Tanzania	255
Thailand	66
Togoloese Republic	228
Tokelan	690
Tonga	676
Trinidad and Tobago	296
Tunisia	216, 217
Turkey	90
Turks and Caicos	1
Tuvalu	688
Uganda	256
United Arab Emirates	971
United Kingdom	44

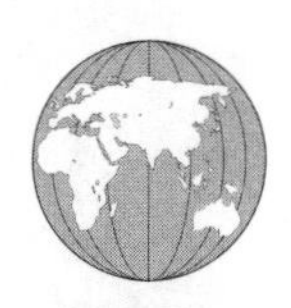

Country	Country Code
United States of America, Puerto Rico, and U.S. Virgin Islands	1
Uruguay	598
USSR (Former Republics of)	7
Vanuatu	678
Venezuela	58
Vietnam	83
Wallis and Futuna Islands	681
Western Samoa	685
Yemen (Arab Republic)	967
Yemen (People's Democratic Republic)	969
Yugoslavian Republics (Former)	38
Zaire	243
Zambia	260
Zanzibar (Tanzania)	259
Zimbabwe	263

Important International Telecommunications Organizations

Commission of the European Communities (CEC)

Directorate for Information, Publications and Documentation
Rue de la Loi 200
B-1049 Brussels
Belgium
Telephone: +32-234-61-11

The CEC has offices in each of its member countries, which include: Belgium, Denmark, France, Germany, Greece, Ireland, Italy, Luxembourg, the Netherlands, Portugal, Spain, and the United Kingdom. Offices are also located in Chile, Japan, Switzerland, the United States, and Venezuela.

International Telecommunications Satellite Organization (Intelsat)

3400 International Drive, N.W.
Washington, D.C. 20008-3098
Telephone: (202) 944-6800
Telex: 89-2707

International Telecommunications Union (ITU)

General Secretariat
Place des Nationes
CH1211 Geneva 20
Switzerland
Telephone: +41-22-730-51-11
Fax: +41-22-730-51-94
Telex: 421000 uit ch
Telegram: Burinterna Geneva

International Telecommunications Users Group (INTUG)

G.G. McKendrick, Executive Director
31 Westminster Palace Gardens
Artillery Row
London SW1P 1RR
England
Telephone: +44-71-799-2446
Fax: +44-71-799-2445
Telex: 927278

While users can participate directly in INTUG, participation occurs more frequently through its member organizations. INTUG currently has member organizations in Australia, Austria, Belgium, Canada, Denmark, Finland, France, Hong Kong, Japan, the Netherlands, New Zealand, Norway, South Africa, Sweden, the United Kingdom, and the United States. You should contact the INTUG office to find out which member organization is the most appropriate for you.

Organisation for Economic Cooperation and Development (OECD)

2, rue Andre-Pascal
75775 Paris CEDEX 16
France
Telephone: +33-1-45-24-82-00
Fax: +33-1-45-24-85-00
Telex: 620 160 OCDE
Telegram: Developeconomie

The OECD also has document sales offices in each of member countries, which include: Australia, Austria, Belgium, Canada, Denmark, Finland, France, Germany, Greece, Iceland, Ireland, Italy, Japan, Luxembourg, the Netherlands, New Zealand, Norway, Portugal, Spain, Sweden, Switzerland, Turkey, the United Kingdom, and the United States. Additional sales offices are located in Argentina, China, Hong Kong, India, Indonesia, Israel, Korea, Malaysia, Pakistan, Singapore, Sri Lanka, Taiwan, Thailand, Venezuela, and Yugoslavia.

Forms

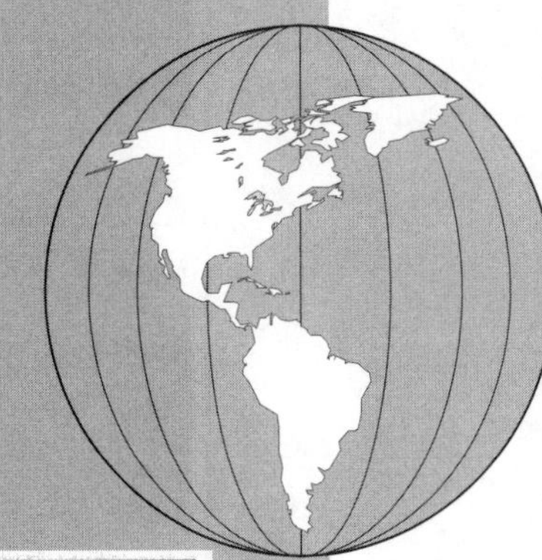

APPENDIX C

Form 1

Quantitative Statistics

Country: ____________
Date: ____________
Completed By: ____________

A) Direct Indicators

1. Compute the Average Telephone Penetration:
   a) Total Number of Telephones ____________

   b) Total Population ____________

   Average Telephone Penetration = (a) divided by (b) = ____________

2. Compute the Average Productivity Per Telephone
   a) Total Gross National Product ____________
   (or Gross Domestic Product)

   b) Total Number of Telephones ____________

   Average GNP per Telephone = (a) divided by (b) = ____________

3. Compute the Public Telephone Ratio
   a) Total Number of Public Telephones ____________
   (or Coin Telephones)

   b) Total Number of Telephones ____________

   Public Telephone Ratio = (b) divided by (a) = ____________

4. Compute Waiting Time for Service per Telephone
   a) Average Waiting Time for new service (months) ____________

   b) Total Number of Telephones ____________

   Waiting time per telephone = (a) divided by (b) = ____________

Form 2

Quantitative Statistics

Country: ____________
Date: ____________
Completed By: ____________

5. Compute Average Annual Growth of Network
   a) Total Number of Telephones 10 years ago ____________

   b) Total Number of Telephones today ____________

   c) Total Change in Telephones = (b) - (a) = ____________

   Annual Average Growth = (c) divided by 10*(b) = ____________

6. Total Calls per Capita and per Telephone
   a) Total Local Calls ____________

   b) Total Long Distance Calls ____________

   c) Total International Calls ____________

   d) Total Calls = (a) + (b) + (c) = ____________

   e) Population ____________

   f) Number of Telephones ____________

   Calls per Capita = (d) divided by (e) = ____________

   Calls per Telephone = (d) divided by (f) = ____________

   Ratio of Domestic calls to Intl. Calls = (a)+(b) divided by (c) = ____________

7. Compute the Number of Telephones per International Circuit
   a) Total number of telephones ____________

   b) Total number of satellite-based international circuits ____________

   c) Total number of cable-based international circuits ____________

   d) Total number of international circuits = (b) + (c) = ____________

   Telephones per International Circuit = (a) divided by (d) = ____________

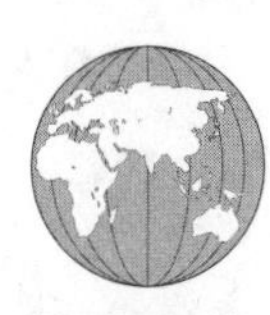

Form 3

Quantitative Statistics

Country: ______________
Date: ______________
Completed By: ______________

---

8. Compute the Number of Calls per International Circuit
   a) Total number of telephones calls ______________

   b) Total number of International circuits ______________

   c) Total number of International calls ______________

   Calls per International Circuit = (a) divided by (d) = ______________

   International Calls per International Circuit = (c) divided by (b) = ______________

9. Compute Average Annual Investment per Telephone
   a) Annual Telecommunications Investment ______________

   b) Total Number of Telephones ______________

   Average Investment per Telephone = (a) divided by (b) = ______________

10. Compute Fiber Optic Transmission Ratios
    a) Total Number of Fiber Optic Cable Route-Kilometers ______________

    b) Total Number of Route-Kilometers ______________

    c) Total Land Area ______________

    d) Total Population ______________

    Percent of Transmission Capacity that is Fiber = (a) divided by (b) * 100 = ______________

    Geographic Fiber Density Ratio = (a) divided by (c) = ______________

    Population Fiber Density Ratio = (a) divided by (d) = ______________

11. Compute the Percentage of Subscriber Lines that are served by Digital Switches
    a) Number of Subscriber Lines Served By Digital Switches ______________

    b) Total Number of Subscriber Lines ______________
       (Total Number of Telephones)

    Percentage Served By Digital Switches = (a) divided by (b) x 100 = ______________

Form 4

Quantitative Statistics

Country: ____________
Date:____________
Completed By:__________

12. Compute the Percentage of Subscriber Lines that are served by Stored Program Control Switches
    a) Number of Subscriber Lines Served By SPC Switches ____________________

    b) Total Number of Subscriber Lines ____________________
    (Total Number of Telephones)

    Percentage Served By Digital Switches = (a) divided by (b) x 100 = ____________

B) Indirect Indicators

1. Compute Electricity Generation per Capita
    a) Total Electricity Generated (kWh) ____________________

    b) Total Population ____________________

    Electricity Generated per Capita = (a) divided by (b) = ____________________

2. Compute Paved Road-km per Square Kilometer
    a) Total Paved Road Kilometers____________________

    b) Total Land Area ____________________

    Square km per Paved Road km = (b) divided by (a) = ____________________

3. Compute Railroad-km per Square Kilometer
    a) Total Railroad Kilometers____________________

    b) Total Land Area ____________________

    Square km per Railroad km = (b) divided by (a) = ____________________

4. Compute Airports with Paved Runways per Square Kilometer
    a) Total Airports with Paved Runways ____________________

    b) Total Land Area ____________________

    Square km per Airport = (b) divided by (a) = ____________________

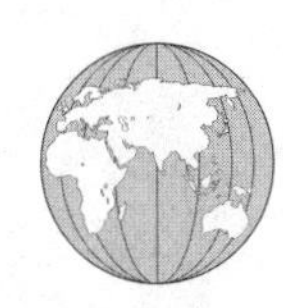

Form 5

Qualitative Information

Country: __________
Date: __________
Completed By: __________

1. What is the current ownership of public telephone provider?
   - Public
   - Private

2. If the ownership situation has changed,

   When did the change occur?

   What was the nature of the change?
   - Nationalization
   - Privatization

   If the change was privatization,
   - How was the ownership transferred?
   - Who now owns the enterprise, and in what proportions?
   - Is the ownership transfer complete?
     - If not, what is the schedule?

3. Is the Customer Premise Equipment (CPE) market competitive?
   - Is Type Approval required?

4. Is there a VAN market?
   - Is this market competitive?
   - What services are available?

5. Are there services that are provided by a private monopoly?
   - What are these services?
     - local
     - domestic long distance
     - international long distance
     - packet switching
     - other

6. What services are regulated?
   - What is the method of regulation?
   - Where in the governmental structure does the regulatory agency reside?
   - Under what law does the agency exist?

Form 6

Qualitative Information

Country: ____________
Date: ____________
Completed By: ____________

7. Are private lines (leased lines) permitted by the carrier?
   Are restrictions placed on their use?
   Do usage-sensitive charges exist?
   Is Resale allowed?

8. Are private networks allowed?
   Can they be connected to the public network?

9. What international organizations is the government/telecommunications provider a member of?
   ITU
   ETSI
   CEPT
   Intelsat
   Eutelsat
   Inmarsat
   Other

10. How is network investment distributed between rural and urban areas?

11. What are the stated telecommunications infrastructure goals with respect to
   Telephone penetration?
   Conversion to digital switching and transmission?
   Installation of fiber optic transmission?
   Conversion to CCITT Signalling System 7?
   Installation of international transmission capacity?
   Privatization, deregulation and liberalization (as applicable)?

12. Are service quality reports available or required?
   If so, what categories are included?

13. What are the pricing relationships among various services?
   business vs. residential
   local vs. long distance
   domestic vs. international
   leased line vs. switched

14. How do the international charges compare with those of other countries?

Glossary

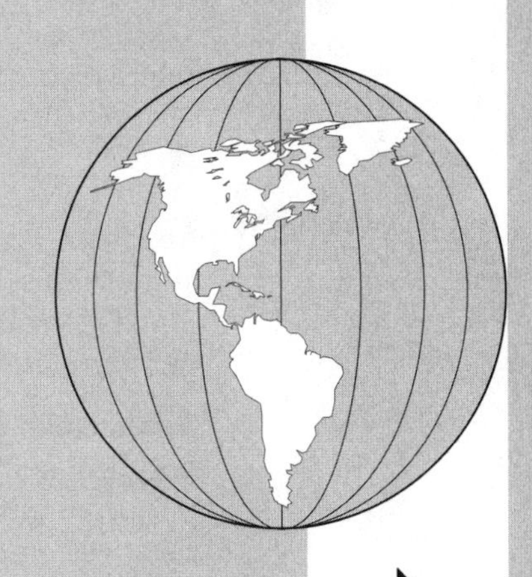

GLOSSARY

µ-Law companding A form of companding used by countries conforming to the North American digital transmission hierarchy (principally, the United States, Canada, and Japan).

A-Law companding A form of companding used by countries following CCITT digital multiplex standards.

ADPCM Adaptive differential pulse code modulation.

Analog telephone A telephone instrument in which the electrical output signal has an analog (as opposed to digital) format.

Anticipatory standards Standards that are developed prior to the existence of a market for products compatible with the standard.

Arabsat Consortium of Arab League nations that provides satellite services to the Middle East.

ARPANET A computer network, sponsored by the U.S. Department of Defense Advanced Research Projects Agency (ARPA), that began in the late 1960s. This network was the predecessor to the U.S. National Science Foundation network (NSFnet).

ASEAN Association of Southeast Asian Nations; an association of nations dealing with issues of mutual economic, technological, trade, and other concerns.

Asynchronous transfer mode (ATM) A technology in which the transmission bandwidth is dynamically allocated among all users by encapsulating the transmitted information in fixed-size cells.

Austro-German Telegraph Union A precursor to the International Telecommunications Union, this organization dealt with issues related to the interconnection of telegraph systems in the Austro-Hungarian Empire and in the German states in the 19th century.

Balanced loading A policy of the U.S. Federal Communications Commission designed to assure the continued economic health of Intelsat, despite the growth of transoceanic fiber-optic cables. This policy requires that U.S. long distance companies divide their international traffic in a balanced manner between Intelsat's satellite facilities and transoceanic fiber-optic cables.

Barrier to entry 1. A technical or economic impediment to market entry. 2. An economic term indicating that some form of impediment exists for new firms to enter a market.

Basic rate interface An ISDN term for the link to the ISDN customer location. The Basic Rate Interface link provides two "bearer channels," at 64 kilobits each, and a "data channel" at 16 kilobits. The bearer channels carry voice, data, or other types of messages. The data channel carries information about those messages.

Bearer channel (B channel) An ISDN term related to the link to the customer location. See Basic rate interface for further details.

Bell Operating Company (BOC) One of the firms of the former Bell System in the United States.

Berne Bureau The administrative body of the International Telegraph Union, which was formed in 1865; the Bureau was named for its location in Berne, Switzerland. In the early 1900s, it was renamed the International Bureau of the Telegraph Union.

Bharat Electronics Ltd. (BEL) An Indian manufacturing company.

Bidirectional line A line that is capable of carrying information in two directions simultaneously.

BT British Telecom, the UK company that was formed by privatizing the formerly state-owned telecommunications administration.

Cable & Wireless PLC A British telecommunications company.

Cable Landing License Act A U.S. law that governs the conditions under which entities may gain permission to "land" or terminate a transoceanic cable on U.S. territory.

Call Pairs Calls between two specific countries.

Call setup The process of setting up a call between two points. This includes the dialing process and the routing process required to establish the call.

CANTV The Venezuelan telephone administration, the official name of which is Compañía Anónima Nacional Teléfonos de Venezuela.

Carrier system A system used within a telephone network to carry information over distances.

CCIF International Telephone Consultative Committee. An ITU committee that merged with its telegraph counterpart in 1956 to form the International Telegraph and Telephone Consultative Committee or CCITT.

CCIR International Radio Consultative Committee. A body of the International Telecommunications Union, which formulates recommendations regarding radio use and services.

CCIT International Telegraph Consultative Committee. An ITU committee that merged with its telephone counterpart in 1956 to form the International Telegraph and Telephone Consultative Committee or CCITT.

CCITT Comite Consultatif Internationale de Telegraphique et Telephonique (or International Telephone and Telegraph Consultative Committee). An organ of the International Telecommunications Union that formulates recommended standards and procedures.

Cellular system A type of mobile telecommunications system in which the service area is divided into several geographically distinct zones (or cells) that serve customers in that area.

Central office (CO) A telephone switching office.

Chaining A technique in which several systems are connected serially, in the manner of a chain.

Channel banks Equipment in the telephone network that converts signals from analog to digital (and vice versa) and then formats them for transmission over a carrier system.

Circuit switches Switches capable of handling circuit switching.

Circuit switching A switching method in which the switching is done at the dedicated circuit level. A dedicated circuit is provided between two points as long as the connection between the two points is maintained.

Coaxial cable The type of cabling most often used to carry data between host computers and dedicated (or "dumb" terminals). Also used to deliver cable television service.

Committed information rate (CIR) The bit rate to which the carrier commits for frame relay service.

Common channel interoffice signaling (CCIS) A signaling system that predated Signaling System 7 in the U.S. telephone network. CCIS was implemented in the 1970s.

Common channel signaling (CCS) An out of band signaling technique in which all signaling information between switches is transmitted over a single, common channel.

Common channel signaling systems Systems that use a single (or common) channel to transmit all signaling information between network entities. An example of such a system is CCITT Signaling System 7.

Companding A technique used in telephone systems that emphasizes low amplitude (quiet) signals so that the dynamic range of the channel is improved.

Compatibility standards Standards concerned with the compatibility between systems, devices, or other entities.

Computer Inquiry I, II, III Investigative initiatives taken by the U.S. Federal Communications Commission during the 1970s and 1980s.

Connectionless network service (CLNS) A form of packet network service in which a virtual circuit need not be established prior to transmitting the data packet.

Coordination Committee on Harmonization (CCH) A committee of CEPT and the EC, formed in 1975, responsible for the harmonization of telecommunications standards.

Copenhagen Memorandum of Understanding (MOU) An agreement between the EC and CEPT establishing a framework for cooperation between the two organizations.

The Council of Europe A precursor to the European Community; formed in 1949, it addressed social issues and human rights.

Council of Ministers The decision-making body of the European Community.

Court of Justice The court of the European Community. The function of the court is to ensure that the laws of the European Community are followed.

CPE Customer premise equipment. Telecommunications equipment that is found at a customer location. CPE includes telephones, modems, and private switching systems. CPE was one of the first parts of the public network opened to competition.

Crossbar technology A type of circuit-switching technology that succeeded the step-by-step technology developed by Almon Strowger, developer of the first automatic switch.

Data channel or D channel An ISDN term related to the link to the customer location. See Basic rate interface for further details.

Data over voice (DOV) A method in which a regular voice line can be subdivided to carry both voice and data.

DCE Data communications equipment.

De facto standards Standards set through the free interplay of market forces.

De jure standards Standards set by force of law.

Digital European cordless telephone standard (DECT) A standard for wireless communications developed in the European Community.

Digital telephone A telephone instrument that produces an electrical signal in digital (as opposed to analog) format.

Directory service A computer network service that provides information services to clients. A typical information service would be to find the electronic mail address for a person or the physical network address from a logical network address.

Directory service agent (DSA) A component of a directory service that responds to inquiries from directory user agents.

Directory user agent (DUA) A component of a directory service that makes inquiries of a directory service agent on behalf of users.

Draft Agreement on Trade in Services See GATS.

Draft international standard (DIS) A stage of a standard in ISO.

Draft proposal A stage of a standard in ISO.

DTE Data terminal equipment.

Duopoly (fixed links) The approach taken by the UK to introduce competition into British telecommunications. In a duopoly there are two competitors. The UK government decided to make it a "fixed links" duopoly by requiring that only the two competitors could provide physical links to the customer. Hence, the term "fixed links."

E-mail A store and forward system concerned with the electronic delivery of messages.

EC European Community. The twelve European nations that, through the Treaty of Rome in 1957, agreed to establish a free trade zone among their nations.

Economic Commission for Europe A body formed after World War II to foster research activities among European nations and to facilitate distribution of relief resources to rebuild after the war.

Economy of scale An economic term indicating that costs per unit decrease as the number of units increase.

Economy of scope An economic term indicating that costs per unit decrease as the number of functions performed by the manufacturer in the development, production, sales, and implementation process increase.

EIA-232 standard An electrical and mechanical interface at the physical layer of the OSI reference model.

Electronic Corporation of India (ECI) A manufacturing firm in India.

Erlang The unit of measurement for telecommunications traffic.

Euratom European organization formed to address issues involving atomic energy. Euratom merged with the European Economic Community in 1965.

European Coal and Steel Community A tariff-free market for coal, steel, and iron formed in 1951 by Belgium, France, Germany, Italy, the Netherlands and Luxembourg. In 1965 it merged with the European Economic Community.

European Conference of Postal and Telecommunications Administrations (CEPT) An organization of European Post, Telephone, and Telegraph (PTT) administrations with the objective of coordinating technical and policy issues within Europe.

European Free Trade Association A group of European nations that didn't join the European Community in the 1960s because it regarded the EC as a customs union, but rather formed its own looser free trade area.

European Parliament The Parliament of the European Community, the members of which are elected by direct election in their respective member nation.

European Telecommunications Standards Institute (ETSI) An organization devoted to the establishment of telecommunications standards for the EC.

Eutelsat European Telecommunications Satellite Organization. A west European consortium of 26 nations providing satellite services to Europe.

Extraordinary Administrative Radio Conference (EARC) A conference of the ITU called in 1963 to deal with newly emerging satellite services and the allocation of frequencies to those services.

FCC Federal Communications Commission. The U.S. regulatory body, established by U.S. law, that has oversight regarding the telephone industry, broadcasting, international telecommunications, and other media.

Fixed links duopoly See duopoly.

Fredericksburg Conference A conference that took place in 1990 to establish the roles of and the coordination between the CCITT and the important regional standards bodies in the process of establishing telecommunications standards.

Frequency division multiplexing (FDM) A form of multiplexing in which a large bandwidth is divided into a collection of smaller distinct frequency bands. The AM and FM radio spectrum is an example of a frequency multiplexed spectrum.

GATS General Agreement on Trade in Services. An outgrowth of the GATT process, the GATS attempts to identify services as an element of trade and to apply GATT principles to services. These principles focus on eliminating trade barriers.

GATT General Agreement on Tariffs and Trade. The GATT process comprises rounds of multilateral trade negotiations among participating countries. Over 100 nations now participate in this process, which is designed to eliminate trade barriers and to facilitate open, nondiscriminatory trading among nations.

General switched telephone network (GSTN) The portion of the telephone network in which calls are established on demand via customer dialing.

Geosynchronous orbit An orbit over the equator and 22,300 miles above the earth's surface. In this orbit, a satellite moves at about the same speed at which the earth rotates and so appears to be stationary from the earth. Also referred to as geostationary orbit.

Green Paper A published list of goals for telecommunications in the European Community. The Green Paper's goals are notable in advocating a substantial move toward liberalizing the provision of telecommunications services and equipment.

Group of negotiators on services (GNS) A group of negotiators formed to develop an overall framework of principles and rules to govern international trade in services.

Half-duplex Refers to a circuit that can transmit data in both directions, but not at the same time.

High speed channel ("H") One of the channels defined for the integrated services digital network (ISDN).

High-level data link control (HDLC) A protocol at the link layer of the OSI reference model.

Hindustan Cable Ltd. (HCL) An Indian cable manufacturing firm.

Hindustan Teleprinters Ltd. (HTL) An Indian terminal equipment manufacturing firm.

HTC or Matav The Hungarian Telecommunications Company. The PTT in Hungary that eventually will be privatized to some extent by the Hungarian government.

In-band signaling A signaling system in which the signaling information is transmitted over the same channel over which the information will subsequently be transmitted.

Indefensible rights of use (IRUs) The permanent right to use a circuit on a cable. The IRU brings with it no rights to manage the cable or to have any rights to salvage value. The holder of an IRU is assured, in return for a one-time charge and monthly rates, the availability of capacity on the cable on a long-term basis for a fixed amount of money.

Indian Telephone Industries (ITI) An Indian manufacturer of telecommunications equipment.

Inmarsat The International Maritime Satellite Organization. A consortium that provides satellite communication for ships and for offshore enterprises.

Institute of Electrical and Electronic Engineers (IEEE) An international organization of electrical and electronic engineers that, among other things, develops standards in telecommunications.

Integrated digital services network (ISDN) A collection of standards defining an end-to-end (telephone-to-telephone) digital telephone network. Service provision is an integral part of the structure.

Intelnet A service, offered by Intelsat, that uses VSAT technology.

Intelsat Business Service (IBS) An integrated digital service, offered by Intelsat, capable of handling voice, data, telex, fax, and videoconferencing.

Interexchange carrier (IXC) A term used in the United States to denote companies that provide long distance services. Examples are AT&T and MCI.

Interexchange Between exchanges. Exchanges are roughly equivalent to local service areas. Interexchange traffic is traffic hauled from one exchange to another.

Interim Communications Satellite Committee (ICSC) The temporary decision-making body that prepared the way for the creation of Intelsat.

International Bureau of the Telegraph Union See Berne Bureau.

International Frequency Registration Board A body of the ITU concerned with the assignment of frequencies and with assignments in the geosynchronous orbit.

International Monetary Fund An international organization concerned with lending money to developing nations.

International standard (IS) The ultimate status of an ISO standards document.

International Telecommunications Union (ITU) A specialized agency of the United Nations with a mission of overseeing global telecommunications. The ITU is open to all nations. Over 160 nations now belong. Membership is at the national government level.

International Telecommunications Union Convention The basic document of the ITU. The Conventions set forth the purposes and goals of the ITU.

International Telecommunications Users Group (INTUG) An organization representing the users of international telecommunications services.

International Telegraph Convention The basic document of the International Telegraph Union.

International Telegraphy Union A precursor of today's ITU. The International Telegraphy Union, formed in 1865, dealt with matters of international telegraphy.

Internet Architecture Board (IAB) The organization responsible for the development of the research Internet and its standards.

Internet Engineering Task Force (IETF) The organization, accountable to the IAB, that develops the standards and practices for the research Internet.

Intersputnik Regional satellite organization owned by the former Soviet republics and now controlled by the Commonwealth of Independent States.

IXC Interexchange carrier. A term used in the United States to denote a long distance company. In the United States, IXCs, such as AT&T and MCI don't provide local service. They only provide long distance services.

Japan Approvals Institute for Telecommunications Equipment The Japanese organization responsible for type approvals.

KBPS Kilobits per second, or 1,000 bits per second.

Kenya External Telecommunications Company (Kenxtel) The Kenyan firm responsible for international telecommunications.

Kenya Posts and Telecommunications Corporation (KPTC) The Kenyan firm responsible for domestic telecommunications.

Kokusai Denskin Denwa Co., Ltd. (KDD) The Japanese firm traditionally responsible for international telecommunications.

Link access procedure for modems (LAPM) A variation of HDLC used to provide error correction in data modems.

Link access protocol-balanced (LAP-B) A subset of HDLC used as the link layer protocol in the X.25 network interface.

Local Access Transport Areas (LATA) A service area defined by the U.S. courts as part of the court settlement that called for the divestiture by the Bell System of its local telephone operations. The LATA denoted the areas within which the local telephone companies formed from the former Bell System could operate. The long distance component of the former Bell System (AT&T) was to provide service between LATAs.

Luxembourg Compromise The agreement formed in 1965 by which the Council of the European Community agreed to require unanimity on all votes dealing with action on significant issues. This unanimity in voting was

replaced by voting by qualified majority with the enactment of the Single European Act in 1987.

Measured service or local measured service A form of billing for local service that involves usage-based pricing. Instead of paying a flat monthly amount regardless of usage, customers pay by the message or by the minute.

Mercury The British company licensed by the UK government to compete with British Telecom in the UK's duopoly approach to the telecommunications industry.

Message store (MS) A component of the X.400 electronic mail standard that stores messages prior to transfer to a user agent or a message transfer agent.

Message transfer agent (MTA) A component of the X.400 electronic mail standard responsible for interacting with a user agent or a message store and the message transfer service element.

Message transfer service element (MTSE) A component of the X.400 electronic mail standard responsible for transferring messages through an electronic mail system.

Metropolitan Fiber Systems A U.S. company that provides an alternative to the public network. Customers can purchase capacity on the company's fiber-optic system rather than purchasing from the telephone company.

Minimum attribute standards Standards that define the minimum attributes of something. For instance, the minimum strength of a material such as steel or aluminum.

Ministry of Posts and Telecommunications (MPT) The Japanese government agency responsible for regulating and overseeing the telecommunications industry.

Modification of Final Judgment (MFJ) The agreement that ended the U.S. Department of Justice antitrust lawsuit against the former Bell System. The MFJ specified that the Bell System would be divested of its local telephone operations, and would retain long distance services and equipment manufacturing.

Most-favored-nation clause (MFN) A basic principle of the GATT process, the MFN specifies that any benefit accorded to one nation that is party to a trade agreement, should also be extended to all nations that are party to the multilateral agreement.

Multicasting Message transmission to multiple destinations simultaneously. This is like broadcasting, except the sender has control over who receives the message.

Multiplexer A device that combines several low bandwidth (low speed) channels into one high bandwidth (high speed) channel. The device that performs the inverse of this function is also called a mulitplexer.

Multivendor compatibility Compatibility across multiple vendors.

Multivintage compatibility Compatibility across multiple versions of a vendor's equipment.

National Institute for Standards and Technology (NIST) A part of the U.S. Department of Commerce responsible for the standards and standardization used by the U.S. government. It also plays a role in private sector standardization.

NCC New common carriers in Japan's telecommunications system.

Nippon Telephone and Telegraph (NTT) The Japanese carrier that had held the monopoly for local service in Japan.

Normes Europeeanes des Telecommunications (NET) Standards developed by CEPT.

OFTEL Office of Telecommunications. The regulatory body in the UK that oversees the telecommunications industry.

Open network architecture (ONA) A U.S. initiative undertaken by the Federal Communications Commission, which requires that the Bell Operating Companies provide enhanced service providers the type of network services and access they need to do business. At the same time, the ONA plan specifies that the Bell Operating Companies provide the same level of services and access to their own enhanced service operations.

Open network provisioning (ONP) An initiative by the European Community, articulated in its Green Paper on Telecommunications, which specifies that telecommunications administrations in the EC countries provide clear definitions to their customers of the requirements for interconnection and access to the telephone network.

Open systems interconnection (OSI) A reference model structuring data communications functions and interfaces.

Open wire A transmission system in which the wires of a transmission system rely on physical separation for insulation. In some cases, a single wire is used for the channel.

Organization for Economic Cooperation and Development (OECD) Paris-based organization that facilitates cooperation on common economic and development issues among its members.

Organization for European Economic Cooperation Formed in 1948, the precursor to the OECD which channeled aid from the United States to European nations in need of post-war relief.

PABX Private automatic branch exchange. Also known as PBX. This is essentially a private telephone switch located on the customer's site. The PBX provides phone service within the customer's location and access to calling outside the customer's location.

Packet assembler and disassembler (PAD) A device capable of breaking a data message into packets of data and reassembling packets of data into the original message format.

Packet data network (PDN) A network over which data is "packetized." In other words, a data message is disassembled into data packets. The packets are sent to their ultimate location in a manner designed to make the most efficient use of transmission facilities. They are reassembled into the original message format at their point of destination.

Packet switches Switches capable of handling packets of data over a packet network.

Pan African Telecommunications Network (Panaftel) A regional microwave network connecting several African countries.

PanAmSat A private satellite company that sought, and received, permission to provide satellite services in competition with Intelsat.

Price cap plans A regulatory approach in which price increases for a telecommunications company's services are tied to the rate of inflation. This approach was first implemented in the UK and has recently been adopted by the FCC in the United States.

Primary rate interface An ISDN interface that operates at 1.536 Mbps or 1.920 Mbps.

Product line compatibility Compatibility across a vendor's product line.

PTT Before privatization, the state agency responsible for providing the postal, telephone, and telegraph services for a country. With privatization, PTTs often become regulatory bodies rather than service providers.

Public switched network (PSN) The basic telephone network consisting of connections between customers and switching offices and the switching offices themselves. The PSN is provided by a regulated company which must offer the PSN services on a nondiscriminatory manner to all customers.

Pulse code modulation (PCM) A specific technique for converting analog voice signals into digital signals, and vice versa.

Purchasing power parities (PPPs) A way of comparing national currencies that relies on baskets of goods that can be purchased instead of exchange rates.

Radiotelegraph Telegraph systems that use wireless techniques instead of cable-based transmission.

Recognized private operating agencies (RPOA) A private operating agency is a nongovernment entity that operates an international telecommunications service. Due to competition, the United States has many such private operating agencies. Many of them encountered difficulties in dealing with the PTTs of other nations. The FCC in the United States can designate these agencies as RPOAs. This designation is meant to assure PTTs that the agencies in question will follow ITU provisions regarding international telecommunications.

Regional African Satellite Communication System (Rascom) An organization planning a regional satellite system for African countries.

Regional Bell Operating Companies (RBOC) Seven companies that resulted from the breakup of the Bell System in the U.S. The RBOCs are only allowed to provide services within their own operating areas. They can't cross over the boundaries of their operating territories (as defined by U.S. regulatory agencies and the U.S. courts) to provide interexchange service.

Requests for comments (RFC) A document produced by the Internet Architecture Board that specifies standards, practices, and other informational items about the research internet.

Research and Development in Advanced Communications in Europe (RACE program) EC initiative to encourage research and development of advanced telecommunications (notably broadband services) in the European Community.

Roamer A user of mobile telephony (including cellular service) who uses that service in more than one serving area.

Secretariat for Communications and Transportation The Mexican PTT prior to privatization.

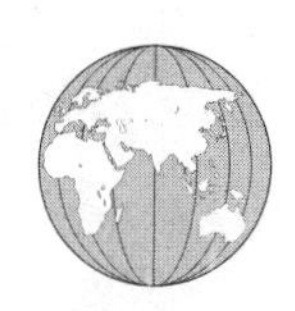

Service control point (SCP) A component of Signaling System 7 that acts as the interface (control point) for a network service provider and the signaling system.

Service switching point (SSP) A component of Signaling System 7 that resides in end office switches and determines the destination of a signaling request.

Signal transfer point (STP) A component of Signaling System 7 that switches signaling messages to other STPs, SCPs, or SSPs.

Signaling System 7 A specific common channel signaling system defined by the CCITT.

Single European Act (SEA) Effective July 1, 1987, this legislation provided the basis for the creation of an internal market in the European Community on January 1, 1993. The SEA also established telecommunications as central to the internal market.

Southeast Asia, Middle East, Western Europe Cable (SEA-ME-WE) An undersea fiber-optic cable under construction interconnecting Southeast Asian, Middle Eastern, and Western European countries.

Special drawing rights (SDR) A unit of exchange used in determining the relative amounts of money a country will get for its share in providing an international call. Because each country has its own currency, the SDR, which is calculated based on the exchange rates for the U.S., German, British, French, and Japanese currencies, is used.

Special Telecommunications Action for Regional Development (the STAR program) EC initiative to improve the telecommunications infrastructure of the developing nations of the European Community.

Spectrum The usable electromagnetic (radio) bandwidth.

Stored program control (SPC) A technique in which a switching system is controlled by a digital computer under software control.

Submit/deliver service element (SDSE) A component of the X.400 electronic mail standard used to deliver messages to user agents.

T1 Carrier System The first carrier system that implemented digital transmission. T1 systems operate at DS1 speeds (1.544 Mbps signal rate, with 1.536 Mbps of usable bandwidth) and use the D4 framing format.

Tandem office A telephone office that connects central offices. Tandem offices may or may not directly serve end users.

Tariff Usually a published document that has been approved by a regulatory body and that sets forth the types of service a telecommunications service provider can provide, the prices which can be charged for those services, and the conditions under which those services can be offered.

Technical Committee of the ISO (TC) The main division in ISO responsible for carrying out the technical work of standards development.

Technical Recommendations Application Committee (TRAC) An organization of CEPT intended to streamline the standards development process.

Telecommunication's Development Bureau A body of the ITU charged with addressing the needs and problems of the developing nations in the ITU membership.

Telecommunications of Jamaica (TOJ) The telecommunications administration of Jamaica.

Telecommunications Technology Council of Japan (TTC) The Japanese standards body responsible for the development of telecommunications standards.

Telekom 2000 Project The German project to bring the telecommunications infrastructure of the former East Germany to parity with the infrastructure of the former West Germany. The end result will be the merger of both telecommunications systems.

Teleport A method of bypassing the public-switched network through the use of fiber-optic cables and satellite uplink and downlink facilities. Teleport providers place fiber-optic cable facilities, usually within a well-populated area. They sell capacity on the fiber-optic facility to customers and then haul their customer's traffic to a satellite uplink site to send the traffic to its destination. Conversely, they collect traffic at a satellite downlink location and then terminate the traffic to their customers over the fiber-optic facility.

Televerket The telephone administration in Sweden.

Telex A switched service provided all over the globe. The telex message is printed on a telex machine and is sent over a global telex network.

Telmex—Telefonos de Mexico S.A. One of the organizations providing telecommunications service in Mexico.

The Six The members of the European Coal and Steel Community.

Time division multiplexing (TDM) A form of multiplexing in which the entire bandwidth of a digital communications channel is dedicated to several slower channels, each for a short period of time.

Time-slot interchanger (TSI) A device that performs the switching function for TDM signals.

Toll center See tandem office.

Trade in Services See GATS.

Transborder data flow (TDF) An international issue concerning the flow of information about a country's citizens across national boundaries.

Transcoder A device that converts one form of PCM into another.

Transponder A system on a satellite that receives transmissions from an earth station and rebroadcasts the transmissions.

Transponder for unrestricted use (TUU) Intelsat service that allows the aggregation of international public-switched traffic, domestic services, and private network traffic over a facility, rather than requiring separate satellite facilities for each type of traffic.

Transport Control Protocol/Internet Protocol (TCP/IP) A collection of protocols used in the research internet.

Treaty of Rome Signed in 1957, this treaty established the European Economic Community; or, as it is now commonly called, the European Community.

Trunk line A telephone line (digital or analog) that interconnects switches.

Twisted pair wire A copper cable that consists of two insulated conductors twisted together.

U2000 DBP Telekom's new management network system, which is in the testing stages and slated to be operational by 1994.

United Nation's International Maritime Organization The group that held meetings resulting in the creation of Inmarsat in 1979.

Universal Electrical Communications Union An organization, which was never formed, proposed by the United States in 1920 to deal with matters of telegraph and radio communications.

UNIX An operating system developed by AT&T Bell Laboratories in widespread use on minicomputers.

User agent (UA) A component of the X.400 mail system that acts as a mail agent for the user.

Value-added network (VAN) A data network that does more than just transmit a message. The simple transmission of a message is regarded as "basic" service by regulatory bodies. VAN networks provide additional value by storing a message, for example.

Very small aperture terminals (VSATs) A small satellite antenna. These small antennae are used to transmit and receive data by retail stores and other operations seeking to transmit information internally without having to incur the costs of using the public-switched network.

Virtual circuit A logical connection across a packet network. A virtual circuit defines the route that packets take. Unlike a circuit-switched network, the bandwidth along this route is shared among many connections.

Virtual private network (VPN) Use of the public-switched network to replicate a private network. This is possible because of more sophisticated switching and network software capability.

Voluntary concensus standards Standards developed in standards committees, such as those administered by the CCITT, ISO, and the IEEE.

Western European Telegraph Union A precursor of the ITU. The Western European Telegraph Union comprised the governments of Western Europe, which in the 19th century, attempted to solve problems of intercountry telegraph service.

Western European Union A European organization which deals with matters of defense. Members include Germany, Italy, Belgium, France, Luxembourg, the Netherlands, and the UK.

Whipsawing A practice the United States wishes to avoid in which the PTTs in other nations play U.S. carriers (MCI, AT&T, and others) against one another in order to "cut a better deal" in dividing the revenues from international calls.

Working Group on Telecommunications Services A subgroup of the GNS convened to resolve areas of dispute involved in applying the GATS approach to telecommunications.

World Administrative Space Conferences (WARCs) ITU conferences at which issues relating to radio communications issues are addressed.

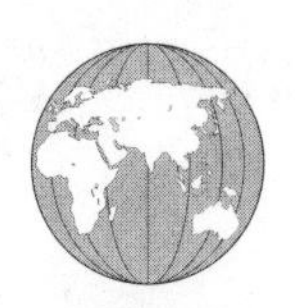

World Administrative Telephone and Telegraph Conferences (WATTCs) ITU conferences at which issues relating to telecommunications services and standards are addressed.

World Telecommunication Advisory Council (WTAC) A council of telecommunications company presidents, chairmen, and representatives from several nations formed to provide the ITU with strategic advice regarding telecommunications.

X.25 An interface between Data Terminal Equipment (DTE) and Data Communications Equipment (DCE) for packet-switched networks.

References

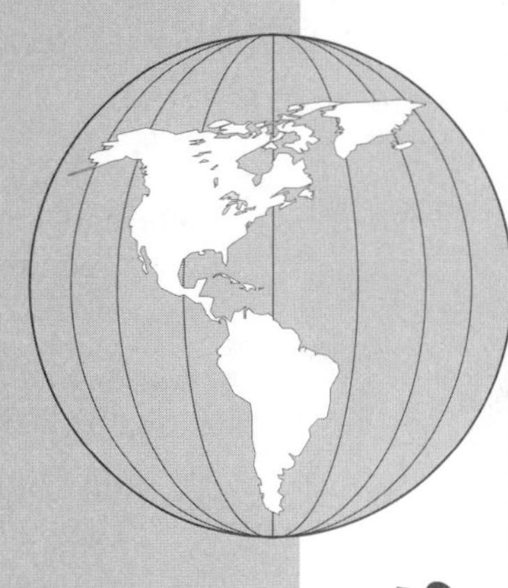

REFERENCES

Agreement Relating to the International Telecommunications Satellite Organization. *23 UST 3814ff.* (August 20, 1971).

Akwule, Raymond U. "Telecommunications in Kenya: Development and Policy Issues." *Telecommunications Policy.* (September/October 1992): pp. 603-611.

Aronson, Jonathan David and Cowhey, Peter F. *When Countries Talk: International Trade in Telecommunications Services.* Cambridge, Mass.: Ballinger Publishing Co., 1988.

Asia-Pacific Economic Cooperation (APEC) Telecommunications Working Group. *The State of Telecommunications Infrastructure and Regulatory Environments of APEC Economies.* (November 1991).

Bell, Alan and Cave, Martin. "Lessons from the UK Duopoly Review." In *Telecommunication: New Signposts to Old Roads.* Ed. Franca Klaver and Paul Slaa. Amsterdam: IOS Press, 1991.

Berrada, Abderrazak. "Telecommunications in the GATT Framework Agreement." *Telecommunication Journal.* Vol. 58, No. VIII (1991): pp. 487-498.

Besen, Stanley M. and Farrell, Joseph. "The Role of the ITU in Standardization: Pre-eminence, Impotence or Rubber Stamp?" *Telecommunications Policy.* Vol. 15, No. 4 (August 1991): pp. 311-321.

Borrego, Jorge and Mody, Bella. "The Morelos Satellite System in Mexico: A Contextual Analysis." *Telecommunications Policy*. (September 1989): pp. 265-276.

Cable Landing License Act. Chapter 12, 42 Stat. 8, 1921, codified as amended at 47 U.S.C. Sections 34-39, 1988.

Cargill, Carl. *Information Technology Standardization: Theory, Process, and Organizations*. Bedford, Mass.: Digital Press, 1989.

Carter, Michael and Wright, Julian. "Optimal Telecommunications Tariffs and the CCITT." *Telecommunications Journal*. Vol. 59, No. III (1992): pp. 125-131.

CCITT Blue Book, Melbourne 1988. *General Tariff Principles: Charging and Accounting in International Telecommunications Services*. Series D Recommendations, Vol. II- Fascicle II.1, 5.1.1.1, p. 354.

Charles, David; Monk, Peter; and Sciberras, Ed. *Technology and Competition in the International Telecommunications Industry*. London: Pinter Publishers, 1989.

Cheong, Ken and Mullins, Mark. "International Telephone Service Imbalances: Accounting Rates and Regulatory Policy." *Telecommunications Policy*. Vol. 15, No. 2 (April 1991): pp. 107-118.

Chowdary, T.H. "Telecommunications Restructuring in Developing Countries." *Telecommunications Policy*. Vol. 16 (September/October 1992): pp. 591-602.

Clarke, Arthur C. "Extra-Terrestrial Relays: Can Rocket Stations Give World-Wide Radio Coverage?" *Wireless World*. (October 1945): pp. 305-308.

Codding, George, A., Jr. "Evolution of the ITU." *Telecommunications Policy*. Vol. 15, No. 4 (August 1991): pp. 271-285.

———"Financing Development Assistance in the ITU." *Telecommunications Policy*. (March 1989): pp. 13-24.

———"Reorganizing the ITU." *Telecommunications Policy*. Vol. 15, No. 4 (August 1991): pp. 267-270.

———"The Nice ITU Plenipotentiary Conference." *Telecommunications Policy*. Vol. 14, No. 2 (April 1990): pp. 139-149.

Codding, George A., Jr. and Rutkowski, Anthony M. *The International Telecommunication Union in a Changing World*. Dedham, Mass.: Artech House, 1982.

Comer, Douglas. *Internetworking with TCP/IP*, 2nd Edition. 1991.

Communications Outlook 1990. Paris: OECD, 1990.

Communications Satellite Act of 1962. Public Law 87-624, 76 Stat. 419. (August 31, 1962).

Cowhey, Peter, and Aronson, Jonathan D. "The ITU in Transition." *Telecommunications Policy*. Vol. 15, No. 4 (August 1991): pp. 298-310.

Crawshay, Grant R. "Recent Commercial Developments in the New Zealand Marketplace." *Proceedings of the 1992 Pacific Telecommunications Conference*, January 1992.

"The Curtain Rises on Telecommunications in Eastern Europe." *Telephony*. (July 6, 1992): pp. 32-33.

Datapro Reports on International Telecommunications. Delran N.J.: Datapro Research.

David, Paul A. "Clio and the Economics of QWERTY." *American Economic Review*. Vol. 75, No. 2 (May 1985): pp. 332-337.

———"Some New Standards for the Economics of Standardization in the Information Age." In *Economic Policy and Technological Performance*. Eds. P. Dasgupta and P. Stoneman. Cambridge: Cambridge University Press, 1987.

Doerrenbaecher, Christoph and Fischer, Oliver. "Telecommunications in the Uruguay Round." *Intereconomics*. (July/August 1990): pp. 185-192.

Doing Business with DBP Telekom in the Center of Europe. Bonn: DBP Directorate, April 1991.

Dow Jones News Retrieval Service. (December 1, 1992).

Draft Final Act Embodying the Results of the Uruguay Round of Multilateral Trade Negotiations, General Agreement on Trade in Services. Trade Negotiations Committee, GATT Secretariat, MTN.TNC/W/FA: December 20, 1991.

Duch, Raymond M. *Privatizing the Economy: Telecommunications Policy in Comparative Perspective*. Ann Arbor, Mich.: University of Michigan Press, 1991.

Dunkel, Arthur. "Telcom Services and the Uruguay Round." *Transnational Data and Communications Report*. (January/February 1992): p. 18.

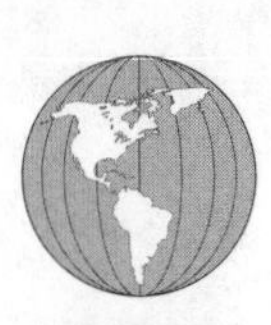

Dyson, Kenneth and Humphreys, Peter. *The Political Economy of Communications: International and European Dimensions*. London: Routledge, 1990.

Eastern European and Former Soviet Telecommunications Report (EESTR). Vol. 3, No. 3-9, 1992.

Elbert, Bruce R. *International Telecommunication Management*. Norwood, Mass.: Artech House, 1990.

Ergas, Henry, and Paterson, Paul. "International Telecommunications Settlement Arrangements." *Telecommunications Policy*. Vol. 15, No. 1 (February 1991): pp. 29-48.

Eward, Ronald. *The Deregulation of International Telecommunications*. Dedham, Mass.: Artech House, 1985.

Federal State Joint Board Staff Report in *CC Docket No. 80-286*. Monitoring Report, *CC Docket No. 87-339*. (January 1991).

Feketekuty, Geza. *International Trade in Services: An Overview and Blueprint for Negotiations*. Cambridge, Mass.: Ballinger Publishing Company, 1988.

Garcia, Renato B. "Telecommunications and Investment Decisions in the Philippines." *Telecommunications Policy*. (March 1984): pp. 51-57.

Garrison, William B., Jr. *Four Case Studies of Structural Alternatives of the Telecommunications Industry*. The Annenberg Washington Program in Communications Policy Studies. (January 1988): pp. 41, 64, 70-71.

General Tariff Principles. 5.1.2.3 iv, p. 356.

Goldberg, Henry. "One-Hundred and Twenty Years of International Communications." *Federal Communications Law Journal*. Vol. 27, No. 1 (January 1985): pp. 131-154.

Griffiths, Sarah and McCarthy, Clare. "The United Kingdom: The Commercial and Regulatory Environment." *Datapro Reports on International Telecommunications*. (May 1992): pp. 803-813, 899.

Halsall, Fred. *Data Communications, Computer Networks, and Open Systems (3rd Edition)*. Reading Mass.: Addison-Wesley, 1992.

Hardy, A.P. "The Role of the Telephone in Economic Development." *Telecommunications Policy*. Vol. 4 (1980): pp. 278-286.

Harrop, Jeffrey. *The Political Economy of Integration in the European Community*. Aldershot, UK: Edward Elgar Publishing, 1989.

Haus, Leah A. *Globalizing the Gatt, The Soviet Union's Successor States, Eastern Europe, and the International Trading System*. Washington D.C.: Brookings Institution, 1992.

Herbst, Jeffrey. "The Politics of Privatization in Africa." *The Political Economy of Public Sector Reform and Privatization*. pp. 88-110.

Heywood, Peter and Saunders, Stephen. "¡Competition, Sí!" *Data Communications*. (September 1991): pp. 70-74.

Hills, Jill. "Universal Service: Liberalization and Privatization of Telecommunications." *Telecommunications Policy*. (June 1989): pp. 129-144.

Hudson, Heather E. "Telecommunications in Africa: The Role of the ITU." *Telecommunications Policy*. (August 1991): pp. 343-350.

———*Communication Satellites, Their Development and Impact*. New York: The Free Press, 1990, pp. 1-38, 131, 148.

———*When Telephones Reach the Village*. Norwood N.J.: Ablex Publishers, 1984.

Hukill, Mark A. and Jussawalla, Meheroo. *Trends in Policies for Telecommunication Infrastructure Development and Investment in ASEAN Countries*. Honolulu: East-West Center, 1991.

"Hungary: A Giant Step Ahead." *Business Week*. (April 15, 1991): pp. 58.

Hush-a-Phone Corp. v. AT&T. 22 FCC 112, 1957.

Hush-a-Phone Corp. v. United States. 238 F.2d 266, D.C. Cir., 1956.

Ikenberry, G. John. "The International Spread of Privatization Polices: Inducement, Learning and 'Policy Bandwagoning.' " In *The Political Economy of Public Sector Reform and Privatization*. Ed. Ezra N. Suleiman and John Waterbury. Boulder, Colo.: Westview Press, 1990.

In the Matter of Regulation of International Accounting Rates. CC Docket No. 90-337, Phase II, 6 FCC RedNo. 12, 3441-3442.

In the Matter of Satellite Systems Providing International Communications, Notice of Inquiry and Proposed Rulemaking. CC Docket No. 84-1299, 100 FCC 2d., 290, 1985.

Intelsat Annual Report. (1988-89).

———(1991-92).

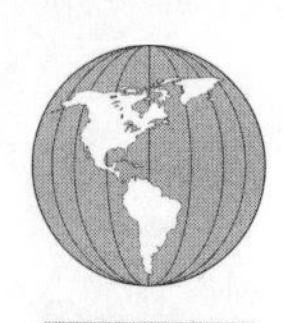

International Telecommunications Convention. "Trade in Telecommunications Services, Notes by the Secretariate." GATT Secretariat, ITU, MTN.GNS/W/52 (May 19, 1989).

International Telecommunications Satellite Organization Headquarters Agreement. *28 UST 2249-2257*. November 22, 24, 1976.

International Telecommunications Union. *Yearbook of Common Carrer Telecommunications Statistics (1981-1990)*. 19th Edition. (1991).

"ITU Structural Reforms Recommended." *Transnational Data and Communications Report*. (May/June 1991): pp 5-6.

———*List of International Telephone Routes*. 32nd Edition. (1992).

Johnson, Leland. *Competition, Pricing, and Regulatory Policy in the International Telephone Industry*. Rand Publication R-3790-NSF/MF. Santa Monica, Calif.: Rand Corporation, July 1989

———"International Telecommunications Regulations." In *New Directions in Telecommunications Policy*. Ed. Paula R. Newberg. Durham N.C.: Duke University Press, 1989, p. 117.

Johnson, Ted. "Caribbean Basin Becomes Large Telecommunications Market." *Telematics and Informatics*. Vol. 7, No. 1 (1990): pp.1-7.

Jussawalla, Meheroo. "Is the Communications Link Still Missing?" *Telecommunications Policy*. (August 1992): pp. 485-503.

Jussawalla, Meheroo and Lamberton, D.M., eds. *Communication Economics and Development*. New York: Pergamon Press, 1982.

Kellogg, Michael K.; Thorne, John; and Huber, Peter W. *Federal Telecommunications Law*. Boston: Little, Brown and Company, 1992.

King, John A.C. "The Privatization of Telecommunications in the United Kingdom." In *Restructuring and Managing the Telecommunications Sector*. Ed. Bjorn Wellenius, Peter A. Stern, Timothy E. Nulty, and Richard D. Stern. Washington D.C.: World Bank, 1989, pp. 55-60.

Kojo, Makoto and Janisch, H.N. "Japanese Telecommunications after the 1985 Regulatory Reform." *Media and Communications Law Review*. pp. 307-340.

Kratvin, Patricia D. "The U.S. Telecommunications Infrastructure and Economic Development." Presented at the 10th Telecommunications Policy Research Conference, Airlie, Va. 1990.

Kudriavtzev, G.G. and Varakin, L.E. "Economic Aspects of Telephone Network Development: The USSR Plan." *Telecommunications Policy*. (February 1990): pp. 7-14.

Legislative Bill 835. Amendment to Revised Statutes of Nebraska, Sections 75-109 and 75-604 and Revised Statutes Supplement, Section 75-609.

Lerner, N.C. "Telecommunications Privatization and Liberalization in Developing Countries." *Telecommunication Journal*. Vol. 58 (1991): pp. 281-282.

Littlechild, Stephen C. *Regulation of British Telecommunications' Profitability: Report to the Secretary of State, February 1983*. Department of Industry, 1983.

Luxner, Larry. "Mexico Reaches for New Telecom Heights." *Telephony*. (February 3, 1992): pp. 22-28.

Martinez, Larry. *Communication Satellites: Power Politics in Space*. Dedham, Mass.: Artech House, 1985.

McCarthy, Clare. "Hungary: The Commercial and Regulatory Environment." *Datapro Reports on International Telecommunications*. (March 1992): pp. 361- 362, 365.

Mendis, Vernon L.B. "Phased Privatization with Proposed Foreign Participation: The Sri Lanka Experience." In *Restructuring and Managing the Telecommunications Sector*. Ed. Bjorn Wellenius, Peter A. Stern, Timothy E. Nulty, and Richard D. Stern. Washington D.C.: World Bank, 1989, pp. 99-106.

Ministry of Posts and Telecommunications Japan. *Telecommunications Market of Japan: OPEN*. (March 1989).

"Multilateral Telecom Trade Regime Prepared." *Transnational Data and Communications Reports*. (January/February 1992): p. 31.

Nelson, Eric G. *Telecommunications Trade Issues: An Overview of the Regulatory Framework and Analyses of Current Events*. Washington D.C.: North American Telecommunications Association, 1989.

Newberg, Paula R., ed. *New Directions in Telecommunications Policy, Volume One—Regulatory Policy: Telephony and Mass Media*. Durham, N.C.: Duke University Press, 1989.

Noam, Eli. *Telecommunications in Europe*. New York: Oxford University Press, 1992.

NTIA Infrastructure Report: Telecommunications in the Age of Information. U.S. Department of Commerce, National Telecommunications and Information Administration, NTIA, Special Publication 91-26, October 1991.

Performance Indicators for Public Telecommunications Operators, OECD ICCP Series Report No. 22. Paris: OECD, 1990.

Policy and Rules Concerning Rates for Dominant Carriers. CC Docket No. 87-313, Memorandum Opinion and Order, DA 91-619, released May 17, 1991.

Poskett, Peter and McDougal, Patrick. "WARC 92: The Case for Mobile Satellites." *Telecommunications Policy.* Vol. 14, No. 5 (October 1990): pp. 355-363.

Pregg, Ernest H. *The American Challenge in World Trade: U.S. Interests in the GATT Multilateral Trading System.* Significant Issues Series, Vol. XI, No. 7. Center for Strategic and International Studies, 1989.

Presidential Determination No. 85-2. November 28, 1964.

"Privatization in Hungary Creates New Opportunities for U.S. Investors." *Eastern Europe Business Bulletin.* (March 1991): pp. 1-2.

Qvortrup, Lars. "Nordic Telecottages: Community Teleservice Centres for Rural Regions." *Telecommunications Policy.* (March 1989): pp. 59-68.

Redwood, John. "A Consultant's Perspective." In *Privatization and Deregulation in Global Perspective.* Ed. Dennis J. Gayle and Jonathan N. Goodrich. Westport, Conn.: Quorum Books, 1990, pp. 55-56.

Reform of the Postal and Telecommunications System in the Federal Republic of Germany: Concept of the Federal Government for the Restructuring of the Telecommunications Market. Heidelberg, Germany: Federal Minister of Posts and Telecommunications, May 1988.

Report and Order. "In the Matter of International Communications Policies Governing Designation of Recognized Private Operating Agencies, Grants of IRUs in International Facilities and Assignment of Data Network Identification Codes." CC Docket 83-1230, 104 FCC 2d, 1986, pp. 214, 256-258.

———"In the Matter of the Regulation of International Accounting Rates." CC Docket No. 90-337, Phase I, 6 FCC Rcd No. 12, 3554, 3555, 3562.

Robinson, Peter. "TDF Issues: Hard Choices for Governments." *Telecommunications Policy.* (February 1989): pp. 64-70.

Rowe, Stanford H., II. *Business Telecommunications, 2nd Edition*. New York: Macmillan, 1991.

Rutkowski, Anthony M. "The ITU at the Cusp of Change." *Telecommunications Policy*. Vol. 15, No. 4 (August 1991): pp. 286-297.

Savage, James G. *The Politics of International Telecommunications Regulation*. Boulder, Colo.: Westview Press, 1989.

Saunders, Robert J.; Warford, Jeremy J.; and Wellenius, Bjorn. *Telecommunications and Economic Development*. Baltimore Md.: The Johns Hopkins University Press for The World Bank, 1983.

Scherer, Peter R. "Perspectives on World Telecom Reform." *Transnational Data and Communications Report*. (May/June 1992): pp. 27-33.

Smith, Anthony. *The Geopolitics of Information*. New York: Oxford University Press, 1980.

Smith, Milton L. *International Regulation of Satellite Communication*. Utrecht Studies in Air and Space Law Number 7. Dordrecht, The Netherlands: Kluwer Academic Publishers, 1990.

Solomon, Jonathan. "Should the UK play the ITU Political Game?" *Telecommunications Policy*. Vol. 14, No. 1 (February 1990): pp. 3-5.

Spiller, Pablo T. and Sampson, Cezley I. "Regulation, Institutions and Commitment: The Jamaican Telecommunications Sector." *Twentieth Annual Telecommunications Policy Research Conference*. Solomons Md.: September 1992.

Stallings, William. *Data and Computer Communications, 3rd Edition*. New York: Macmillan, 1991.

Statement of FCC Chairman Alfred C. Sikes Regarding the Commission's Companion International Settlements & Accounting Rates Decisions CC Dockets 90-337, Phases I and II. Washington D.C.: Government Printing Office, 1991.

Stern, Peter A. "The International Telecom Accounting and Settlements Debate." *Transnational Data and Communications Report*. (July/August 1991): pp. 25-33.

Suleiman, Ezra. "The Politics of Privatization in Britain and France." In *The Political Economy of Public Sector Reform and Privatization*. Ed. Ezra N. Suleiman and John Waterbury. Boulder, Colo.: Westview Press, 1990, p. 115.

Taylor, Leslie. "Depoliticizing Space WARC." *Satellite Communications*. (January 1989).

Tedeschi, Michael. *Live Via Satellite: The Story of COMSAT and the Technology that Changed World Communication*. Washington D.C.: Acropolis Books, 1989.

"Telecom Important to Restarted Services Negotiations." *Transnational Data and Communications Reports*. (March/April 1991): p. 5.

"Telecommunications Annex to Draft General Agreement on Trade in Services, Section 2.1." *Transnational Data and Communications Reports*. (January/February 1992): pp. 31-33.

Telecommunications Rules and Regulations. Nebraska Public Service Commission.

"Tomorrow's ITU: The Challenges of Change." *Telecommunications Policy*. (August 1991): pp. 269-270.

"Towards a Competitive Community-Wide Telecommunications Market in 1992: Implementing the Green Paper on the Development of the Common Market for Telecommunications Services and Equipment." Brussels: Commision of the European Communities, COM (88), 1988.

United States Central Intelligence Agency. *1991 World Factbook*, available through the US National Technical Information Administration (NTIA), Springfield VA, USA.

Universal Service and Rate Restructuring in Telecommunications. ICCP Series Report No. 23 Paris: OECD, 1991, pp. 70-82.

"U.S. Formally Exempts Basic Telecom from Uruguay Round." *Transnational Data and Communications Reports*. (March/April 1991): p. 5.

"U.S. Spectrum Management Policy: Agenda for the Future." *NTIA Special Publication 91-23*. U.S. Department of Commerce, 1991.

U.S.C. 47 Section 34, 1988.

U.S.C. 47 Section 35, 1988.

Van Slageren, Steven and Gooskens, Pit. *Connected with the Future: Telecommunications and Economic Development, India as an Example*. Eindhoven, the Netherlands: Stichting Onderzoek Bedrijfstak Elektrotechniek, 1990.

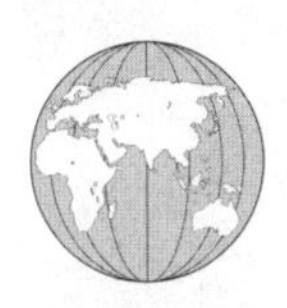

Vercruysse, Jean-Pierre. "Telecommunications in India: 'Deregulation' vs. Self Reliance." *Telematics and Informatics*. Vol. 7, No. 2, pp. 109-121.

"WARCs, WATTCs, & PLENIPOTs." *Via Satellite*. May 1989.

Weinhaus, Carol and Oettinger, Anthony. *Behind the Telephone Debates*. Norwood, N.J.: Ablex, 1988.

Weiss, Martin B.H. "Compatibility Standards and Product Development Strategies: A Review of Data Modem Developments." *Computer Standards and Interfaces*. Vol. 12, 1991a. pp. 109-121.

Weiss, Martin B.H. and Cargill, Carl . "Consortia in the Standards Setting Process." *Journal of the American Society for Information Science* (JASIS). (September 1992).

Wellenius, Bjorn. "Beginnings of Sector Reform in the Developing World." In *Restructuring and Managing the Telecommunications Sector*. Ed. Bjorn Wellenius, Peter A. Stern, Timothy E. Nulty, and Richard D. Stern. Washington D.C.: World Bank, 1989, pp. 89-98.

———"On the Role of Telecommunications in Development." *Telecommunications Policy*. (March 1984): pp. 59-66.

Wenders, John T. *The Economics of Telecommunications: Theory and Practice*. Cambridge, Mass.: Ballinger Publishing Company, 1987.

"What the Brochures Don't Tell You." *Business Week*. (April 15, 1991): pp. 58.

White, Rita Lauria and White, Harold M., Jr. *The Law and Regulation of International Space Communication*. Norwood Mass.: Artech House, 1988.

Whitlock, Erik and Nyevrikel, Emilia. "The Evolution of Hungarian Telecommunications Policy." *Telecommunications Policy*. (April 1992): pp. 249, 253- 254.

Willey, Bruce. "A Latin America Telecommunications Primer." *Telecommunications*. North American Edition. (March 1992): pp. 45-50.

Williamson, John; Titch, Steven; and Purton, Peter. "The Curtain Rises on Telecommunications in Eastern Europe." *Telephony*. (July 6, 1992): pp. 27-28, 32-33.

Woodrow, R. Brian. "Tilting Towards a Trade Regime." *Telecommunications Policy*. Vol. 15, No. 4. (April 1991): pp. 323-342.

Index

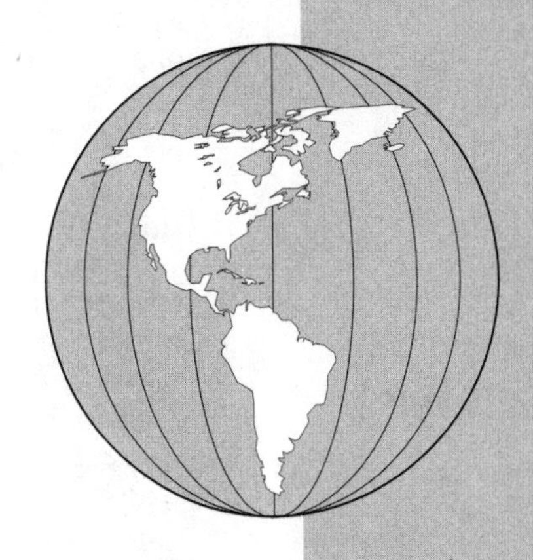

Symbols

A

B

C

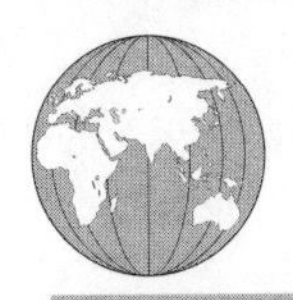

D

E

F

G

H

I

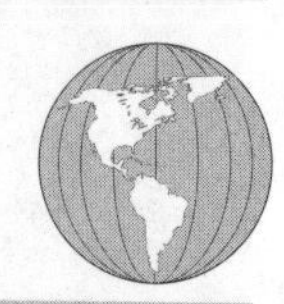

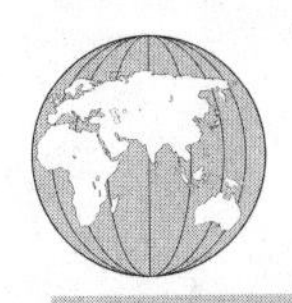

J

K

L

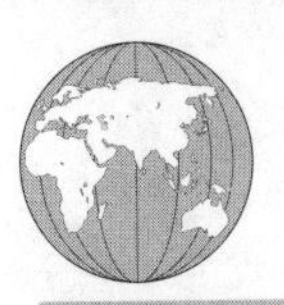

M

N

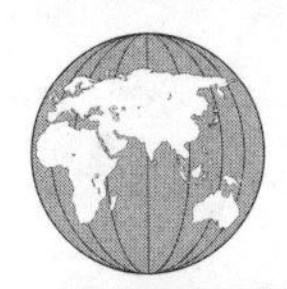

O

P

S

T

U

V

W

X-Y-Z